MATHEMATICAL PHYSICS
(An In-Depth Study)

R. H. ATKIN

Published 2010 by arima publishing

www.arimapublishing.com

ISBN 978 1 84549 466 7

© R. H. Atkin 2010

All rights reserved

This book is copyright. Subject to statutory exception and to provisions of relevant collective licensing agreements, no part of this publication may be reproduced, stored in a retrieval system, or transmitted in any form or by any means, without the prior written permission of the author.

Printed and bound in the United Kingdom

This book is sold subject to the conditions that it shall not, by way of trade or otherwise, be lent, re-sold, hired out, or otherwise circulated without the publisher's prior consent in any form of binding or cover other than that which it is published and without a similar condition including this condition being imposed on the subsequent purchaser.

Abramis is an imprint of arima publishing.

arima publishing
ASK House, Northgate Avenue
Bury St Edmunds, Suffolk IP32 6BB
t: (+44) 01284 700321

www.arimapublishing.com

List of Contents for PART-1

Prologue for **PART-1**

[0] **Chapter-0** : Why Projective Geometry ?
  [0.1] Using algebras $\Re^{\#}, \mathbb{C}, \mathbb{Q}$
  [0.2] Axiomatic basis for P^n
  [0.3] Cross-Ratio as Invariant
  [0.4] P^2 as an Algebraic Geometry
  [0.5] Homographies, $\Gamma(1\text{-}1)$, Harmonic ranges, and Involutions
  [0.6] Introducing a metric into P^2
  [0.7] Euclidean metric space.

[1] **Chapter-1** : Scales and Base Elements
  [1.0] Examples
  [1.1] Generating Elements
  [1.2] Role of Homographies, $\Gamma(1\text{-}1)$
  [1.3] Some Geometrical Optics
  [1.4] Involutions, Reflexive Measures
  [1.5] Geometrical Optics as entirely Reflexive.
  [1.6] Further examples of reflexive measures
  [1.7] Defining a **Physics, Theoretical Physics**
  [1.8] Metrical and non-metrical geometries
  [1.9] Using dimensional Analysis
  [1.10] Defining Physics-1
    Projective, Affine, Metric geometries
    Co-ordinatisation, **Eulerian** angles
    Duality ; Envelope and Point curves
  Figs. [1.1] ... [1.12]

[2] **Chapter-2** : Physics-2 ; Rigid Bodies and Statics
  [2.1] Centre of Gravity
  [2.2] Weights and the Law of the Lever (**Archimedes**)
  [2.3] Dual of the Lever; Triangle of Forces
  [2.4] Bow's notation ; (torque) moment of force
    Significance of Quaternion algebra ; vector algebra
  [2.5] Force Field and Work Function
    Conservative field ; Potential Function
  Figs. 2.1, 2.2a, 2.2b, 2.3

[3] **Chapter-3** : Physics-3 ; Particle Dynamics
  [3.1] Ordering time, motion
  [3.2] **Poincare-Lorentz-Einstein** transformations
  [3.3] **Newton's** law of motion
  [3.4] Gradient of "momentum" v. "time" curve
    Definition of Impulse
  [3.5] Gravitation re **Newton** and **Einstein**
  [3.6] S.H.M and dual systems
  Figs. 3.1, 3.2

[4] **Chapter-4** : Physics-4 ; Dynamical Systems
  [4.1] **Huyghens** and **d'Alembert**
  [4.2] Radius of Gyration
  [4.3] Polhode and herpolhode cones
  [4.4] Equations of **Lagrange** ; $L = T - V$
  [4.5] Equations of **Hamilton** ; Hamiltonian H
    Holonomic systems ; $H = T + V$
  [4.6] The Principle of Least Action $\delta \int_C T dt = 0$

Hamilton's Principle via $\delta \int_C L \, dt = 0$
- [4.7] **Pfaff's** differential form
  Covariant derivatives
- [4.8] Illustrating via planetary orbit
- [4.9] **Hamilton's** Contact Transformations
  Poisson Brackets
- [4.10] Small oscillations about a stable configuration
  Eigenvalues and eigenvectors

[5] **chapter-5**   : Physics-5 ; Heat and Gas Laws
- [5.1] Heat and Gas Laws ; Boyle, Charles, Berthelot, etc.
  Van der Waals equation;
- [5.2] Kinetic theories ; Use of **Physics-3**
- [5.3] First Law of Thermodynamics
- [5.4] Specific Heats, C_p, C_v ; Adiabatic changes

[6] **Chapter-6**   : Physics-6 ; Electrostatics
- [6.1] **Kelvin's** theory of images
- [6.2] Force field ; Inverse-square Law
- [6.3] Work of **Coulomb, Gauss, Poisson**
  $\nabla^2 V = 0$
- [6.4] Homography and the Potential barrier
- [6.5] Conformal mappings in Complex algebra $\mathbb{C}$.
- [6.6] Schwarz-Christoffel transformations
- [6.7] Other solutions to $\nabla^2 V = 0$
  Figs. 6.1 ... 6.4

[7] **Chapter-7**   : Physics-7 ; Magnetostatics
- [7.1] Magnetic "poles" and "dipoles"
- [7.2] Magnetic fields in Quaternion algebra, $\mathbb{Q}$
- [7.3] Magnetic field $\underline{\mathbf{H}}$ from the Quaternion potential
- [7.4] Magnetic shells ; $\Phi_P = \xi \Omega_P$
- [7.5] **Laplace's** equation ; Complex analytic functions
  Fig. 7.1

[8] **Chapter-8**   : Physics-8 ; Electro-Magnetic Dynamics
- [8.1] Work of **Ampere, Oersted, Faraday**
- [8.2] Quaternion algebra $\mathbb{Q}$
  Field quaternion $\underline{\mathbf{F}} = \underline{\mathbf{E}} + i\underline{\mathbf{H}}$
  Maxwell field equations
- [8.3] Invariance under **Lorentz/Einstein** laws
  The **Lorentz** force on a charged particle
- [8.4] Electromagnetic waves in Free Space
- [8.5] Plane E-M waves

[9] **Chapter-9**   : Physics-9 ; Fluid Mechanics
- [9.1] Compressible and Incompressible fluids
- [9.2] Hydrostatics
- [9.3] Principle of **Archimedes**
- [9.4] Centre of Buoyancy, Metacentre & Stability
- [9.5] Equation of Continuity in Fluid Flow
- [9.6] Stream Lines ; velocity potential
- [9.7] **Bernoulli's** equation
- [9.8] **Laplace's** equation ; c.f. Electrostatics
- [9.10] Comparison with Electrostatics
- [9.10] Allowing for Viscosity

[10] **Chapter-10**   : Physics-10 ; Wave Motion
- [10.1] Basic Equation
  d'Alembert's solution

- [10.2] Waves on a stretched string
  Progressive and Standing Waves
- [10.3] Sound waves ; Pitch and Timbre
  Fourier analysis ; Fundamentals & Harmonics
- [10.4] Surface waves on (eg) water
- [10.5] **Huyghens' Principle** ; Wave Theory of Light
  Fresnel and **Hughens**
- [10.6] Interference & Diffraction
- [10.7] Electro-Magnetic waves
Fig. 10.1

[11] **Chapter-11** : Physics-11; Invariants
- [11.1] Algebraic invariants w.r.t groundforms
- [11.2] Quadratic differential form
- [11.3] Projective invariants w.r.t groundform
- [11.4] Absolute Invariants
- [11.5] Covariants, contravariants, mixed concomitants
- [11.6] Differential operators ; **Aronhold** operator
- [11.7] Cogredient and contragredient Tensors
- [11.8] Differential forms in General Relativity Theory

[12] **Chapter-12** : Physics-12; General Relativity
- [12.1] **Einstein's** use of **Riemannian** metric
- [12.2] **Riemann/Christoffel** tensor
- [12.3] **Schwarzchild** metric
  Gravitational law of force ; c.f **Newton's** law
- [12.4] **Forsyth's** derivation of invariants
- [12.5] **Weyl's** attempt at Unified field Theory

[13] **Chapter-13** : Products of Algebras
- [13.1] Outer product of algebras **A** and **B**
- [13.2] Matrix representation
  Trace as an Invariant
- [13.3] The mapping $\pi : Q^2 \to Q$
- [13.4] Electromagnetic field $E + iH$
  Poynting vector
- [13.5] E-M field ; Quaternions over $\mathbb{C}$ viz. $Q(\mathbb{C})$
- [13.6] E-M field energy in $\mathbb{C} \times Q \times Q$
- [13.7] Homographies in **Clifford** algebra, $\mathbb{G} \times \mathbb{G}$
- [13.8] **Eddington's** Sedenion algebra
Fig. 13.1

[14] **Chapter-14** : Quantum Matters
- [14.1] Quantisation (various)
- [14.2] **Pythagoras'** black sheep
- [14.3] **Young's** slits via algebras $\mathfrak{R}^\#$ and $\mathbb{C}$
- [14.4] An experimental model of quantisation
- [14.5] **Planck's** quantisation of Action
- [14.6] **Bohr's** theory of hydrogen atom
- [14.7] Spectral series
- [14.8] Particle or Wave ?
Figs. 14.1, .1a, .1b, .1c, .1d, 14.2, 14.3

[15] **Chapter-15** : **Heisenberg's** Matrix Mechanics
- [15.1] **Hilbert Space**
  Eigenvalues/Eigenvectors
  Hermitian matrices
- [15.2] **Born, Jordan, Dirac** and **Poisson Brackets**
- [15.3] $PQ - QP = -ih/2\pi$

[16] **Chapter-16** : Quantisation in a Continuum
- [16.1] Uncountable numbers of axes
- [16.2] Derivation of $\mathbf{p}_{op}$ and $\mathbf{E}_{op}$
- [16.3] **Dirac's Space**
  - **bra** and **ket** vectors
  - Linear operators
- [16.4] Eigenkets in a continuum
- [16.5] **L. de Broglie**'s postulate
- [16.6] **Schrödinger**'s Wave equation
  - Solution for H-atom in (r,θ,φ)
- [16.7] **Schrödinger**'s solution for the Harmonic Oscillator

[17] **Chapter-17** : Cross-Ratios in a Field Theory
- [17.1] Extending the Projective definition of metric
  - Defining **poins**
- [17.2] Consequences for a Quantum Mechanics
  - Involutio and the Harmonic Oscillator
  - Metaphysics : question **MQ** and answer **MA**
- [17.3] Localised cross-ratio viz., $(d\xi\ 0\ d\eta\ \infty)$
  - Examples in Physics :
  - **Bernoulli's** equation
  - Energy Equation
  - **Hamilton's** equations
  - Equation of continuity in Fluid Mechanics
- [17.4] Applying **Planck's Postulate** to **action** variables
  - Energy operator $\mathbf{E}_{op} = (ih/2\pi)\ \partial/\partial t$
  - Momentum operator $\mathbf{p}_{op} = (h/2\pi i)\ \partial/\partial q$
  - Quantised energy values as $h\nu$
- [17.5] The operator **crop**
  - Applied to $d\mathbf{A}$ and $\mathbf{pdq}$
  - Momentum operator (again)
- [17.6] **Schrödinger's** wave equation
- [17.7] **Newton's** Law of Motion

. . . .

List of Contents for PART-2

Prologue for **PART-2**

[1] **Chapter-1** : Holes (Objects) in Fields
- [1.1] Base Elements - seeing objects
- [1.2] Example - Magnetic field
- [1.3] Observing generating elements
- [1.4] Values attached to pinholes
- [1.5] Analysis procedure
- [1.6] Notes on topology
- Figs. 1.1, 1.2

[2] **Chapter-2** : Chain Complexes & Homology Groups
- [2.1] Simplices $\{\sigma_p\}$, Chain Cmplexes
- [2.2] Convex Polyhedra in E^n
- [2.3] **Dowker** complexes
- [2.4] Simpices in Exterior algebra
- [2.5] ÇW-complexes
- [2.6] Cech complexes
- [2.7] Triangulation

[2.8] Homology groups in a complex
Figs. 2.1, 2.2, 2.2a, 2.3, 2.3a, 2.4, 2.5

[3] **Chapter-3** : Cohomology and Measures/Observations
    [3.1] Classical Acyclic backcloth
    [3.2] Mesh of Observations
    [3.3] Mappings on a Chain Complex
    [3.4] Duality between C^* and C_*
    [3.5] Examples from Vector Field theory
    [3.6] The Ring of Cochains
    Figs. 3.1, 3.2

[4] **Chapter-4** : The Cocycle law in Physics
    [4.1] **Heisenberg** Uncertainty Principle & Measures in a Čech complex
    [4.2] **Non-Hausdorf** space in Physics Measures
    [4.3] **Laurent** expansion in $\mathbb{C}$
    [4.4] **Laplace's** equation ; a Cocycle in a Potential Field $V(r)$
    [4.5] **Poisson's** equation as a Coboundary statement
    [4.6] **Cocycles** via the use of the **Calculus of Variations**
    [4.7] **Lagrange's Equations for a general dynamical system**
    [4.8] **Newton's Law of Motion**
    [4.9] **Hamilton's equations** for a dynamical system
    Figs. 4.1, 4.2

[5] **Chapter-5** : Cocycle laws in Exterior Algebra
    [5.1] One-to-one correspondences, $\Gamma(1\text{-}1)$
    [5.2] Cocycle Laws in **De Rham** cohomology
    [5.3] **Hamilton's** equations for a dynamical system
    [5.4] **Maxwell's** equations for the electromagnetic field
    [5.5] **Hodge** *-operator in Λ^n space
    [5.6] Equations of Wave Motion in **Lorentz** 4-space
    [5.7] **Schrödinger's** wave equation
    Fig. 5.1

. . . .

List of Contents for PART-3

Prologue for **PART-3**

[1] **Chapter-1** : **Gravity and the Cohomology Ring**
    [1.1] Direct product of complexes
    [1.2] Gravity defined by two finite bodies

[2] **Chapter-2** : **Relativistic Hamiltonian - Dirac's treatment**
    [2.1] Relativistic Hamiltonian for a Particle
    [2.2] **Hamiltonian** linear operator in **Dirac's** treatment
    [2.3] Quantum numbers for **spin** in **Dirac's** treatment
    [2.4] **Spin** and an **Involution** in **Projective Space**

[3] **Chapter-3** : **Lorentz-Einstein** 4-space : $\Gamma(1\text{-}1)$ in $\mathbb{C}$
    [3.1] Using Magnetic Field iH as Base Element 3_L

[3.2] Hypothesis-1
[3.3] Hypothesis-2

[4] **Chapter-4** : **Dirac's** Theory of Anti-Particles
[4.1] Matter and Anti-matter
[4.2] Calculating masses of particles/anti-particles
[4.3] Particle/Anti-particle Involution properties

[5] **Chapter-5** : **The Spectrum of Particles**
[5.1] Experimental Basis (Only) for Observing Particles
[5.2] Categories of Observed Particles
[5.3] Safeguarding Conservation Laws
[5.4] Websites for further information
[5.5] Some ideas to explore in the spirit of this thesis

. . . .

List of Contents for PART-4

Prologue for **PART-4**

[1] **Chapter-1** : **Basic Set Properties**
[1.1] Mathematical and Other Sets
[1.2] Subsets, Union, Intersection
[1.3] Hierarchy of Sets
[1.4] The Russell Paradox
[1.5] An Illustrative example
[1.6] Equivalence relation defined on a set
[1.7] Partitions and Quotient Sets
[1.8] Cartesian products of Sets

[2] **Chapter-2** : **The World of a Complex**
[2.1] The **Dowker** complex
[2.2] Embedding in **Euclidean** space
[2.3] A q-Connection relation
[2.4] Partition of a Complex via Q-Analysis
[2.5] Defining a p-Event
[2.6] Complexes representing **Time-Moments** and **Time-Intervals**
[2.7] **Birth**, **Death**, and the bit in-between
[2.8] **Komplex-Time** and **Clock-time**
[2.9] The relevance of t-forces

[3] **Chapter-3** : **Various Examples**
[3.1] A few of many **Dowker** arrays
[3.2] Suggestions for patterns of traffic on complexes
[3.3] Anne Chamberlain : A Study of Behcet's Disease
[3.4] J.H.Johnson : The Q-Analysis of Road Intersections
[3.5] J.H.Johnson : The Q-Analysis of Road Traffic Systems
[3.6] Beaumont & Gatrell : Examples in Dynamic Geography
[3.7] J.H.Johnson : A Contribution to Artifical Intelligence

[4] **Chapter-4** : **Dynamics** and **Complexes**
[4.1] Building a Complex
[4.2] Destroying a Complex - the Game of Nim
[4.3] NIM and Q-Analysis

[5] **Chapter-5** : **Tactical & Positional** play in **Chess**
        [5.1] Complexes arising in the game of Chess
        [5.2] The Immortal Game
        [5.3] **Steinitz** v **Mongredien** positions

[6] **Chapter-6** : **Probabilities** and **Events**
        [6.1] A Graded Pattern of Probabilities on Events in a Complex
        [6.2] Probabilities in the game of CRAPS
        [6.3] Loaded Dice in CRAPS
        [6.4] The Time Significance of the Grading
        [6.5] An Unexpected Event versus a Surprise Event

Appendices

[A] **Appendix-A** : Introduction to various algebras
        [A.0] Fields, Rings, Groups (examples of)
        [A.1] Integers **J**, Rational numbers $\Re$
             Real numbers $\Re^{\#}$
             Non-singular n x n matrices
        [A.2] Vector spaces and Modules
        [A.3] Algebras, Division algebras
             Elementary algebra
             Complex algebra $\mathbb{C}$
        [A.4] Quaternion algebra $\mathbb{Q}$
        [A.5] Exterior algebra Λ
        [A.6] **Clifford** algebras $\mathbb{C}$
        [A.7] Function spaces, field equations
        [A.8] Exterior derivative of p-Forms in Λ-space

[B] **Appendix-B** : Discussion of various geometries
        [B.0] Role of rational and irrational numbers
        [B.1] Projective geometry
             Duality
        [B.2] Perspectivities, Cross-Ratios & Conics
        [B.3] Algebraic geometry
             Homogeneous co-ordinates
             Line at Infinity, Λ_{∞}
        [B.4] Homographies
             Harmonic ranges
             Involutions
        [B.5] Introducing a Metric in Projective geometry
             Referring to an absolute conic
        [B.6] **Euclidean** geometry
        [B.7] Deducing the **Pythagorean** metric
        [B.8] **Riemannian** geometry
        [B.9] **Hilbert Space**
      Figs. B.1 ... B.10

[C] **Appendix-C** : Remarks re **Q-Software**
        [C.0] Introduction
        [C.1] Simple form of a Data File
        [C.2] Other useful Properties
        [C.3] Concept of a Pattern Π^* on a Complex K
        [C.4] Generating a Π^* from Π^0
        [C.7] When using a weighted matrix

[C.8] Concept of an Anti-Complex $\overline{K}$
[C.9] Data files for chessboard positions

[D] Appendix-D : Representing a **Dowker** complex in **Exterior Algebra**
    [D.1] The basis of vectors
    [D.2] Wedge product
    [D.3] General case
    [D.4] Representing a **Dowker** complex

[E] Appendix-E   References and Bibliography

Index

PART 1

Physics is Multi-Layered

PROLOGUE to Part-1

The Moving Finger writes; and having writ,
Moves on; nor all thy piety nor wit
Shall move it back to cancel half a Line,
Nor all thy Tears wash out a Word of it.

(We quote from Fitzgerald's translation of the "Rubaiyat" of **Omar Khayyam**)

Throughout this book we shall pursue the development of Mathematical Physics from its early days to modern times, bearing in mind the discipline has been based on the following paradigm, viz.,

(1) the geometric ideas inherited from the ancient Greeks, via the scholarship of **Pythagoras**, **Euclid**, et al - the geometry being known as **Euclidean Geometry**, and

(2) the algebraic system which was later introduced by Arabic mathematicians (**Al Jebr** in particular).

It has produced many positive results, as we shall see, but it has also become rather heavily laden with intricate functional thinking - which has required the invention of various branches of mathematics, particularly the inspired birth of the Differential and Integral Calculus with all its various offsprings.

We shall herein regard Mathematics as a **language** and its different parts as **dialects**, such dialects having grown out of the need to represent various new and specialist studies in Physics.

It is obvious that the development of Physics and of Mathematics have grown side-by-side ; that of Mathematics provided by an intellectual collaboration between those who pursue the language for its own sake (the Pure Mathematicians) and those who pursue the study of using the language for its applications (the Applied Mathematicians).

And now we come to the **major theme** in Part-1 of this book viz., that Physicists have overlooked that **elegant dialect** that grew up in Western Europe during the period of the 18th to 20th centuries, viz. that of

Projective Geometry.

In this book we shall attempt to remedy this - without any lack of respect for the traditional approach - by demonstrating the relative ease with which the projective approach can produce the basic Laws of Physics.

In Chapter-0 we shall begin by outlining the basic concepts of Projective Geometry - which will NOT involve a collection of abstract theorems, but which will lay out only the simplest basic ideas necessary for our analysis of physics.

Let us hope that there are probably a few academics who still remember studying the subject while at school and of needing it to pass their exams at university !

A gentle reading and reflection on the subject will see the reader through any initial anxieties he/she might have. Motivation will follow by noticing how many hard-won "Laws of Physics" fall out very naturally via this Projective approach.

We shall exhibit these matters, throughout the whole range of theoretical physics, by setting them side-by-side with the conventional/traditional approach.

But we cannot study Projective Geometry by simply looking at pictures : we must also use some suitable **algebra** as an accompanying dialogue, and this is when it becomes inovative and revolutionary.

For the moment let us anticipate just a few things from Appendix-A.

The algebra we learned at school, called **elementary algebra**, is an echo of the properties passesssed by **real numbers**. These we shall denote by $\Re^{\#}$ and these consist of all the naural numbers (integers), say **J**, as well as all the rational numbers (of the form p/q where p and q are members of **J** (q > 0, p any number in **J**), as well as all the so-called irrational numbers.

When it seems helpful the rational numbers can be represented by $\Re$.

The irrational numbers are things like $\sqrt{2}$, π, or the exponential number e.

We mention these matters because, although mathematicians have a legitimate right to use the irrationals - since they needed the solution of equations like $x^2 = 2$, to make the system more complete - nevertheless it can be proved that, for example, π and e are not the solutions of any such algebraic equations, and so are designated as **transcendental numbers** (since they transcend the others !).

The other thing that needs to be pointed out is that <u>in the laboratory</u> these irrationals cannot be precisely measured (observed). The only thing a physicist can honestly do is to approximate their values by ever closer rational numbers.

(This is the basis of **Dedekind's** definition of a real number)

However this need not stop physicists using all the symbols to be found in $\Re^{\#}$ and so we shall assume that $\Re^{\#}$ can be used as a proper illustration of Elementary Algebra.

Here the reader may benefit from looking at the Appendix-A where a more formal definition of Elementary Algebra can be found, as well as a discussion of other algebras.

In particular Appendix-A explains why Elementary Algebra, among others, is what is called a **division algreba**. Briefly this means that the algebra contains the solution of an equation like $\qquad ax = b$
since we can "divide" b by a and write $x = b/a$ --- not as trivial a matter as one might think. It is explained by noting that the "a" possesses an **inverse** (relative to the operation of multiplication) which is written a^{-1}, or $1/a$, and called a **multiplicative inverse**. Such an inverse is a mate of the member "a" of the algebra satisfying the relation $a^{-1}a = aa^{-1} = 1$, and so the solution of the above equation is formally obtained by writing

$$ax = b \longrightarrow a^{-1}ax = a^{-1}b \longrightarrow 1x = a^{-1}b \longrightarrow x = b/a$$

Seems a bit involved (?) but neat (?).

The significance of such an inverse becomes more apparent whenever we move to consider other stra algebras - which we certainly need to do.

Of these we shall find that Projective Geometry makes use of the following three algebras,

(a) Elementary algebra $\Re^{\#}$, of course,

(b) the algebra of **Complex numbers**, $\mathbb{C}$,
being of the form $z = x + iy$; $i^2 = -1$, x and $y \in \Re^{\#}$

(c) the algebra of the **Quaternions**, $\mathbb{Q}$, usually taken over the reals $\Re^{\#}$
but sometimes taken over the complex numbers $\mathbb{C}$.
This algebra (which is a division algebra) contains within it a subring which is known as **vector algebra** - although this has limitations.

A division algebra plays a dominant role in Physics, as do the real numbers $\Re^{\#}$, and it can be shown that there are only three associative division algebras over the reals, viz., $\Re^{\#}$, $\mathbb{C}$, and $\mathbb{Q}$. It is hardly surprising therefore that these will dominate our discussion. If in any doubt a reading of Appendix-A is meant to be helpful.

We shall also, in Chapter-0, try to explain the profound differences between Projective Geometry and Euclidean Geometry - the latter being what we have all been (more or less) schooled in.

By looking at the List of Contents the reader will see the extensive list of the topics covered in the book. He/She may be surprised to see that there seems to be a number of specialist Physics (plural). But this is how the author sees the subject, and it roughly corresponds to the historical development of the discipline, though the topics are by no means in a serial order - most of them overlapping with one or other of them, and each of them corresponding to a **general definition** of what a **Physics** really is. We hope that it diminishes the tendency to be confused by trying to understand it all at once.

The book is also written in four parts, there being an introductory Prologue for each part. This enables us to see into a possible future for the mathematics of theoretical Physics via insights provided by modern topological ideas ; the fourth part goes beyond the laboratory and sees the world as a **digital** structure.

This also gives the reader the chance to sit back and change gear, if he/she so wishes.

Enjoy !

CHAPTER-0 / PART-1   Why Projective Geometry?

[0.1] Using algebras $\Re^\#$, $\mathbb{C}$, $\mathbb{Q}$

As a part of the **language** of mathematics, geometry is an **abstract exercise** which exists independently of any physical representation of the concepts. For this reason, as mathematicians commonly require, drawing pictures to represent these concepts is merely a tool (albeit often a very useful one) to help the imagination in its perambulations. It has therefore been extremely helpful to use an algebra as a dialogue to discuss the properties of that geometry. Consequently we shall see how the introduction of **algebraic geometry** is an essential step in the development of the subject, and in this connection we need to say something about the introduction of what are called **co-ordinates** ; and these must be symbols of, and expressions in, the dialogue (the algebra). An essential step is that of providing co-ordinates to points, lines, curves, etc..

Naturally in P^n we use any one of the three division algebras already mentioned - and more carefully described in Appendix-B. In studying the geometry itself $\Re^\#$ is the usual choice but when it comes to Physics we shall see that the choice of algebra not only varies but also dominates the the observations.

It was **Von Staudt** who successfuly showed that setting up co-ordinates in P^2 does not require the use of a metric in the Space and the frame of reference can be imagined to be a so-called **triangle of reference** (see below) : compared with the fixed sytem of axes required by **Euclidean Geometry**, and where the algebraic geometry is that introduced by **D'Escartes**.

[0.2] Axiomatic basis for P^n

After the great achievements by the Ancient Greeks (typified by **Pythagoras**, **Euclid** and **Appollonius** etc) there was another startling advance by the Europeans in the 18th and 19th centuries. Pionering work by (eg) **Poncelet** provided the **Propositions of Incidence** as basic axioms for Projective Geometry, viz.,

> (i) the projective plane, P^2, consists of relations between two undefined entities - **points** and **lines**.
>
> (ii) two distinct points define a unique line.
>
> (iii) two distinct lines define a unique point.

and in 3-dimensions, P^3, the entities are called **points**, **lines**, and **planes** and these satisfy the extended propositions of incidence, viz.,

(i) two distinct lines define a unique point

(ii) two distinct points define a unique line

(iii) a line and a plane define a unique point

(iv) a line and point define a unique plane

(v) two distinct planes define a unique line.

It follows, for example, that any three non-collinear points define a plane.

We shall find most results, but not all, which are relevant to this study can be expressed in $\mathbf{P}^2$.

A noticeable absence in a projective geometry is the concept of **parallelism** - so <u>any</u> two distinct lines meet in a point.

Notice, too, that if we complain about this - because of what we think we observe - then we have really drifted into the realm of **Physics** (where <u>observation</u> is supposed to be experienced). In spite of this, and because the business of **observing** has yet to be analysed, we shall see that the properties of $\mathbf{P}^n$ are highly relevant to a **Physics**.

An equally important feature of these basic axioms of projective geometry is the absence of the idea of **distance** (or **metric**). This does not mean that a metric cannot be introduced into the geometry, as indeed it can, but then it is more useful and contains within it a degenerate form, viz., that of the **Euclidean** case - in which **distance** is defined via the theorem of **Pythagoras**.

It also follows from the propositions of incidence that there is a natural **duality** to be found among theorems. For example, in $\mathbf{P}^2$, a theorem about lines and points gives rise to a **dual theorem** about points and lines : that is to say, we need only replace "point" with "line" - and "line" with "point".

In a similar way, in $\mathbf{P}^3$, we need only replace "point" with "plane" and "plane" with "point" to obtain a **dual result** : in this case "line" is **self-dual** (v. figs B.1, B.2).

[0.3] Cross-Ratio as Invariant

As its name implies **Projective Geometry** is centred on the idea of **projection**, and this is a generalisation of the idea of **perspectivity** - one figure being in **perspective** with respect to another if the joins of corresponding points are concurrent at, say **V**, the **vertex** or **centre of perspective**. The dual of this being when the meets of corresponding lines are collinear on, say the line v; this being the **axis of perspective**.

A series of perspectives (via distinct centres V_r) is called a **projectivity** (v. fig B.3).

In Physics we are particularly interested in any **invariants** in P^2 which are preserved under this process of projection, and the most important of these is that of **cross-ratio** of any four points (or lines) related by a projectivity.

The reason for this is that what are to become the Laws of Physics must be independent of the different ways of observing events. Since we shall assume that any observer is to be identified with the centre of projection, the vertex V, in a typical diagram - we can the better interpret such a "V" as the **Viewer** of the events. Then the rays (lines) centred on V denote the paths of whatever signal is involved in the observation.

For example (v. fig B.3) the four points A_1, B_1, C_1, D_1 on the line p_1 have an associated cross-ratio written as $(A_1\ B_1\ C_1\ D_1)$ and, however we characterise it algebraically, this is to be equal to the cross-ratio of (eg) $(A_2\ B_2\ C_2\ D_2)$ on the line p_2, via a perspectivity from V_1, and thence equal to the cross-ratio of the range $(A_3\ B_3\ C_3\ D_3)$, via the perspectivity with vertex V_2.

When we define parameters with which to co-ordinatise the lines, or **rays**, of a pencil we also equate their cross-ratios with the above - so the duality holds.

A <u>significant property</u> of **projectively related** pencils of lines, that is to say those which have equal cross-ratio values, is the following.

Given two vertices S_1 and S_2 with associated pencils of lines, (a_1,b_1,c_1,d_1) and (a_2,b_2,c_2,d_2), which are projectively related, then the intersection points of corresponding lines (rays) viz., P_a, P_b, P_c, and P_d all lie on a **point-conic** which also contains the vertices S_1 and S_2. [ref Appendix-E [2] [3] [7] [43] [44]].

This suggests that the observation of the conic sections (mostly ellipses) will be a frequent experimental experience.

[0.4] P^2 as an Algebraic Geometry

Moving into **algebraic geometry** we introduce "co-ordinates" as follows. [v. Fig B.4]

Let X, Y, Z constitute a **triangle of reference** and let co-ordinates be such that the point X has co-ordinates $(1,0,0)$, the point Y has co-ordinates $(0,1,0)$ and the point Z has co-ordinates $(0,0,1)$. It is also postulated that, for any non-zero $k \in \Re^{\#}$, the symbolic points **kP** and **P** are the same point. This means that it is to be the **ratios** of the co-ordinates, viz., x/z and y/z which will be significant. So that for example the points (x,y,z) and (kx,ky,kz) represent the same point - often expressed as $kP = P$ for any point P. The triad $(0,0,0)$ is excluded from the scheme - since $0/0$ cannot be interpreted. And when we allow **z** to be the number 1 we arrive at our "normal" **Cartesian** system used in the **Euclidian** geometry.

When we do not use this ratio the algebric geometry is said to be in the **homogeneous** mode. For example, in this mode the equation of a line referred to the usual axes becomes

$$ax + by + cz = 0 \quad \text{the symbols being in } \Re^{\#}.$$

and shifting to the **Euclidean** case this relates to

$$ax/z + by/z + c = 0 \quad \text{or} \quad ax + by + c = 0.$$

Any **line**, λ say, in $\mathbf{P}^2$ is defined by the **homogeneous equation** $ax + by + cz = 0$ and is said to have **line co-ordinates** (a, b, c) - which contains the **point** (x,y,z).

Lines which, in traditional language, are called **parallel** will have eqations like $ax + by + cz = 0$ and $ax + by + c^*z = 0$. Consequently they will meet where these equations are simultaneously true - which is where $(c - c^*)z = 0$. If the lines are distinct it follows that $z = 0$. Since in Cartesian terms this means that they meet at $(x \div 0, y \div 0)$ we call this a **point at infinity** and the line $z = 0$ as the **line at infinity** (denoting it by Λ_∞). [These concepts are required by the propositions of incidence, since <u>any</u> two lines must meet in a point]. We also see that by projection any point on a line can be projected onto Λ_∞ - [v. Fig B.5 following] - so such a point is accessible in a commonsense way, since **if we can project a finite point to infinity then we can reverse the procedure and bring the point at infinity back to a finite position.**

[These matters are fully discussed in any of the following references in Appendix-E : [2], [3], [4], [43], [44]]

Following the work of **von Staudt**, in particular, we can demonstrate how the introduction of algebraic "co-ordinates" in the projective plane $\mathbf{P}^2$ can give rise, by simple constructions which do not involve any concept of distance, representations of all the usual algebraic operations. A full discussion of this point will be found in **H.F.Baker's** excellent work [v. Appendix-E Ref. [2]] - and we demonstrate this in Figs. B.01 & B.02.

Returning to our need for invariants what is called the **cross-ratio** of four distinct points

A,B,C,D and which are such that no three of them lie on a line is written as (A B C D).
If "co-ordinates" a, b, c, d are attributed to these points then the cross-ratio of this range
is to have the value $\quad$ (a-b)(c-d) ÷ (a-d)(c-b)

Clearly we must ensure that the co-ordinate values are taken from a **Field** - so using $\mathfrak{R}^{\#}$
becomes almost (but not quite) essential, $\mathfrak{R}$ would serve equally well.

In **Euclidean** geometry it will accur as directed lengths AB... and be calculated as

$$(A\ B\ C\ D) = AB.CD \div AD.CB$$

[To remember: arrange the letters around a circle, then go round it twice, first clockwise and
secondly anti-clockwise - each time starting with A (or **a**)].

For example : $\quad$ (2 5 1 0) = (2-5).(1-0) ÷ (2-0).(1-5) = -3 ÷ -8 = +3/8

and another : $\quad$ (2 0 1 ∞) = (2-0).(1-∞) ÷ (2-∞).(1-0) = (2-0) ÷ (1-0) = 2

since adding or subtracting anything to/from ∞ has no effect on the latter.

In a similar way this discussion equally applies to a pencil of four lines, say λ_1, λ_2,
λ_3, and λ_4 with vertex V : assuming they have "co-ordinates" c_1, c_2, c_3, c_4
(such as the four possible values of a suitable parameter), then the pencil has a cross-ratio
($c_1\ c_2\ c_3\ c_4$) defined as above.

[0.5] Homographies, Γ(1-1), Harmonic ranges, and Involutions

These play an important part in their appearances throughout the Physics to be discussed in
the following chapters.

If one range of points on a line λ_1 are projectively related to another range of points
on a line λ_2 then it is clear that the separate points are in a **one-to-one correspondence**.
We denote such a correspondence by the symbol Γ(1-1) and in the algebraic geometry this becomes
an algebraic expression which is linear in each variable. Thus if **x** denotes the co-ordinate
of any point on λ_1 whilst **y** denotes the same thing for the **corresponding point**, on λ_2, in
the projectivity, the expression (called a **homography**) is

$$axy + bx + cy + d = 0 \quad \text{the symbols being in } \mathfrak{R}^{\#}$$

where ac-bd ≠ 0 (otherwise the expression factorises and is degenerate).

The points **A,C** are called **mates** in the homography - as are the points **B,D**.

It is not difficult to prove that, by simple substitution from the Γ(1-1), the cross-ratios
$\quad (x_1 x_2 x_3 x_4) \quad$ and $\quad (y_1 y_2 y_3 y_4) \quad$ are equal in $\mathfrak{R}^{\#}$.

So, in **algebraic geometry** we have an **invariant** (viz., cross-ratio) under such a Γ(1-1).

In a given $\Gamma(1\text{-}1)$ although an **x** might correspond to a particular **y** it will not necessarily follow that the particular **y** corresponds to the original **x**. In other words it is not always a **symmetric** homography. This will occur whenever $b \neq c$.

In the event that (i) the ranges lie on the same line λ
(or, dually, rays lie on the same point **V**)

and (ii) the homography is symmetrical, i.e. $b = c$,

the homography is called an **involution** (on the range of points, or on the pencil of rays).

When the cross-ratio (**abcd**) = -1 we call the range (or pencil) **harmonic** so that, under a projectivity a harmonic range (or pencil of rays) maps into another harmonic range.

[The name "harmonic" derives from the following fact :
Taking **A** as an origin and mapping points **B,C,D** into co-ordinates b,c,d we can show that the harmonic condition (**ABCD**) = -1 gives the relation $2/c = 1/b + 1/d$ - showing that b,c,d are members of a harmonic series].

In an **involution** we require that the **interchange of mates** AC → CA (BD → DB) leaves the cross-ratio unaltered - because of the symmetry. That is to say that, in an **involution**

$$(ABCD) = (CBAD) = (ADCB).$$

But it is shown in Appendix-B that if (**ABCD**) = λ then (**CBAD**) = λ^{-1} giving $\lambda^2 = 1$

Since $\lambda = +1$ corresponds to a degeneracy we must have $\lambda = -1$.

When A B C D is an involution then its cross-ratio (ABCD) = -1, an harmonic range.

The homography then looks like

$$axy + b(x + y) + c = 0 \quad \text{the symbols being in (say) } \Re^\#.$$

Special cases arise by taking various values for the coefficients **a,b,c** (v. later chapters).

In an **involution** an important role is played by the existence of its **double points**, that is to say those points for which $x = y$. These will be given by the equation

$$ax^2 + 2bx + c = 0$$

and if we are given its roots, as α and β, then the equations

$$a\alpha^2 + 2b\alpha + c = 0 \quad \text{and} \quad a\beta^2 + 2b\beta + c = 0$$

enable us to solve for the ratios b/a and c/a - thus defining the involution.

So an involution is determined by its two double points.

Also, if we take an origin O such that points **A,C** have co-ordinates $-a, +a$ whilst **B,D** have co-ordinates **b,d**, it follows from (-a b a d) = -1 that $bd = a^2$ or in terms of metrical geometry $OB.OD = OC^2$.

This situation is illustrated in the case of **inverse points** with respect to a circle.

Figures B.8 and B.9 illustrate some of these ideas.

B and **D** are defined as inverse points with respect to the circle σ if

OB.OD = OC² = (radius)², and consequently **B**,**D** are mates in the involution defined by the double points **C** and **C***. It is also a theorem of Euclidean Geometry that any circle through the points **B** and **D** cut the circle σ orthogonally.

[Note: All these concepts and definitions (except the ideas of parallelism and the line at infinity, Λ_∞) can still apply in a metrical geometry such as **Euclidean geometry**. But in that case the factors in the cross-ratio expression, viz., $(x_1 - x_2)$ etc. will be interpreted as **distances** or **angles**. In many textbooks this is how they are introduced. It is only in ,eg, P^2 or P^3, that "distance" is not a basic need.]

[0.6] Introducing a metric into P^2

We can show that P^2 (and P^3) contains the well-known **Euclidean Geometry** by showing how a **metric** (concept of **distance** or **interval**) can be introduced.

First we need to list the properties of such a thing. It is in fact a mapping of **P**×**P** into an algebra **A** : usually **A** may be the complex numbers $\mathbb{C}$ (of which $\Re^\#$ is a sub-algebra).

So the **metric** (distance) function d : **P**×**P** → **A** satisfies the following conditions.

(i) d(**P**,**P**) = 0

(ii) d(**P**,**Q**) > 0 and d(**P**,**Q**) = - d(**Q**,**P**) whenever **P** ≠ **Q**

(iii) d(**P**,**R**) ≤ d(**P**,**Q**) + d(**Q**,**R**) for all triads **P**,**Q**,**R**.
   equality occurring when the points are collinear.

If **O**,**U** are points on a line λ and **P**,**Q** are any two other points on λ we can define the distance d(**P**,**Q**) by

d(**P**,**Q**) as (1/2i)ln(**POQU**).

[Note: This was first introduced by the French mathematician **Laguerre** (Nouv. Annal. xii, 1853)]

Since it can easily be shown that, generally,

(**POQU**) * (**QORU**) = (**PORU**)

we deduce that this gives d(**P**,**Q**) + d(**Q**,**R**) = d(**P**,**R**) + nπ, n being a positive integer.

[Note: In $\mathbb{C}$ the function $\ln(re^{i\theta}) = \ln(r) + i(\theta + 2\pi n)$, n being a +ve integer]

We must appreciate that "distance" between points **P**,**Q** in the plane is no longer an absolute property of the two points, as in Euclidean Geometry, but depends on the choice of what are usually called **absolute points** (following **Cayley**) .

These points of reference are found by asserting a **basic**, or absolute, **conic** (often a circle) whose intercepts on a line joining **P**,**Q**, say **I** and **J**, and which allow us to specify the cross-ratio needed in the above definition of distance. [It is sometimes convenient to take the points **I**,**J** as the two points intercepted (by every circle) on the line at infinity, Λ_∞.]

In a similar way, if **V** is a vertex, **o,u** two arbitrarily chosen rays through **V** and **p,q** two other rays through **V**, the **angle** between **p** and **q** can be defined by the expression $\angle(\mathbf{p},\mathbf{q}) = (1/2i) \ln (\mathbf{poqu})$. The lines **o,u** are identified as the two tangents from **V** to the chosen **absolute conic**.

[Further discussion of this topic can be found in Appendix-E [2],[43],[44],[45]]

But suffice it to say that, following the work of **Laguerre** and **Cayley**, we can show that metric notions can be introduced into $\mathbf{P}^2$ (and $\mathbf{P}^3$) which result in a geometry transcending the well-known **Euclidean Geometry**.

[A fuller discussion of the this topic can be found in Appendix-E [2], [4], [43] [44]]

[0.7] Euclidean metric space

Euclidean geometry possesses a **metric** which is usually called the **Pythagorean Metric** - since it imposes the famous **Pythagoras' Theorem** on the space as a "natural" measure of distance in the plane. In this space the incremental distance between the point (x,y) and $(x+dx, y+dy)$ is written as

$$ds^2 = dx^2 + dy^2$$

where eg dx^2 means $(dx)^2$.

This naturally assumes that the concept of a "right-angled Δ" is admitted and that the real numbers $\Re^{\#}$ are to be used in the geometry.

This metric, together with the notion of **parallelism**, constitutes the essence of **Euclidean Geometry**. Other concepts and properties of $\mathbf{P}^2$ (or $\mathbf{P}^3$) are accepted which do not involve any of these special additives.

In this geometry **distance, angle**, and **congruence** are **invariants** under the general group of **linear transformations**.

[Note : We assume that the reader is knowledgeable about the common theorems of this Euclidean Geometry].

Following the notation and argument found in Ref. Appendix-D [2] we can show that this **Euclidean** metric can be contained within the structure of projective geometry.

Chapter-1 /PART-1   Scales & Base Elements

[1.0] Examples

In the realms of **Physics** we shall regard all observations as being defined by members of a **Scale**, say $\mathfrak{C}$ (or $\{\mathfrak{C}_r\}$, when there are more than one). And associated with such a Scale, and being an essential part of it, we shall require a **Base Element**, say $\mathfrak{Z}$ (the cyrillic zed), or, again the set $\{\mathfrak{Z}_r\}$, whenever it is appropriate to consider more than one.

Without a **Base element**, or **seeing/observing agent**, we cannot observe anything.

Thus a **Useable Scale** will be the combination

$$U\mathfrak{C} \equiv \{\mathfrak{C} \,;\, \mathfrak{Z}\}$$

and often abbreviated to $\mathfrak{C}(\mathfrak{Z})$ in simpler cases.

Some examples of **Base elements** are :

   (0) Light signal, $\mathfrak{Z}_L$

   (1) Sense of Ordering, or of Time, $\mathfrak{Z}_T$

   (2) Electric Charge, $\mathfrak{Z}_Q$

   (3) Electric Field, $\mathfrak{Z}_E$

   (4) Magnetic Field, $\mathfrak{Z}_H$

   (5) Sound Frequency, $\mathfrak{Z}_F$

   (6) Gravitational Field, $\mathfrak{Z}_G$

   (7) X-rays

which do not exhaust the possibilties.

Some examples of **Scales** are :

   (a) Linear range/marker eg, Ruler, Thermometer

   (b) Marked Dial (rotatory)

   (c) Digital Dial

   (d) Video Screen

   (e) Musical scale

   (f) Spectrum (eg of colours)

   (g) Photographic Plate

   (h) Spectrometer

   (j) Clock (various)

In practice **Useable Scales** will involve more than one physical structure and become an

experimental set-up.

Also we primarily expect Scales to offer measures of elements, say E_r, <u>other than</u> those of the Base Element(s). In this normal kind of experimental observation we are trying to capture and measure our **intuitive sense** of the phenomena - because the experimental association of the scale measures with this sense is the only way we can "identify" the phenomena. This means that, in reality, we are trying to **define the phenomena** by way of the **measures** of them. This is why it is essential for the experimentalist to try and isolate the phenomena, as far as possible, and so to restrict the scale action that he can produce a suitable mapping

$$\gamma : E_r \to A \quad : A \text{ being some suitable Algebra. [v.Appendix-A]}$$

This process of **observation**, and its expression in mathematical terms, has usually been interpreted via an **injective mapping** into a **Field, F**, - where F has usually been the set of **real numbers**, $\Re^{\#}$.

A Scale which is dedicated to observations of, say E_0, should therefore be specified as

$$\{\mathfrak{C} \; ; \; \mathfrak{Z} \; ; \; E_0\} \quad \text{or} \quad \{\mathfrak{C}(\mathfrak{Z}) \; ; \; E_0\}$$

[Note: An injective mapping, say $\gamma: A \to B$, is such that, for every $a \in A$ there
exists a unique $b \in B$ where $\gamma(a) = b$: but it does not guarantee that for any
$b \in B$ there is an $a \in A$ for which $\gamma(a) = b$. Such a mapping is therefore an
<u>into</u> mapping, and not necessarily an <u>onto</u> mapping]

In this we have already distinguished the **real numbers,** $\Re^{\#}$ from the **rationals,** $\Re$, because the experimentalist can never directly observe any of the **irrational numbers**.

This is why we can only speak of an **injective mapping** (not a **surjective mapping**) into the reals $\Re^{\#}$. Only with this **strong reservation** can we allow the physicist to continue to use the real numbers $\Re^{\#}$ - but we do so allow!

With this in mind we can describe a traditional **useable scale U$\mathfrak{C}$** as

$$U\mathfrak{C} \equiv \{\mathfrak{C}(\mathfrak{Z}) \; ; \; E_r \; ; \; \Re^{\#}\}$$

meaning also that we assume there exists an **injective mapping**, say γ, from the measures of E_r <u>into</u> the field $\Re^{\#}$.

[1.1] The Generating Elements

The results of any experimental observations on a scale $\{\mathfrak{C}(\mathfrak{Z}) \; ; \; E \; ; \; A\}$ we will call the **measures** of E and these are located in the **algebraic structure, A** ; E being called a **generating element**.

We are therefore inclined (or unwittingly forced) to attribute the structure of **A** to the measures of the generating element **E**.

This may or may not be a good thing, but it plays an essential part in constructing what we like to think of as the "Laws of Nature".

During our discussion of these matters we shall find it natural to speak of **a Physics** rather than just "Physics".

It will hardly be anticipating the discussion to suggest at this stage that the **content** will be defined entirely by the Base Elements $\{\mathfrak{Z}\}$ whilst the the **form** (or "structure" of the Physics) will depend solely on the above Scales $\{\mathfrak{C}\}$ peculiar to the observations together with the nature of the chosen algebra **A**.

[1.2] Role of Homographies, the $\Gamma(1\text{-}1)$

If we are to agree on

   (i) the recognition of at least one Base Element $\mathfrak{Z}$, and

   (ii) the search for phenomena associated with our intuitive element **E** and the setting up of Scale(s) $\{\mathfrak{C}(\mathfrak{Z})\ ;\ \mathbf{E}\ ;\ \mathbf{A}\}$ to observe this element then we must look for the essential features of such observations.

Reflecting on one of the simplest of physical experiments - and assuming that we are in a "pre-Newton" and a "pre-Hooke" environment - viz., trying to understand our intuitive idea of "mass", by using a <u>spring balance</u>, then we can see that **observing** this **E** on a scale $\mathfrak{C}(\mathfrak{Z}_L)$ means :

(a) regarding $\mathfrak{C}$ as ofering us a whole set of **states**, or conditions ;

(b) demanding that these states be associated with our **E** ;

(c) setting up the algebraic/arithmetical symbols which correspond to these states

   via an injective mapping (v. supra).

Then we map these states into a geometrical linear space and attribute numbers to them. Hanging nothing on the spring gives us a measure we map into the zero element of $\mathfrak{R}^{\#}$; increasing the "mass" eventually gives us a limiting measure which corresponds to the loss of elasticity of the spring (it fails to return to its neutral condition). This corresponds to ∞ on the linear space. Otherwise each thing possessing "mass" occurs with co-ordinates (on the line) $\pm x$. Taking the $+x$ as a correlation with our supposed "mass" we are able to compare, say, $\mathbf{m}_1$ with $\mathbf{m}_2$. We can get no further with this ; no **absolute** sense of what "mass" is. We would therefore naturally proceed to map some particular "mass" into the state with co-ordinate = 1 and then be able to identify the measures of "mass" with the cross-ratio $(-x, 0, +x, \infty)$.

In Physics we can only hope to find "laws of nature" if we can observe **invariants**, as we move from one Scale of measures to another. This means that if our measures are those of cross-ratios then the invariant properties observed (among a set of Scales) will always lead us to observe **conic sections** [v. Appendix-B §[B.2]]. This might well suggest that we embrace what we might call the **ubiquitous conic section** as a sine qua non in all our observations [see the rest of this thesis]. It also suggests that **identifying a (1-1)-correspondence, Γ(1-1)**, is already a pointer to a "Law" in that physics since it encompasses the invariance of the **cross-ratios** involved in the observations. Thus we shall notice that a discovered Γ(1-1) between two measures of, say, "x" and "y" - whatever they may be - and where **we have found the cross-ratios of their respective quadruple occurrences to be equal** then that Γ(1-1) will itself be eligible to be caste as in the role of a **Law of Physics**.

This also means that we need to consider measures associated with FOUR points (or lines) as discussed at some length in the reference [Appendix-E [5]].

We notice, for example, that changing from one Scale $\mathfrak{C}_1$ to another $\mathfrak{C}_2$, but measuring the same "mass" was later catered for by Hooke's introduction of "Hooke's Law" - which characterised each Scale by a multiplicative factor (a constant of "elasticity", λ). It then appeared that the measures attributed to the "mass", viz. $\mathbf{m}_1$ and $\mathbf{m}_2$, become related by a simple factor, giving us something like $\mathbf{m}_1 = k\mathbf{m}_2$, where $k \in \Re$.

Although this is a particularly simple case it is significant to notice that this is a special case of a (1-1)-correspondence, Γ(1-1), which, as we have already introduced, is of the general form

$$\Gamma(1\text{-}1) \equiv axy + bx + cy + d = 0 \quad \text{the symbols being in the field } \Re^{\#}$$

by taking $a=0, d=0, c/b = -k$ $(b \neq 0)$ - which therefore already contains **Hooke's** Law.

The measures found in the observations are referred to as extensions and the ordering of their values (in $\Re^{\#}$) allows us to distinguish one "mass" from another.

After Newton's work we would now say that by this experiment we say that we are observing "weight" rather than "mass".

[1.3] Some Geometrical Optics

The basic laws of **geometrical optics** can be found in the involutions of the **reflexive measures** associated with the Base Element $\mathfrak{Z}_L$.

Example-3.1

A source of light can be observed through a plane sheet of glass or through a crystal of calcite. If the correspondence in the first case is a $\Gamma(1\text{-}1)$ then it is definitely a $\Gamma(1\text{-}2)$ in the second.

Example-3.2

If the light source, in Example-1, is observed through **Young's Slits** then the correspondence becomes a $\Gamma(1\text{-}n)$, for large n.

This strongly suggests that the role of a base element, $\mathcal{B}$, in a scale $\mathfrak{C}$ is so fundamental that we should specify it when describing the scale. Then we should keep any measures of another element, say **E**, separate and dependent on $\mathcal{B}_0$. So we specify such a scale by

$$U\mathfrak{C} \equiv \{\mathfrak{C}(\mathcal{B}) \,;\, E \,;\, A\}$$

when speaking of **measures of E** in an algebra, say $A \equiv \mathfrak{R}$ or $\mathfrak{R}^\#$.

[1.4] Involutions, Reflexive measures

Certainly a very important property of a $\Gamma(1\text{-}1)$, in a division algebra **A**, is the **invariance** of the **cross-ratio** of a projectively related range of four elements, whether they be points on a line (or some other one-dimensional set, such as the points of a conic) or rays (lines) on a point - or tangents defining an envelope curve.

The characteristics of a scale $\mathfrak{C}$ will be decided by the role which is played by its **Base element**, $\mathcal{B}$. It will be the $\mathcal{B}$ which gives **structure** to the scale - for the observable element, **E**, on it.

It would follow that if we insist on **observing $\mathcal{B}$ itself** on the scale $\mathfrak{C}(\mathcal{B})$ we must expect to find something "incestuous" about the measures - and the first thing we would expect to find is an apparent **lack of ordering** of the observations.

Definition-D1 : The measures of a base element $\mathcal{B}$ on a scale $\mathfrak{C}(\mathcal{B})$ will be called

Reflexive Measures - associated with the scale $\mathfrak{C}(\mathcal{B})$.

Now we can see that, when observing these reflexive measures (of $\mathcal{B}$), there will be observables **X,Y** which we cannot distinguish from **Y,X**. So in any invariant cross-ratio **(PXQY)** we shall equally well "see" **(PYQX)**. But this means that, in a representative division algebra (where we map the things **P,X,Q,Y** into **A**) we must observe that

$$(PXQY) = (PYQX)$$

and this means that $(\mathbf{PXQY}) = -1$ a **harmonic range** (v.Appendix-B).
and the measures of the **Base Element** $\mathcal{3}$ appear as **mates** in the **involution**
defined by the **double points**, say, **I** and **J**.

Definition-D2 : If **I,J** are self-corresponding (double) elements among the pairs of
$\mathcal{3}$-measures on the scale $\mathfrak{C}(\mathcal{3} \,;\, \mathcal{3})$ then all other pairs occur as
mates in the involution defined by **I,J**.

[1.5] Geometrical Optics as entirely Reflexive

These definitions find a ready expression in the basic laws of **geometrical optics**.
Indeed it is not an exaggeration to say that **geometricl optics** is entirely defined in terms
of **reflexive measures** on scales $\mathfrak{C}(\mathcal{3}_i)$.
The following examples illustrate this.

<u>Example-5.1</u> If the algebra is the field $\mathfrak{R}^{\#}$ and if we assign co-ordinates $0, \infty$ to
I,J then observations **P,Q** occur with co-ordinates $+x$ and $-x$.
This occurs when a plane mirror is used to observe pairs of light measures.
If the **x** is regarded as a distance (as in Euclidean geometry) we deduce
the "Law", viz., object and image are equally distant from the mirror.
The **dual** scale gives the simple **law of reflection** at a surface.
[Refer to Fig.1.1 and its **Dual** Fig.1.2].
If we take account of experimental data, we notice that the co-ordinates u, v
in the above example need to be associated with characteristics of the **medium** in
which the measures are taken. This is achieved by attributing some **scaling factor**,
say μ, to the co-ordinates. So if, for example, the measures of **v** in Fig. 1.1,
are found in a region-2, this characteristic factor is assumed to be some constant
dependent upon the interaction between the light rays and the medium.
This means that we should replace **v** by the (projectively equivalent) variable
μv. The involution then gives us $(\mu v \, 0 \, \mathbf{u} \, \infty) = -1$, which results
in $\mathbf{u} = -\mu v$. Simple experiments give examples where $\mu > 1$.
[v. Fig 1.5]

<u>Example-5.2</u> If **I,J** are defined by $x=0$ and $x=r$ (r finite in $\mathfrak{R}^{\#}$) then

$P(x=u)$ and $Q(x=v)$ occur where $(u0vr) = -1$, and a metrical interpretation gives the "Law" $1/u + 1/v = 2/r$ the formula for the object/image with a spherical mirror scale.

This case includes that of Ex.3.1, for as $r \to \infty$ the reflecting surface (circular arc) becomes a straight line, and the involution becomes $u + v = 0$. The case of parallel rays striking the mirror is illustrated in Fig.1.4. In each case the "laws" are given by the involution range/pencil. [Refer to Fig.1.3 and Fig.1.4].

<u>Example-5.3</u>   The **refraction** of light rays on passing from a **region-1** to a more dense **region-2** was successfully explained by **Huyghens** by his postulate of light as a **wavelike** phenomenon [Further consequences of this are discussed in Chapter-10]. But it may also be described in geometrical optics by considering the two **dual** situations (v. Fig. 1.6) - the one being the reflection in a plane mirror when the image appears in a region-2 (Ex-4.2 points $P, \Omega, \mu Q, \Sigma$ on a line λ), the other being the dual of rays $(p, \omega, q, \sigma$ on a point $\Lambda)$.

We need only assume that rays entering region-2 are **constrained along the normal** (i.e line ω). This constraint will be equivalent to the projection of a **circle** in region-1 into an **ellipse** (axes **a** and **b**) and we can interpret this as requiring $\cos i : \cos r :: a : b$.

The harmonic pencil $(p \; \omega \; \overline{q} \; \sigma) = -1$, evaluated via the gradients of the rays gives $(-\sin i/\cos i \; 0 \; \sin r/\cos r \; \infty) = -1$

which means $(-(\sin i)/a \; 0 \; (\sin r)/b \; \infty) = -1$

giving $(\sin i)/(\sin r) = a/b$ (a constant μ)

<u>Example-5.4</u>   We shall see in Chapter-10 that the **Huyghens** postulate of Light as Wave Motion not only produces an explanation (in **Euclidean Geometry**) of refraction at the bounding surface between region-1 and region-2, as we have seen in Fig.1.7, but has an elegant description of other phenomena, viz., **interference** and **diffraction**. [v. Appendix-E [10],[22],[29],[52]]

The diagram in Fig. 1.7 shows the simple geometry behind the refraction of a plain wave-front (of light) when it passes from region-1 to region-2. This only required that the velocity of the wave should be different in the two regions - say v_1 in region-1 and v_2 in region-2 - the resulting relation being $(\sin i)/(\sin r) = v_1/v_2$.

Taking account of this **refractive index** factor, μ, we can demonstrate the action of **convex** and **concave** lenses in **geometrical optics** - including theories and practical applications of **microscopes** and **telescopes**. Other properties of such instruments (such as **chromatic dispersion** and **resolving power**) require the use of **Huyghens** postulate. [Appendix-E [10],[52] for standard formulae in these and other problems].

Example-5.5 If the circular arc in Ex.5.2 is replaced by that of a parabola the double points, on the axis, become **O** (the vertex) and **N** (where the normal at a point **P** meets the axis) (v. Fig.1.8).

[The parabola is a special case of the circle as both the centre and focus → ∞].

If the focus of the parabola is **F(a,0)** then for parallel rays along the axis $u = \infty$ and since **N** is $(2a + at^2, 0)$ the involution gives (as in Ex-4.2 above) **(TVNU)** = -1 and this results in the image appearing at $v = a$, the Focus of the parabola.

In other words all parallel rays are brought to the Focus **F**.

Equally well, a source at **F** outputs a truly parallel beam.

Hence the practical use of paraboloidal mirrors.

[1.6] Further examples of reflexive measures.

Example-6.1 The reflexive measures with **sound waves** occur when with two such frequencies f_1 and f_2 we observe the **beat frequencies**, viz.

$f_1 + f_2$ and $f_1 - f_2$.

These are the mates in an involution whose double points occur when $f_1 = f_2 = 0$ and when $f_1 = f_2 =$ "out of range" (i.e. at ∞).

Example-6.2 The ±x of Ex-5.1 occur in the observation of **electric charges**. Thus electric charge, with double elements 0 and ∞, will be manifest as **positive** and **negative** charges. This is an exact parallel with the object/image situation in the geometrical optics, of Ex-5.1.

Example-6.3 If the double elements in the scale $\mathfrak{C}(\underline{3} ; \underline{3})$ coincide then the involution is degenerate, and every pair of observable mates coincide. This is illustrated on scales which use **magnetic base elements**, and the **magnetic field** is indistinguishable from the "electric/magnetic dipole" field - in which the $\pm$ elements coincide.

Example-6.4 The **gravitational field**, $\mathcal{B}_G$, is a Base Element which is manifest whenever two (or more) masses are in the same universe. A single mass, in an otherwise empty universe does not experience a **gravitational field**, even though it might well generate one. Thus the reflexive measures, associated with points in a line, λ, in $\mathbf{P}^1$ are mates in the case where double points are O (when the two masses coalesce, the attraction being then infinite) and $\mathbf{U}$ ($= \infty$) (when the attraction is effectively zero). The involution is the $(\mathbf{Q,O,P,U}) = (q\infty p0) = \mathbf{-1}$.

These measures, in $\mathbf{P}^1$, denote the mutual attractions $\pm\mathbf{F}$ [v. Fig.1.9].

Example-6.5 Anticipating the discussion in Chapter-3 [Particle Dynamics] it is appropriate here to describe the dynamics (motion) of a particle, **P**, in the presence of a second such particle, **Q**. This is contained in the **dual** geometry to that of Ex-4.8. This is because when a point **P** is moving on any curve we must associate its **velocity** with the **tangent** to that curve - and this means that the velocity is a Scale measure on the **envelope curve** [v. Appendix-B].

Now the **dual** of Fig.1.9 is shown in Fig.1.10 and correponds to the idea of four rays (lines) meeting at a point Λ. These give us the cross-ratio $(qopu) = (q\infty p0) = -1$; in **Euclidean** space the lines o and u are angle bisectors of the lines q,p (on the vertex Λ).

Moving into $\mathbf{P}^3$, this is equivalent to the **Cone** on vertex Λ.

It follows that a **plane motion in this field** must correspond to a **plane section of the cone** - that is to say, the motion will be a geometrical **conic section** (ellipse, parabola, hyperbola etc).

We shall see, in Ch-3, that **Newtonian dynamics** explained the motion of a planet in an ellipse (compatible with **Kepler's Laws**) by postulating the **inverse square** law of force - directed towards one of the foci of an ellipse (the Sun).

Example-6.6 An earthbound projectile describes a parabola because :-

A small particle-mass (relative to the earth) moving near to the surface "sees" the earth as a plane (tangential plane) - and the constant gravity **g** is acting vertically downwards [v. Fig.1.11]. These parallel forces meet at ∞ and so the apparent "other particle-mass" is at ∞. The **conic section** which the projectile describes is therefore one whose centre, and the important **focus**, is at ∞.

Such a conic is a parabola and the **Newtonian** centre of attraction is therefore at ∞.

[1.7] Defining a Physics

From what we have said above it seems that we cannot have **a Physics** without accepting four ideas.

(a) The idea of a set of indefinable intuitive concepts (or elements E_r)

(b) The idea that these elements can be associated with observations (measures) on a Scale (or Scales) - with specified Base elements, $\{\mathfrak{C}(\mathfrak{Z}) \,;\, E_r\}$

(c) The measures on a Scale can be injectively mapped into an algebraic structure, **A**, certainly a division algebra, and this algebra brings with it a suitable **algebraic geometry** [v. Appendix-B].

(d) The apparent "Laws of Nature" will be expressed in terms of suitable **invariants** of the algebraic structures. These will be invariants under either **changes of Scale**, **changes of Algebra**, or **both**.

Definition-D3 :
Given a set of elements $\{E_1, E_2, ...\} \equiv E^n$ and a set of Scales $\{\mathfrak{C}_\alpha(\mathfrak{Z})\}$, then the complete set of measures $\{\mathfrak{C}(\mathfrak{Z}) \,;\, E^n\}$ is said to define a **Physics**.

Definition-D4 :
Given a Physics $\{\mathfrak{C}(\mathfrak{Z}) \,;\, E^n\}$ and an algebra **A** containing the images of the measures of the E^n via injective mappings $\{\sigma_r : \mathfrak{C} \to \mathbf{A}\}$, then the set
$$[\mathfrak{C}_\alpha(\mathfrak{Z}) \,;\, E^n \,;\, \mathbf{A} \,;\, \{\sigma_r\}]$$
will define a **Theoretical Physics**.

Definition-D5 :
Given a Theoretical Physics and a set of mappings **M** between the elements $\{\mathfrak{C}(\mathfrak{Z}) \,;\, E_r\}$ and $\{\mathfrak{C}(\mathfrak{Z}) \,;\, E_s\}$ for all relevant **r,s**, then the algebraic invariants under **M** will be called the **Laws of Nature** in that Physics.

[1.8] Metrical and Non-Metrical Geometries

Although the ideas of **Euclidean Geometry** have been extremely useful in the development of **Physics** it does not follow that this is an essential tool for that study. Indeed we shall see that many of the concepts of the **non-metrical Projective Geometry** are not only very relevant to that discipline but also can be seen to underly various searches for the **Natural Laws** of that science.

And we should emphasise the distinction between any **abstract** mathematical language we care to use and the results of **observation** of nature's pheneomena. Perhaps the word "abstract" is here being misused since it conjures ideas about things which are "not real".

This completely misses the point about the use of a "language". In mathematics all its concepts are "real" ; it is only in "translation" into some other language that its words can be questioned as "real" or "not real". This is why students of mathematics (who are secretly students of Physics) frequently ask for a translation of the terms - thereby never quite mastering the language. A mathematician is a scholar who does not need to "translate" his language - he is fluent in it. If he "applies" the language to his everday experiences then it is a valuable tool in his vocabulary - for communicating with non-mathematicians.

[1.9] Using Dimensional Analysis

Some acknowledgement of the role of the Base Elements in Physics has been the use of what is known as **dimensional analysis**. This is the selection of some basic qualities in terms of which other (defined) elements (our **generating elements**) can be expressed.
The method is wellknown and, although it is transparently simple, has proved to be very helpful. A few examples follow.

If we select the base elements "mass", **M**, "length", **L**, "time", **T**, and write their "dimensions" as [M], [L], [T] we can introduce the "dimensions" of

Force as $[F] = MLT^{-2}$ (via the **Newtonian** laws)

Velocity as $[V] = LT^{-1}$ (length per unit of time)

Momentum as [mom] = [Mass x Velocity] = MLT^{-1}

Energy as $[E] = [F][L] = ML^2T^{-2}$ (force times distance)

Action as [A] = [Energy x Time] = ML^2T^{-1}

Frequency as [freq] = [number per sec] = T^{-1}

and so on. The reader is no doubt familiar with such analysis.

[1.10] Definingg Physics-1

We begin at the beginning, with the simplest observations, viz. what we regard as physical manifestations of **geometry**, but which are in fact the projections of geometrical concepts into some agreeable algebraic geometry. The early geometry must have been a two-way traffic between observation and abstraction. So this **Physics-1** is really "abstract geometry" masquerading as a Physics - in all innocence!

Thus, what we can say is that **Physics-1** [v. §[1.4] above] is to be defined by way of

$$\textbf{Physics-1} \equiv \{\mathfrak{C}(\mathfrak{Z}_L) ; E_p\}$$

and the associated **Theoretical Physics** has traditionally been defined by

$$\textbf{Physics-1 (theoretical)} \equiv \{\mathfrak{C}(\mathfrak{Z}_L) ; E_p ; A ; \sigma\}$$

where E_p denotes "position" and **A** is a suitable algebra - usually assumed to be $\mathfrak{R}^\#$ and its associated Euclidean Geometry - σ being an injection into $\mathfrak{R}^\#$, since we can only admit actual measures being in $\mathfrak{R}$.

We shall see that most of the **Laws** have an underlying expression in **Projective Space(s)**

viz., P^1, P^2 or P^3.

The properties of **Physics-1** are obtained by considering mappings between the **subsets** of $\{E_p\}$, such as **point**, **line** and **plane** whence the set **M** is commonly the **full Linear Group, LG** and the "natural laws" are the geometrical theorems under that group (being more or less verified). The basic invariants among the scales form the **Propositions of Incidence** for Projective Geometry. An interesting example of our earlier ramarks about structure is that associativity and commutativity in **A** correspond to the theorems of **Desargue** and **Pappus** respectively [v. Appendix-B] It is therefore possible for the Physics measures to define either a Pappian projective geometry (when **A** may be taken as $\mathfrak{R}^\#$) or a non-Pappian geometry (when **A** may be taken as a matrix algebra - where commutativity need not apply).

If the additional concept of "parallelism" is available on the scales then the geometry becomes **affine geometry** and the invariants must be taken under the affine group of mappings. This latter permits the physical concept of **displacement** in the Physics. In **Physics-1** the $\Gamma(1\text{-}1)$ is an equivalence relation and, of course, contains in it the invariance of cross-ratio. Also, of course, these invariant mappings cannot alter the basic involution properties of the reflexive measures.

But **Physics-1** does not actually require the concept of "metric", the introduction of which generates a modified Physics, viz.

$$\textbf{Physics-1}^* : \{\mathfrak{C}(\mathfrak{Z}_L) ; E_p, E_D ; A ; \sigma_r\}$$

E_D denoting this idea of metric/distance/interval. Since measures of E_D are mapped into $\mathfrak{R}^\#$ this latter must be a subalgebra of **A**, or should be a field over which the additive vector space of **A** is defined.

The metric which the projective geometry of **Physics-1** carries [v. Appendix-B] is defined by the introduction of an **absolute conic/quadric** and this conic, say, may be taken as an ellipse, hyperbola, parabola, or even degenerate forms of these - this last case being the one in which the Pythagorean metric can be found, and which is inherent in **Euclidean Geometry**. [Appendix-B]. In this **Physics-1*** the natural laws are invariants under those mappings **M** which leave the absolute conic/quadric invariant. These are the **Euler-Rodrigues Transformations** and under this set an equivalence relation is that of **congruence**.

In this Physics our intuitive ideas of "substance", or "matter", are associated with the concept of **particle** - and this is attached to the geometrical notion of **point** (however large the "particle" actually is). A **particle** is an entity **without any internal structure**. It is entirely **external** in its effect.

When we begin to think that the spatial extension of a "body" cannot be ignored then we try to think of it as a collection of these particles - and often with some satisfactory theories to go with it - but see Part-3 for an alternative.

The Physics then concerns itself with bodies of various shapes, particles of various sizes, and curves and surfaces associated with them. Thus we shall be concerned with "particles" which appear on "straight" lines, on curves (in a plane or in 3-space); and with other bodies embedded in similar geometrical structures.

The choice of the relevant algebra, **A**, provides us with a mathematical language we can use in the Theoretical Physics - and the interpretation of experimental measures.

Whereas particles can be described by position co-ordinates (in Projective or Euclidean space) - things like (**x,y,z**) - more complex bodies need more variables to identify them. An acceptable set in **Euclidean** space are the **Eulerian Angles** which fix the orientation of such a body in 3-space. [v.Fig.1.12].

These are usually denoted by the angles (θ, φ, ψ) as shown in the diagram - where it is helpful to imagine an ordinary conical top as the body. The point of the Eulerian angles is to be able to fix the orientation of the body (cone) relative to a set of fixed axes in 3-space (centred on the origin **O**).

So if $(\ell, \mathbf{m}, \mathbf{n})$ are axes fixed in the cone (**n** being along the axis of the cone) we can find them by   (i) taking **n** inclined at angle θ to **OZ**,

       (ii) the plane **OZ,** ∪ making an angle φ with the plane **OZ,OX**

and   (iii) axes **Ox,Oy** (which form a triad in the cone with the axis **Oz** (≡ **n**)) being rotated through angle ψ about **n**

these giving the cone's orientation via its axes - which become $\ell, \mathbf{m}, \mathbf{n}$. [v. Fig: 1.12].

But we also know that, in Projective Geometry, every curve which is described by a set of **points** (a point-curve) also exists as an **envelope-curve** described by the tangents to that curve. This is because of the **duality** found in that geometry, and either view is found in the algebra being used to describe the "curve".

It is notable that a **conic section** possesses the property that **the envelope** of the envelope-curve is identical with the original **point-curve**.

It was the problem of finding the **tangent to a curve** which led earlier mathematicians, notably **Newton** and **Leibnitz**, to invent the **Differential/Integral Calculus** ; a discipline which allows us to examine a curve in a **vanishing neighbourhood** - in other words "at a point". This has naturally become a very powerful tool in the development of Theoretical Physics - which will become apparent in the following chapters.

Of course it only applies at points of those curves which possess tangents. Such curves are described by functions, say $f(x,y)$, which are called **differentiable**, and there are plenty functions which are not. Certainly differentiability must not be confused with continuity.

[Indeed the mathematician Weierstrass discovered a function which is continuous
everywhere but differentiable nowhere (mathematicians enjoy such sports)]

For example, if we are given a family of curves, say $f(x,y,\alpha) = 0$, where α is a continuous parameter (and f is differentiable w.r.t α), neighbouring curves will share a common point (be part of the envelope of something) whenever $f(x,y,\alpha+\delta\alpha) = 0$ as well as $f(x,y,\alpha) = 0$. This means that we need only eliminate α from the simultaneous equations

$$f(x,y,\alpha) = 0 \quad \text{and} \quad \partial_\alpha f(x,y,\alpha) = 0.$$

[It is easy to check, eg, that the family of lines of the form $x\cos\alpha + y\sin\alpha = p$
envelopes the circle $x^2 + y^2 = p^2$].

In general algebraic geometry it is well known that the point-conic

$$ax^2 + by^2 + cz^2 + 2fyz + 2gzx + 2hxy = 0$$

is equivalent to the envelope-conic

$$A\ell^2 + Bm^2 + Cn^2 + 2Fmn + 2Gn\ell + 2H\ell m = 0$$

where A,B,C,F,G,H are the minors of a,b,c,f,g,h in the determinant Δ, being

$$\Delta = \det \begin{bmatrix} a & h & g \\ h & b & f \\ g & f & c \end{bmatrix}$$

When the Pythagorean metric is introduced into the geometry it is commonly quoted in terms of differentials in the calculus dx, dy etc. : the distance between two neighbouring points P(x,y,z) and Q(x+dx, y+dy, z+dz) being ds where

$$(ds)^2 = (dx)^2 + (dy)^2 + (dz)^2 \quad \text{(rectangular co-ordinates)}$$

Choosing other co-ordinates, for example, plane polars, spherical polars, cylindrical polars, give corresponding equivalent metric forms (in Euclidean metric geometry).

[v. Apendix-B for the generalised Riemannian metric].

But we shall also see that **Laguerre** showed that **projective metric** can be introduced which has a large and profitable range of applications.

In such an extension we use a generic term for what can generalise the words "point" "line" and we have chosen the word "poin" for this purpose in our later discussions.

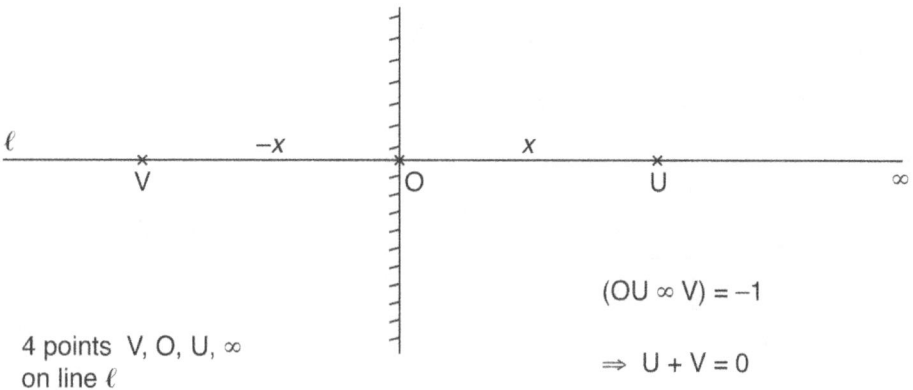

4 points V, O, U, ∞
on line ℓ

$(OU \infty V) = -1$

$\Rightarrow U + V = 0$

Fig 1.1

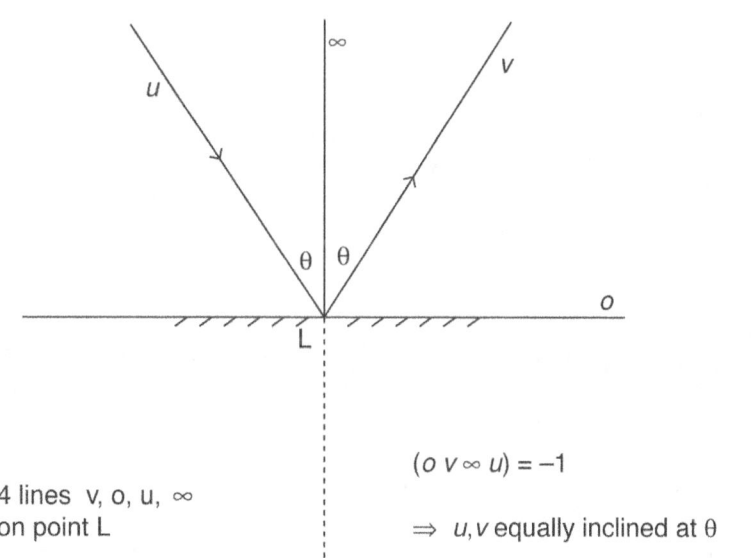

4 lines v, o, u, ∞
on point L

$(o\, v \infty u) = -1$

$\Rightarrow u, v$ equally inclined at θ

Fig 1.2

Note: The 2 scales show DUALITY
Point sources become Rays Cross Ratio invariant

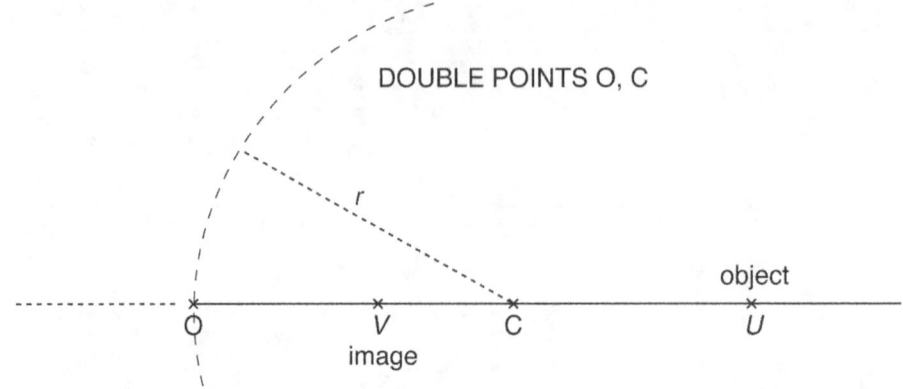

Fig 1.3

$(O\ V\ C\ U) = (o\ v\ r\ u) = -1$

$$\Rightarrow \frac{1}{u} + \frac{1}{v} = \frac{2}{r}$$

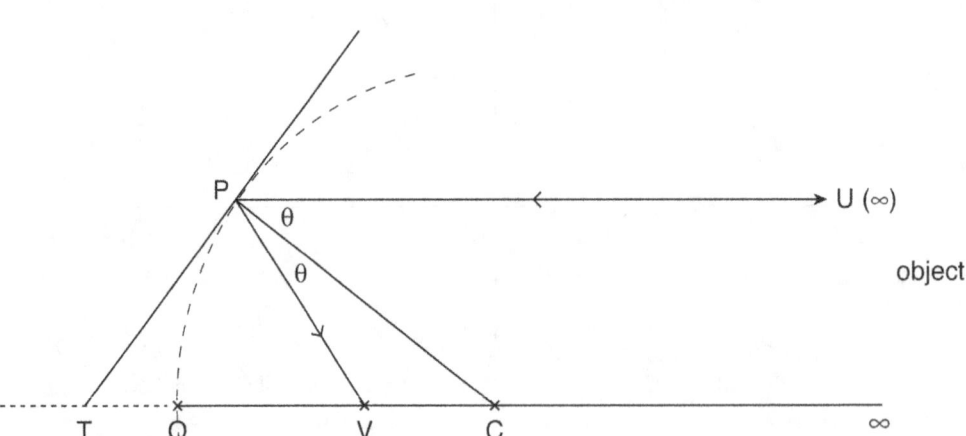

$(T\ V\ C\ \infty) = -1$

Fig 1.4

Because Angles θ are Equal and PT is ⊥ to PC

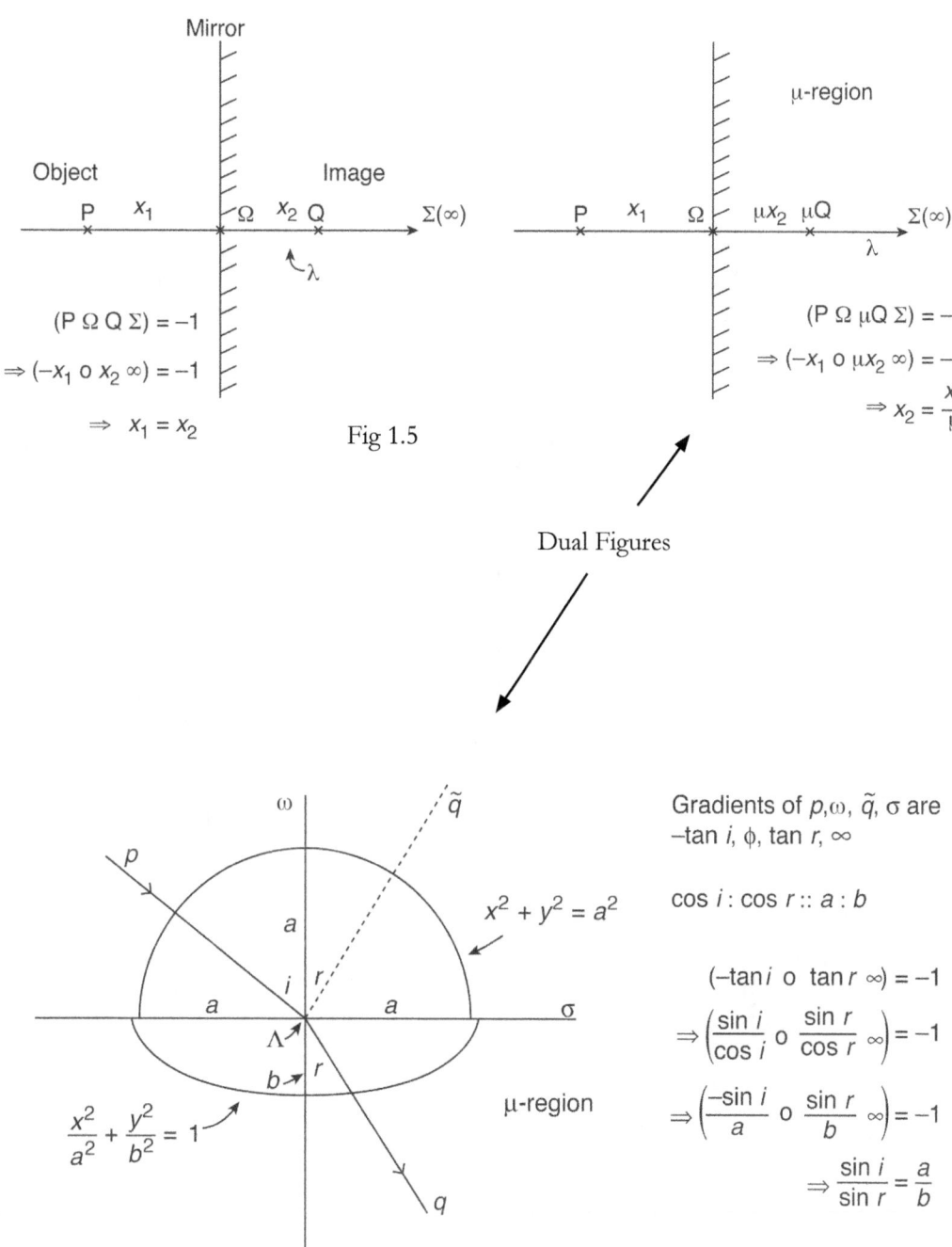

Fig 1.5

Fig 1.6

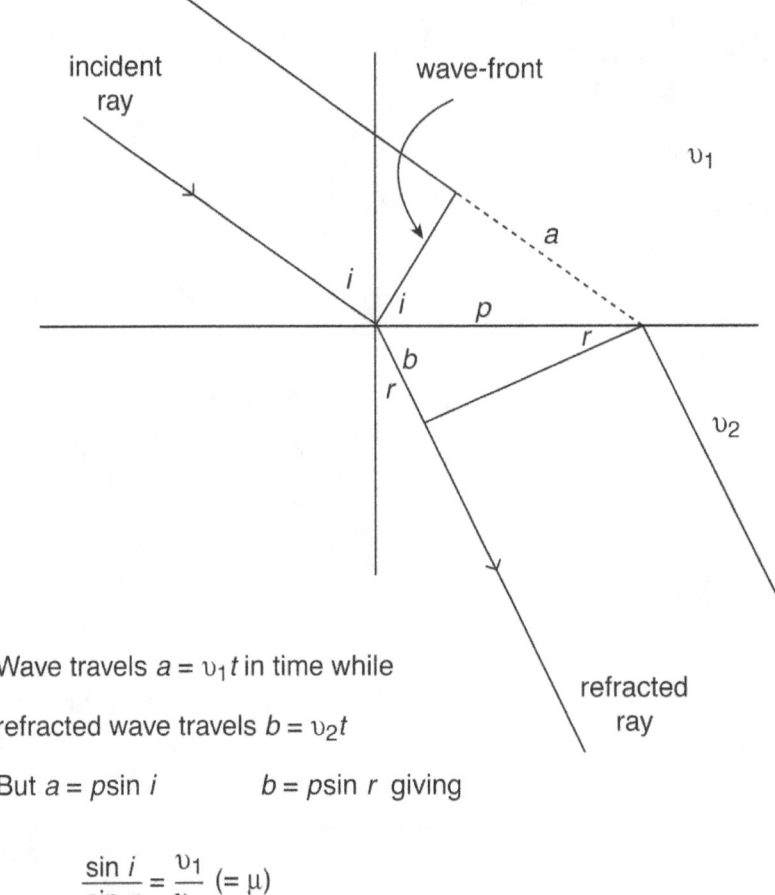

Wave travels $a = v_1 t$ in time while refracted wave travels $b = v_2 t$

But $a = p\sin i$ $\qquad b = p\sin r$ giving

$$\frac{\sin i}{\sin r} = \frac{v_1}{v_2} \,(= \mu)$$

Fig 1.7

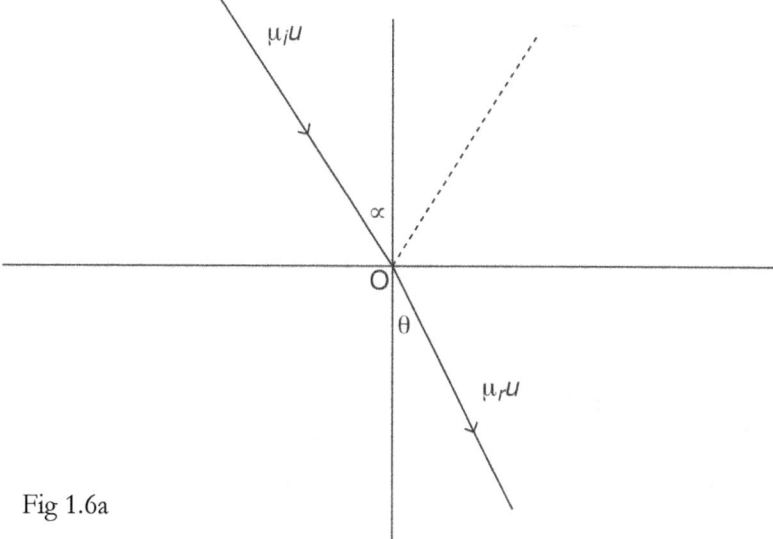

Fig 1.6a

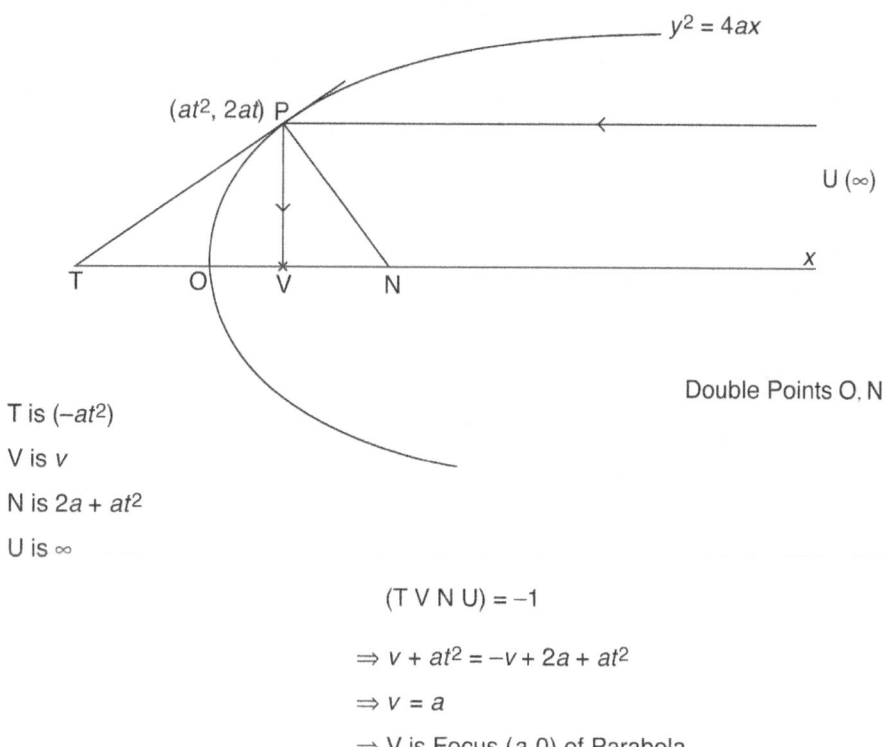

T is $(-at^2)$
V is v
N is $2a + at^2$
U is ∞

$(T\ V\ N\ U) = -1$

$\Rightarrow v + at^2 = -v + 2a + at^2$

$\Rightarrow v = a$

$\Rightarrow$ V is Focus $(a, 0)$ of Parabola

Fig 1.8

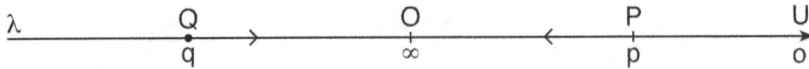

Fig 1.9

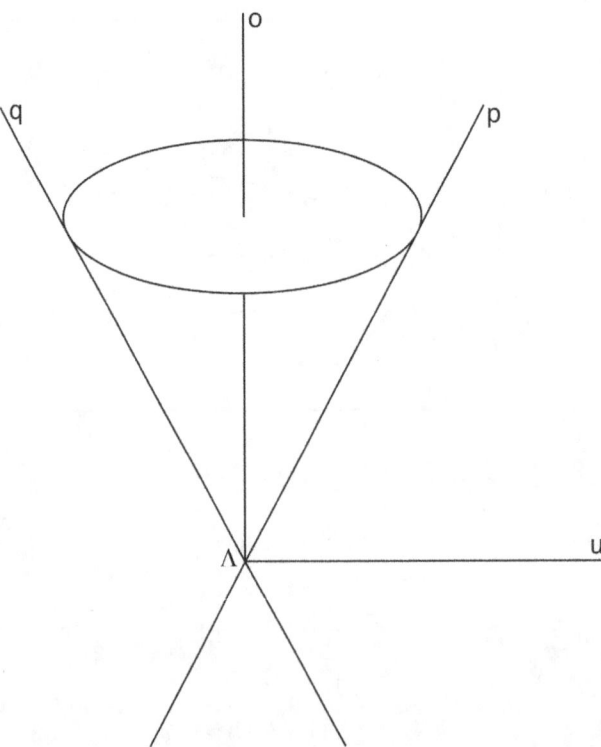

Fig 1.10

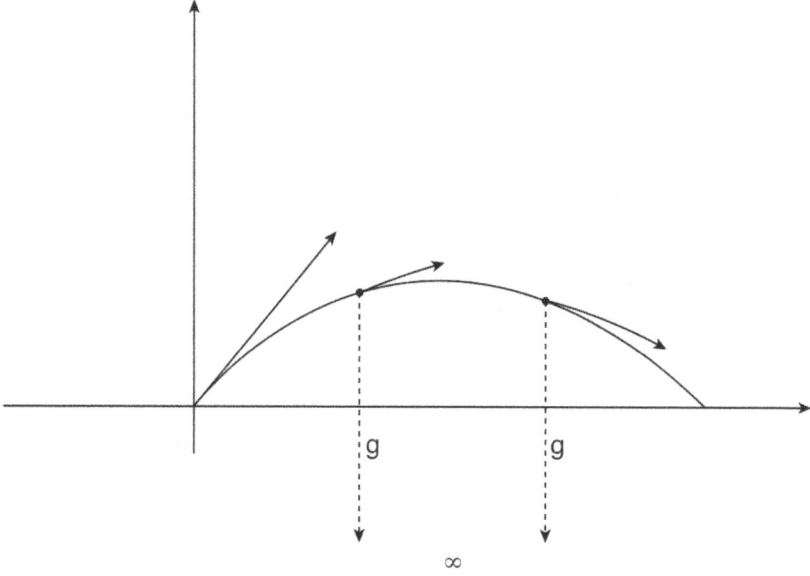

Fig 1.11

36 MATHEMATICAL PHYSICS

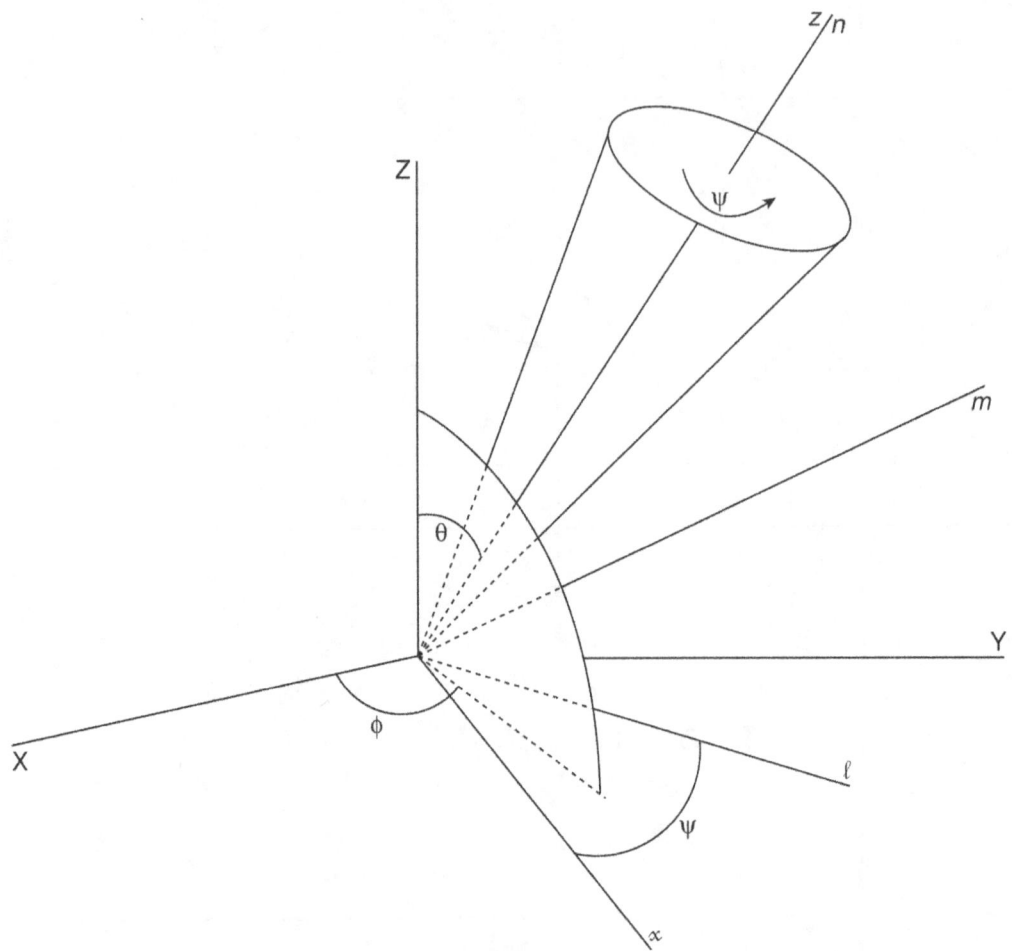

Eulerian angles (θ, φ, ψ)

Fig 1.12

Chapter-2 /PART-1   Physics-2

[2.1] **Centre of Gravity**

We must distinguish between the **geometry** of **Physics-1*** - which may legitimately form the framework for the "particle" of classical Physics - and that geometry/Physics which contains in it the idea of **rigid body**. This idea involves the additional assumption that **length** is **independent of position** (the traditional way in which metric has been introduced).

We can express this difference by saying that **rigid bodies** have been introduced into the measures, so the Physics becomes

$$\textbf{Physics-2} \equiv \{\textbf{Physics-1}^* \cup (\text{rigid bodies})\}$$
$$\equiv \{\mathfrak{C}(\mathfrak{Z}_L) \; ; \; \textbf{E}_p, \; \textbf{E}_F \; ; \; \mathfrak{R}^\# \; ; \; \sigma_r\}$$

where $\textbf{E}_p$ is a position measure, $\textbf{E}_F$ is a force measure.

The **Physics-2** can equally well be identified with **Statics** and in this Physics "rigidity" is itself an equivalence relation which decomposes the measures into disjoint sets which define the **rigid bodies**. The geometry is metrical **Euclidean** and this allows us to contemplate the metric-dependent ideas of **area** and of **volume**.

Traditionally an extended body has been regarded as a collection of points - as in **Pysics-1** - and this has enabled us to define, eg, the idea of a **centroid**. In 3-dimensions we assume we have an axis in the body and that this body is made up of an innumerable number of **points**, as in **Physics-1***. If we assume that each such point **P** has co-ordinates (x_p, y_p, z_p) then centroid, **G**, has 3-dimensional Cartesian co-ordinates $(\overline{x}, \overline{y}, \overline{z})$ given by, eg

$$(\text{total volume,etc}) * \overline{x} = \Sigma_p(x_p \delta_p)$$

(where δ_p denotes an element of volume (or whatever) at the point **P**) and so on for $\overline{y}$ and $\overline{z}$. Whenever this **centroid, G** is taken as an origin in the Cartesian system it follows that ,eg, $\Sigma_p(x_p) = 0$.

When we associate the points $\{P\}$ with particles of apparent "mass" m_p then **G** becomes the **centre of mass** of that total set.

When we associate the points $\{P\}$ with "weights" w_p then **G** becomes the **centre of gravity** of that total set.

Also the sense of ordering being absent from the observations we have no need of any "clocks" in the measures - the Physics is **Statics** [v. Appendix-E [46]].

[2.2] Weights and the Law of the Lever (**Archimedes**)

In this Physics we introduce the idea of **force** - which makes its appearance by way of measures of **weight**, **tension** & **thrust** in things like strings & rods. In this way we need to refine the definition via

$$\text{Physics-2}^* \equiv \text{Statics} \equiv \{\text{Physics-1} \cup \{\mathbb{C}(3_L)\ ;\ E_F\ ;\ A\}$$

where **A** allows an **affine geometry** to carry the measures of the forces E_F.

Affine geometry allows a representation of **displacement** as well as **parallelism** and thereby we can introduce the concept of **vector** - a representation via **magnitude** and **direction** associated with a **line** of definite **length**. These ideas dominate the usual **Euclidean geometry** which Physicists have traditionally used.

Weights can be compared (but not measured as absolute elements) by, eg, comparative extensions in elastic strings (or springs).

By balancing two weights, (w_1, w_2), on the arms (x_1, x_2) of an **Archimedean Lever**. (Ref. Fig.(2.1)) we can expect a $\Gamma(1\text{-}1)$ between the ratios x_1/x_2 and w_1/w_2.

[Note: Notice the ratios are effectively homogeneous co-ordinates on a projective line $\mathbf{P}^1$]

Writing $\quad \Gamma(1\text{-}1) \equiv A x_1 w_1 / x_2 w_2 + B x_1/x_2 + C w_1/w_2 + D = 0$

we obtain $\quad A x_1 w_1 + B x_1 w_2 + C w_1 x_2 + D x_2 w_2 = 0$

this being a $\Gamma(1\text{-}1)$ in homogeneous co-ordinates.

By taking $A = 1$ and shifting the origin via $x \to (x - C)$ and $w \to (w - B)$ this reduces to

$$x_1 w_1 = x_2 w_2$$

this being **Archimedes' Law of the Lever**. Since no sense of time is involved in the measures this must correspond to the **statical equilibrium** situation.

Another view of the matter follows easily from the idea of centre of gravity. For the parallel forces w_1 and w_2 must be balanced, in equilibrium, by the upward force exerted by the fulcrum, and so the centre of gravity of the weights, viz., P_3, lies at the fulcrum and that is where $\quad w_2 x_2 + w_1(-x_1) = 0$.

An observed "natural law" for line forces acting at a point is to be found in what is called the **parallelogram rule**.

We normally represent linear forces, F_1, F_2, by lengths and directions of lines and we do this by mapping the measures into an **affine geometry** $\cup$ **Euclidean metric**.

The invariance we depend on is that associated with the projective properties of ranges of points, or pencils of lines. Referring to Fig.(2.2a) and Fig.(2.2b) we see that the move from a representation of "force-measures" F_1 & F_2 in a projective P^2 can be associated with ranges (eg) (A,P,B,Y) and (A,R,D,Z). These are both projectively related (the same cross-ratios) to the range (A,X,C,Q) - which becomes thereby the equivalent of these two forces. In **Statics** this is regarded as the **resultant** (or equivalent single force) of F_1 and F_2 acting at the point "A".

When we insist on viewing the geometry as **affine/Euclidean** this amounts to

(i) projecting the line YZ to Λ_∞

(ii) representing the magnitude of the forces by lengths of lines

and then we obtain the **parallelogram of forces** - wherein the "resultant" of F_1 and F_2 is F_3 and represented by the diagonal AC.

[2.3] Duality in Statics

We can also see the "duality", which is inherent in projective geometry, [v. Appendix-B] exhibited in this Physics when we compare the above two problems ; for if we look at Fig.2.3 we can see that whereas the "lever problem" is concerned with

three forces w_1, w_2, $(w_1 + w_2)$ acting in equilibrium at three points P_1, P_2, P_3 on a line ℓ the dual problem consists in

three forces F_1, F_2, F_3 acting in lines p_1, p_2, p_3 at a point L.

The conditions of equilibrium have a natural parallel form, viz.,

$w_1 x_1 = w_2 x_2$ in the case of the lever, and

$F_1 \sin\alpha_1 = F_2 \sin\alpha_2$ in the dual case.

Since we operate in an affine space the force(vector) F_1 can be considered as the side HK in the diagram whence $\angle$LKH becomes α_1 and the diagram becomes the parallelogram LJKN. Then the diagonal is given as $F_3 = F_1 \cos\alpha_1 + F_2 \cos\alpha_2$.

We then see that the idea of the **triangle of forces** <u>precedes</u> the idea of the **parallelogram of forces** (as opposed to the usual teaching in these matters).

[2.4] The significance of Quaternions $\mathbb{Q}$

The Statical cases of frameworks - consisting of **struts** and **ties** - is well illustrated by the analysis contained in **Bow's notation** and leads to the analysis of the forces contained therein.

The Archimedean lever experiment strongly suggests that, if equilibrium is disturbed, then the "weights" have a **turning effect** about the fulcrum. This turning effect is then identified with the product (eg) x_1*w_1 - which is given the name of **torque** or **moment** about the fulcrum. This torque exists even in the Statical situation (of no motion) and so it must be taken into account via the Algebra used for the measures.

Since rotation is associated with an "axis of rotation" it would suggest that any vector in the affine geometry which is to represent such a rotation must have a direction which is perpendicular to the "plane of rotation". and this requires an extension to the algebra (and its associated geometry).

This can only be achieved by changing the algebra **A** to $\mathcal{Q}$, the algebra of **Quaternions** $\mathcal{Q}$, over the reals. This is because we need a division algebra if we wish to "solve" a $\Gamma(1-1)$ in the algebra. This is not, however, what has been traditionally done - ever since the mathematician **Heaviside** carved up the algebra $\mathcal{Q}$ into its separate parts of **scalar** and **vector**. The result of this was to produce a sub-ring - called "vector algebra" - without either associativity or multiplicative inverses. In this "vector algebra" we certainly obtain a sensible "vector product", viz., $\wedge$, under which (eg) the vector $\underline{a} \wedge \underline{b}$ is a vector perpendicular to the plane of $\underline{a}$ and $\underline{b}$. But although $\underline{a} \wedge \underline{b} \neq \underline{b} \wedge \underline{a}$ is even a desirable property, $\underline{a} \wedge (\underline{b} \wedge \underline{c}) \neq (\underline{a} \wedge \underline{b}) \wedge \underline{c}$ is not, and we cannot solve an equation of the form $\underline{x} \wedge \underline{a} = \underline{b}$ for x -since we need an inverse $\mathbf{a}^{-1}$.

On the other hand we need only use the quaternion algebra $\mathcal{Q}$ to put all these ills right. Of course we need the non-commutative property, $\underline{a} \wedge \underline{b} \neq \underline{b} \wedge \underline{a}$, because rotations about succesive axes through finite angles are not commutative.

But we then reap the advantage of being able to unravel $\Gamma(1-1)$ correspondences in the algebra.

In particular, the "torque" of a force about a point in a plane - say **F** acting at a point whose position vector is $\underline{r}$ (relative to an origin **O**) is given by the quaternion product   rF   when we write **r** & **F** as quaternions

$\mathbf{r} = 01 + x\mathbf{i} + y\mathbf{j} + z\mathbf{k}$ and $\mathbf{F} = 01 + X\mathbf{i} + Y\mathbf{j} + Z\mathbf{k}$

or as **r** = (scalar r) + (vec r) and similarly for **F**.

The resultant quaternion is then

$$-(xX + yY + zZ)\mathbf{1} + (yZ-zY)\mathbf{i} + (zX-xZ)\mathbf{j} + (xY-yX)\mathbf{k}$$

the scalar part being the **scalar product** of the two "vectors" (vec **r**) and (vec **F**), whilst the vector part gives the **vector product** (vec **r**) $\wedge$ (vec **F**).

[2.5] Force Field and Work Function

These notions find their place both here in Statics as well as in Dynamics (v.later).

The idea is based on a consideration of the consequences of a **virtual displacement** of the points at which Forces are acting, and occurs where there is a Force **F** defined at all points of some space. This is called a **field of force** - gravity is a good example of such a field. In our terrestrial experience it is found "everywhere".

The displacement is best described in terms of the differentials of the Calculus and, if we take the simple case of a linear Force **F** moving its point of application through a change of co-ordinate, say d**r**, we define the change in the Work Function by

$$\delta W = (\text{scalar product } \mathbf{F} \cdot d\mathbf{r})$$

or

$$\delta W = X dx + Y dy + Z dz$$

where we are assuming that the components **X,Y,Z** are individually functions of **r** (x,y,z). This means that **W** is a function of **r** (or x,y,z) - **W(x,y,z)** - and when this δW is itself a perfect differential, giving $\delta W \equiv dW(r)$, we call this function **W(x,y,z)** the **Work Function** associated with the field of Force **F(r)**. In this case we say that the field of Force is a **conservative field** [v. Appendix-E [46]].

The characteristic of a "conservative field" is that the total Work Done in moving from point $\mathbf{P}_1$ to point $\mathbf{P}_2$, along a curve **C** is independent of the choice of **C**.

That is to say that the integral $\int_C \mathbf{F}(r) d\mathbf{r}$ is independent of **C**.

Its role in Statics occurs because it can be shown that

$$\delta(\Sigma_p \mathbf{F}(r_p)) \text{ must equal } \delta(\text{resultant Force})$$

and therefore must be zero in an equilibrium configuration.

Thus we have an equilibrium condition, in a conservative field, given by $dW = 0$ and, furthermore, since this means that the function **W(r)** has a **turning value** in the equilibrium condition, a **minimum** value will correspond to **stability** of the equilibrium condition whilst a **maximum** value will mean that the condition is **unstable**.

The **potential energy** (or "potential function") **V** is simply defined by

$$dV = -dW$$

representing the "work" which is **potentially available** in the configuration. Thus a conservative system is one in which we can find a potenrial function **V**.

[Note: The reader will find a few worked examples of these ideas in Appendix-E [46]].

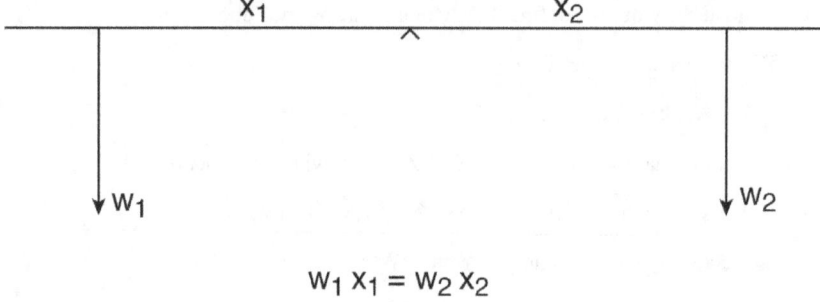

Fig 2.1 Archimedes' Law of the Lever

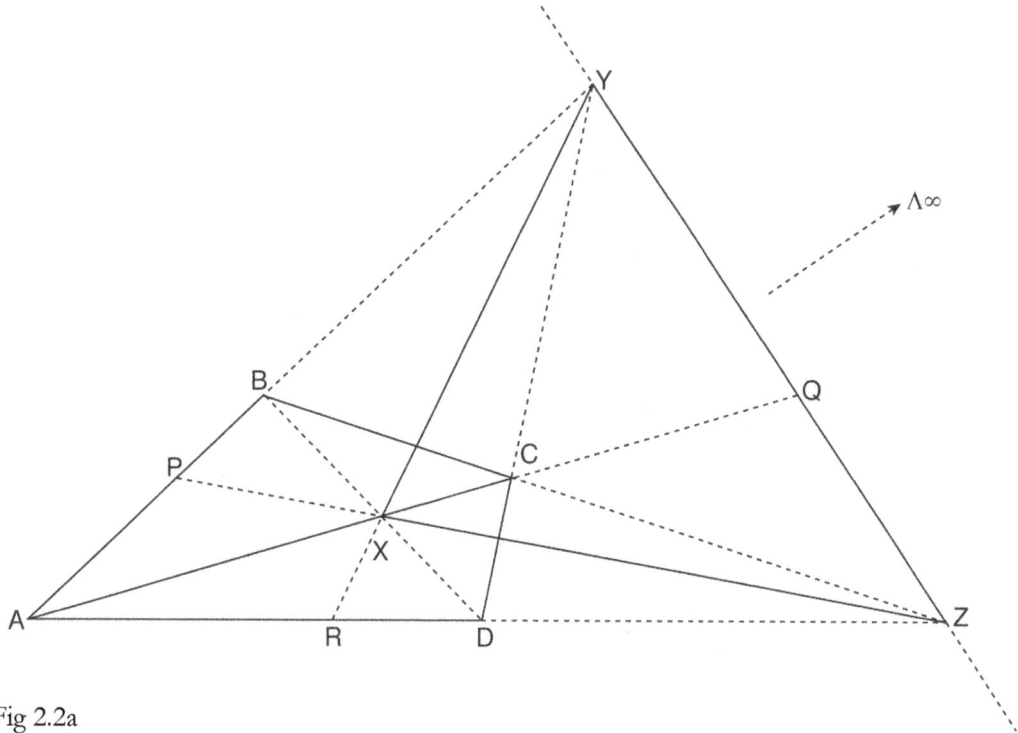

Fig 2.2a

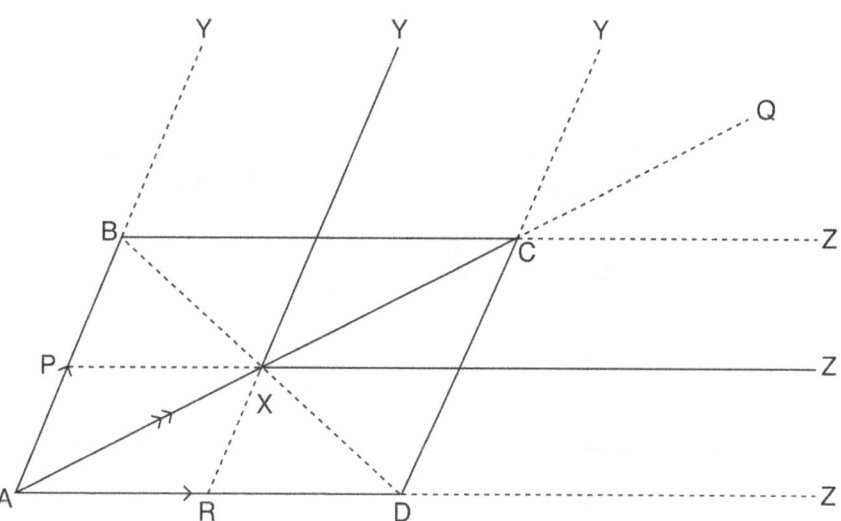

Fig 2.2b

44 MATHEMATICAL PHYSICS

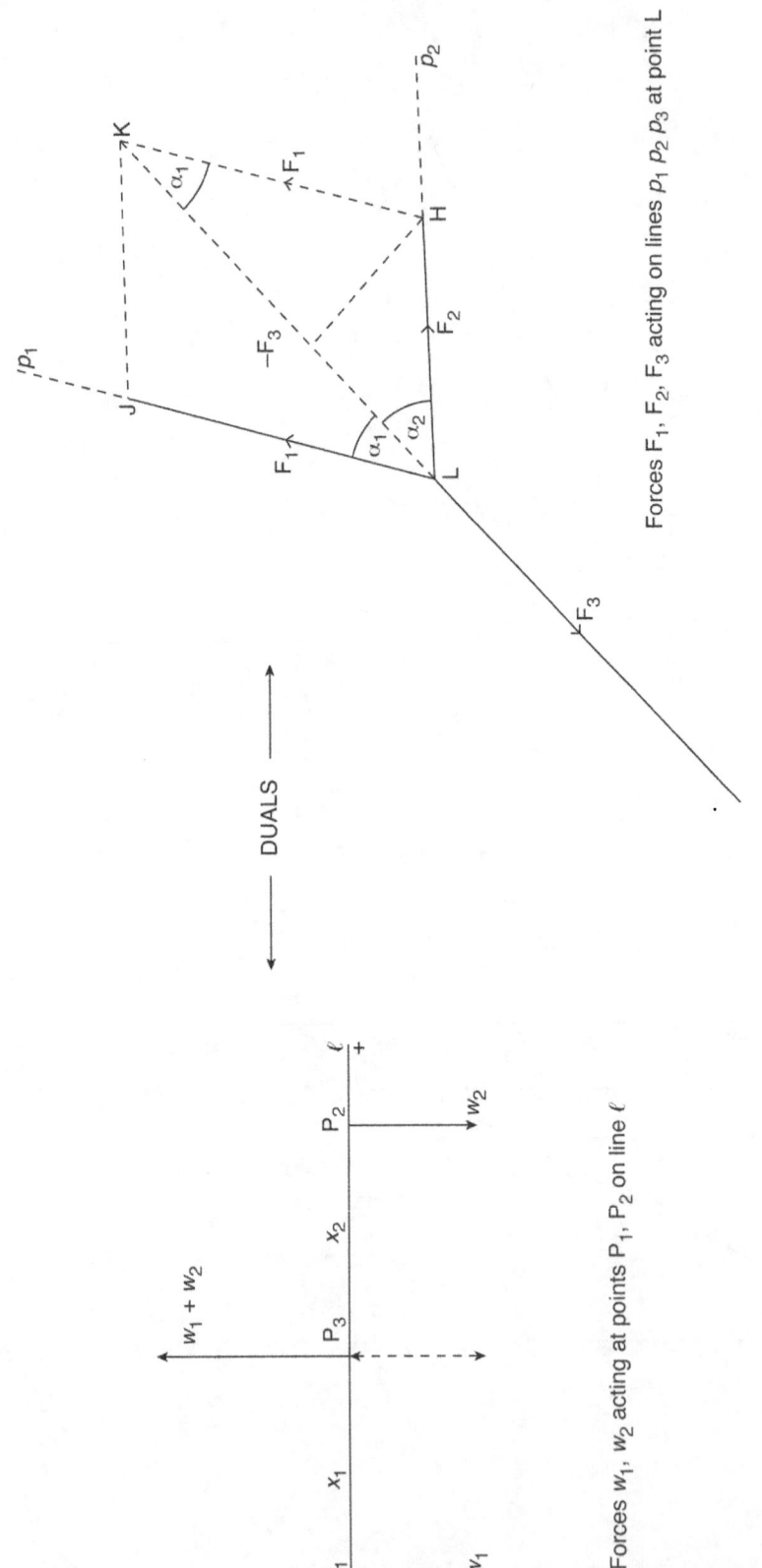

Fig 2.3 Duality in Statics

Chapter-3 /PART-1   Physics-3 : Particle Dynamics

[3.1] Ordering time, motion

We can now introduce the simplest dynamical Physics, viz., **Particle Dynamics**. This **Physics** [v. §[1.4] in Chapter-1] will be defined by Scales like :

$$\{\mathfrak{C}(\mathfrak{Z}_L, \mathfrak{Z}_T) \, ; \, E_m, \, E_p\}$$

where the Base Elements are "light signal" and "timing", and the associated **Theoretical Phyics** has traditionally been defined by

$$\{\mathfrak{C}(\mathfrak{Z}_L, \mathfrak{Z}_T) \, ; \, E_m, \, E_p \, ; \, \mathfrak{R}^\# \, ; \, \{\sigma_r\}\}$$

where E_m denotes "mass", E_p denotes "position" in a **Euclidean Space**, using the algebra of the reals $\mathfrak{R}^\#$. The elements "mass" and "point" are observed as an item called **particle**. We shall, however, prefer the algebra to be that of **quaternions**, $\mathbb{Q}(\mathfrak{R}^\#)$. We shall also see that the **Laws** can be just as easily expressed in **Projective Space(s)** viz., P^2 or P^3.

The concepts which underly the sense of **dynamics** (that of **movement**, either through changes of "value" or through changes of "direction") require the use of the Base Element $\mathfrak{Z}_T$. This **ordering** of the observational phenomena is an expression of our intuitive feeling of **time** - or **timing**. Thus, [v. Appendix-A], the algebraic sign " > " becomes the idea of **later than**, whilst " < " becomes **earlier than**. We naturally call the appropriate scales "clocks".

A physical **particle** in the Theoretical Physics becomes represented by the geometrical **point** (even if it is the size of the Earth !) and its "position" in the geometry is naturally defined by its co-ordinates - either (x,y) in Cartesian/Euclidean terms, or (x,y,z) in the homogeneous co-ordinates of a Projective Space.

When the motion is observed to be in 1-dimension (linear) then only one co-ordinate (x) or (the homogeneous) (x,z) - which often becomes (x,t), t being the timing. Parallels between Cartesian and Projective geometries are obtained by writing (other things being equal) a z (or t) co-ordinate as the unit 1. Points "at infinity" are obtained by writing (eg) $z=0$.

When the co-ordinate x changes in harmony with a change in t then we interpret that as a **velocity v**, or **speed**, and identify it as a "property" of the "particle". This means that we can define velocity by way of a $\Gamma(1\text{-}1)$ between x and t, viz.,

$$\Gamma(1\text{-}1) \equiv x - vt = 0$$

In a similar way we can define **acceleration**, **f**, by the homography

$$\Gamma(1\text{-}1) \equiv v - ft = 0$$

This naturally means that the Scale we are using must use $\mathfrak{Z}_L$ and $\mathfrak{Z}_T$ - something we think of as the combination of a "ruler" and a "clock".

When we wish to observe these variables in vanishing neighbourhoods (of **x** and/or **t**) then we naturally use the differential forms :-

$$\Gamma(1\text{-}1) \equiv dx - vdt = 0 \quad \text{and} \quad \Gamma(1\text{-}1) \equiv dv - fdt = 0$$

In higher dimensions we replace the scalars by vectors, whence

$$\underline{dr} - \underline{v}dt = 0 \quad \text{and} \quad \underline{dv} - \underline{f}dt = 0$$

where we notice that **t** is always regarded as a scalar. The other variables may be found either in **Heaviside's** vector algebra or in (our preferred) quaternions $\mathfrak{Q}$ (over $\mathfrak{R}^\#$).

[3.2] Einstein's Special Relativity

This emerges as part of the scale properties in this Physics and, in that sense, it should be regarded as part of the **kinematics** of the study.

[Note: **kinematics** is the description of the motion, ignoring its causes ; when we introduce the causes (forces) it becomes **kinetics**]

When we observe the linear velocities of a particle on two projectively related scales, say $\mathfrak{C}, \mathfrak{C}^*$, we expect to find a $\Gamma(1\text{-}1)$ of the form

$$(1) \qquad \alpha v v^* + \beta v + \gamma v^* + \delta = 0 \qquad \alpha, \beta, \gamma, \delta \in \mathfrak{R}^\#$$

In **Galileo-Newton** physics it was assumed, in effect, that the double points of (1) must be taken as $v = v^* = \infty$ and this requires $\alpha = 0$ and (1) becomes

$$(2) \qquad \beta v + \gamma v^* + \delta = 0$$

Now if the scale $\mathfrak{C}^*$ has an observed velocity **V** relative to $\mathfrak{C}$ two pairs in (2) will correspond to $v = V, v^* = 0$ and $v = 0, V^* = -V$

whence $\beta = 1, \gamma = -1$ and $\delta = -V$.

Now (2) becomes $\quad (3) \quad v = v^* + V$

which is how the Physics of **Galileo/Newton** viewed relative velocities.

But then **Poincare-Lorentz-Einstein** suggested that the double points should represent the limiting **velocity of light** (our Base Element $\mathfrak{Z}_L$), viz., **c**.

The $\Gamma(1\text{-}1)$ (1) then gives us, with $v = v^* = c$

$$(4) \qquad Vvv^* + c^2(v - v^*) - c^2 V = 0$$

It is clear that if we take **c** large enough it will be difficult to distinguish between

(3) and (4).

From (4) we obtain the **Lorentz** transformation laws as follows.

Writing $v=x/t$ (or dx/dt) and $v^*=x^*/t^*$ (4) becomes the homogeneous form

(5) $\quad Vxx^* + c^2(xt^* - x^*t) - c^2Vtt^* = 0$

or $\quad$ (5a) $\quad x^*(xV-c^2t) + c^2t^*(x-Vt) = 0$

Since this holds for all ratios x^*/t^* we can introduce a separation constant, say β, such that

(6) $\quad x^* = \beta(x-Vt) \quad$ and $\quad ct^* = \beta(ct-xV/c)$

and we can add to these the other co-ordinates via

(7) $\quad y^* = y \quad$ and $\quad z^* = z$

(6) and (7) are the **Lorentz transformations** for $\mathfrak{C}$ and $\mathfrak{C}^*$.

They can be expressed in the matrix algebra of linear transformations as

(8) $\quad \mathbf{r}^* = \mathscr{L}\mathbf{r}$

where $\mathscr{L}$ is the matrix

$$\mathscr{L} = \begin{bmatrix} \beta & 0 & 0 & H \\ 0 & 1 & 0 & 0 \\ 0 & 0 & 1 & 0 \\ H & 0 & 0 & \beta \end{bmatrix} \quad \text{where } H = -\beta V/c,$$

and now the vector $\mathbf{r}$ has rectangular co-ordinates x,y,z,ct.

Again we notice that this derivation does not require a metrical geometry even though metrics can always be used in it.

From the way we have derived $\mathscr{L}$ we expect the roles of $\mathfrak{C}$ and $\mathfrak{C}^*$ can be interchanged by replacing V by $-V$ and this will give us the **inverse operator**

$\mathscr{L}^{-1} \quad$ as the matrix

$$\begin{bmatrix} \beta & 0 & 0 & K \\ 0 & 1 & 0 & 0 \\ 0 & 0 & 1 & 0 \\ K & 0 & 0 & \beta \end{bmatrix} \quad \text{where } K = -H,$$

and since $\mathscr{L}\mathscr{L}^{-1} = \mathbf{I}$, the unit matrix we get

(9) $\quad \beta^2 = (1-v^2/c^2)^{-1}$

identifying the separation constant β.

We also notice that (8) and (9) give us

$$x^2 - c^2t^2 = x^{*2} - c^2t^{*2}$$

and so we may regard the **Lorentz** transformations as a set of collineations in the (**x,ct**) space which preserves (leaves invariant) a certain degenerate conic. This conic may be used as an absolute conic to define a metric and in this case we shall obtain a **hyperbolic geometry**. In this geometry there is a distinction between **interior** and **exterior** points and this distinguishes **past** and **future**.

To incorporate this absolute conic we notice that an invariant metric is given by
$$ds^2 = d(ct)^2 - dx^2 - dy^2 - dz^2$$
which puts our observations into the **space-time continuum**, after **Einstein** and **Minkowski**.

And it is also clear that the operator does not preserve "distance"; so **interval** is not preserved in this space (c.f. **Fitzgerald contraction** and **time dilation**).

[3.3] Newton's Law of Motion

This Law links **Physics-2** with **Physics-3** by introducing the concept of **force** into the kinematics of motion and thus creating **kinetics**. It was a profound development in Physics. The experience of **force** is taken from that of **Statics** (v. **Physics-2**) and **Newton's** contribution was to encompass this idea into Dynamics.

His **Law of Motion** is expressed in terms of Acceleration and amounts to
$$(\text{mass}) * (\text{acceleration}) = (\text{applied force})$$
or using **Newton's** notation for derivatives with respect to time **t** this is
$$m\ddot{x} = F$$
In a field of force, say $F(x)$, with a work function $dW(x) = Fdx$ and writing the acceleration in the form vdv/dx the law gives, on integration, the **Kinetic Energy** of a particle as $K.E = \frac{1}{2}mv^2$.

Prior to **Newton's** time there had been much speculation as to how forces affect changes of position and/or velocity. **Newton's** law identified that effect via the **change in velocity** (**acceleration**) produced by the force - the commonest example of force being, of course, that of **weight**.

Thus it became possible to distinguish between "mass" and "weight", the latter being the "force" due to "gravity" acting on the "mass". Thus the Physics was able to move from kinematics to kinetics whilst changes (motions) from the equilibrium configurations of Statics to Dynamical phenomena became understandable, and the "weight" of a mass **m** is then the familiar **mg**, **g** being a constant acceleration vector in local neighbourhoods on the surface of the earth.

Newton's Law is more comprehensively contained by identifying **momentum** (the product **mv**)

when it becomes $\quad d\mathbf{p}/dt = \underline{\mathbf{F}}$, $\mathbf{p}$ being the momentum vector $m\underline{\mathbf{v}}$,
and both variables being regarded as vectors in $\mathcal{Q}$ (over $\mathfrak{R}^{\#}$).

None of this gets us out of the dilemma of trying to identify "mass" as an <u>absolute</u> measure since out of the three variables $\mathbf{m,p,F}$ the Law can only define one of them in terms of the other two. However this is hardly a serious obstacle in **Physics-3**.

The equation $\quad d^2\mathbf{x}/dt^2 = \mathbf{f}$ (eg, a constant) immediately gives us many elementary results in **Physics-3**.

In a **conservative field of force** with work Function $W(x)$ the introduction of the **Potential Function** $V(x) = -W(x)$, and since $F(x) = -dV/dx$, the princple known as the **conservation of energy**, viz. K.E + P.E = constant, follows.

Considering measures in the **Einstein** world we need to replace the above momentum by an expression which is invariant under the **Lorentz** operator $\mathcal{L}$, and this turns out to be $\quad \mathbf{p} = \beta_0 m\underline{\mathbf{v}} \quad$ where $\beta_0 = \{1 - |\underline{\mathbf{v}}|^2/c^2\}^{-\frac{1}{2}}$.

The apparent "mass" of the particle will then be $\beta_0 m_0$, m_0 being the "mass" on a scale where the particle is at rest.

[In the literature of Relativity these "scales" are referred to as "inertial frames of reference"]

In a field of force, with Work Function $W(\underline{\mathbf{r}})$ we have grad $W = \underline{\mathbf{F}}(\underline{\mathbf{r}})$ and taking $\underline{\mathbf{F}} = d(\beta_0 m\underline{\mathbf{v}})/dt \quad$ by **Newton's** Law we can show that

$$W = m\beta_0 c^2$$

and this leads to some well known results, particularly the appearance of the **intrinsic energy** to be found in a mass at rest - $\mathbf{m_0 c^2}$. [v. Appendix-E [13]]

[3.4] Momentum v Time urve

Newton's law relating force and momentum expects to find (eg) the curve of $\mathbf{mv}$-$\mathbf{t}$ (or $\mathbf{mv}$-$\mathbf{x}$) as a **differentiable curve**, possessing a tangent at the selected point. But common experience has led to the realisation that there can occur sudden **finite discontinuities** in that curve. [v. Fig. 3.1]. When this occurs we interpret the finite jump in $\mathbf{mv}$ as an **impulse** and from this we establish the conservation of momentum when inelastic objects collide [v. Appendix-E [12 In mathematical terms this means that the momentum mv is a possible function of position and of time - and that it is not necessarily either continuous or differentiable at all points in the geometry.

[3.5] <u>Newtonian and Einsteinian Gravitation</u>

No doubt the most notable of **Newton's** achievements was the law of **universal gravitation**, viz., point masses m_1, m_2 at a vector distance $\underline{r}$ apart attract each other with a force which is proportional to $m_1 * m_2/|\underline{r}|^2$ and which acts along the direction of the vector $\underline{r}$ - the **inverse square law**. He was then able to show that the planetary orbits are **ellipses**, with Sun at a focus, compatible with observations obtained by **Tycho Brahe** at an earlier date, as well as being a verification of the remarkable deductions made by **Kepler** from these observations - and which are referred to as **Kepler's Laws**. [v. eg Ref D-12]

The $\Gamma(1\text{-}1)$ which embodies the gravitational field, with potential function **V**, is the simple

$$\Gamma(1\text{-}1) \equiv Vr + \mu = 0$$

and this gives the **Newtonian** gravitational force as $F = -dV/dr = -\mu/r^2$.

In $\Re^\#$ **V** and **r** are scalars, but if we place them in **Q** (over $\Re^\#$) we must write

$$V + \underline{U} \quad \text{in place of V} \quad \text{and} \quad r_0 + \underline{r} \text{ in place of } r.$$

The $\Gamma(1\text{-}1)$ then becomes

$$\Gamma(1\text{-}1) \equiv (V + \underline{U})(r_0 + \underline{r}) + \mu = 0$$

The gravitational results are then given by

$$(V + \underline{U})(r_0^2 - |\underline{r}|^2) = -\mu(r_0 - \underline{r})$$

Equating the scalar and vector parts of the quaternions give us

$$V = -\mu r_0/(r_0^2 - |\underline{r}|^2) \quad \text{and}$$

$$\underline{U} = \mu\underline{r}/(r_0^2 - |\underline{r}|^2)$$

If **h** = plane angular momentum/unit mass we write $h/c = r = |\underline{r}|$ whence, expanding **V**, gives

$$V = -\mu/r_0 \{1 + (h/cr_0)^2 + \text{higher powers}\}$$

and a first approximation gives us $V = -\mu/r_0 - \mu h^2/c^2 r_0^3$

and the gravitational force via $F = -dV/dr_0$ becomes

$$F = -\mu/r_0^2 - 3\mu h^2/c^2 r_0^4$$

this being the accepted **Einstein correction term** to the **Newtonian** force.

[Note: See Chapter-12 in this PART-1 for the **Einstein** treatment of gravitation - in which he derives the result for a 4-space with a **Riemannian** metric.]

But we notice that the vector component $\underline{U}$ has the dimensions of **angular momentum** or **spin**, which we therefore expect the mass-particle to possess. It is highly suggestive that this should refer to the spin of the planets in the solar system, being part of, if not all, of the planet's angular momentum.

The derivation of a planet's orbit, under the inverse-square law of force, is discussed in Chapter-4, via the work of **Lagrange** and **Hamilton**.

[3.6] Duality in **Physics-3**

A common and significant example in particle dynamics is that of repetitive and oscillatory motion, viz., **simple harmonic motion**, SHM. This occurs (eg) linearly when the applied force is proportional to the displacement and directed always towards the centre (origin) **O**. It also occurs in the familiar case of the simple pendulum.
These two cases can be seen to be **duals** of each other (in the sense of projective geometry).

Referring to Fig.3.2, the upper diagram shows the oscillations produced by a particle suspended by a spiral spring (or elastic string) ; being disturbed from its static position and then released under gravity then it moves in SHM. This is illustrated by the horizontal line A'A and as it moves backwards and forwards along this line we can see that by orthogonal projection [The centre of perspective being at ∞] the motion is equivalent to the motion of a point around the circle on AA' as diameter. So the x-co-ordinate on the line is given by $x = \cos\theta$, and $\theta = \omega t$, the **Newton's** equation for the motion being $d^2x/dt^2 = -\omega^2 x$. Here ω is a constant real number and the angle θ in the diagram is then ωt.

But this motion is exactly the **dual** of the simple pendulum motion, also shown in the diagram. For whereas the first example can be described as

"the motion of a point **P** on a line ℓ between extremes **A** and **A'**"

so the second example can be described as

"the motion of a line **p** on a point **L** between extremes **a** and **a'**".

The algebraic description of the motion is the same in each case - being a trigonometric (periodic) function of the time.

For any parameter of a dynamical system which moves under a law of motion like

$$d^2\theta/dt^2 + \omega^2\theta = 0$$

the solution of the problem is of the form $\theta = a\cos\omega t + b\sin\omega t$. In mathematical terms it explains the close relation between the trig functions and the exponential number e. This is because, in the latter case, $d\theta/dt = k\theta$, which gives $\theta = ke^{kt}$, and thence

$$d^2\theta/dt^2 = k^2 e^{kt} = k\theta.$$

[Note : **Euler** first pointed out that $e^{i\pi} = -1$ - alarming many mathematicians]

52 MATHEMATICAL PHYSICS

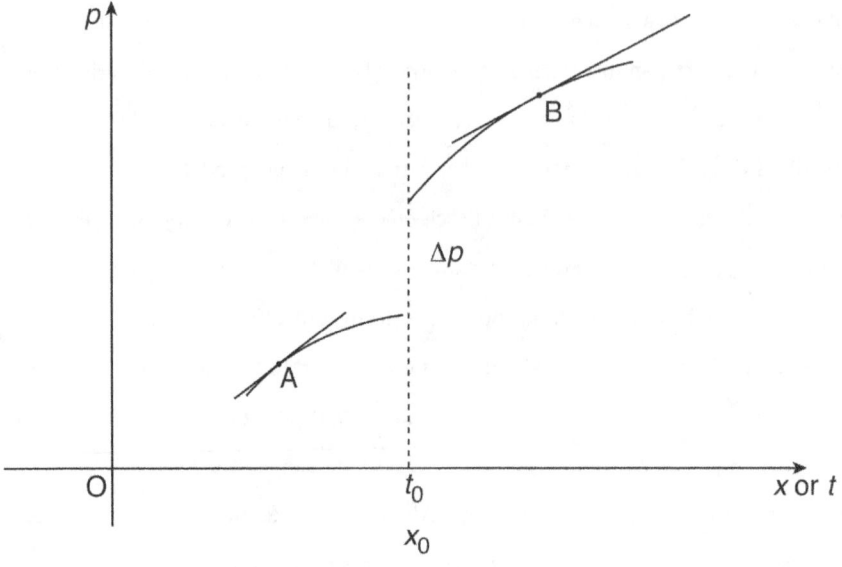

"Force" at A or B   "Impulse" at t_0

Fig 3.1

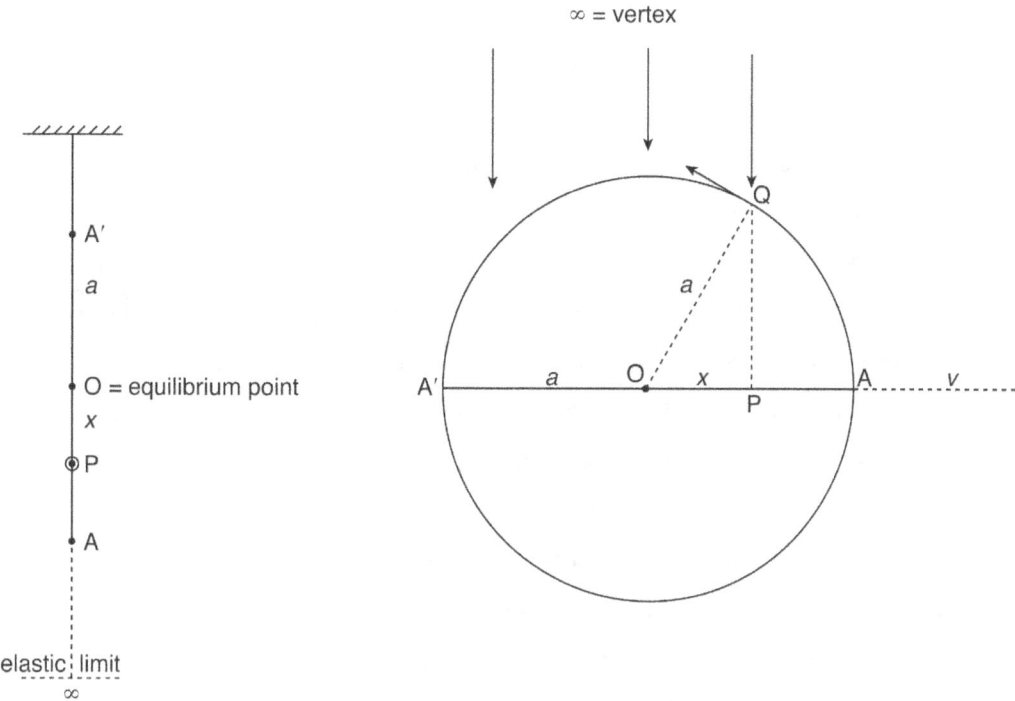

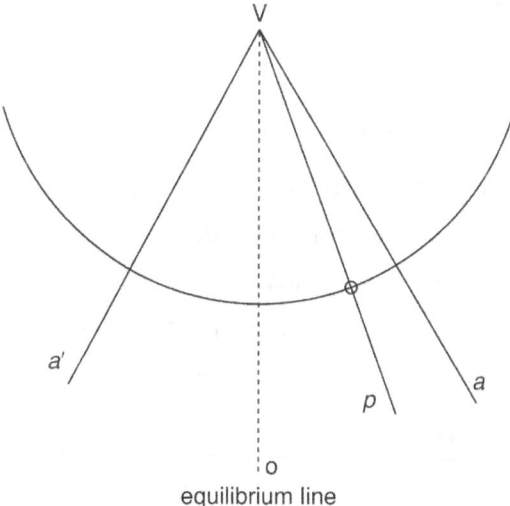

Fig 3.2

Chapter-4 / PART-1   Physics-4

[4.1] Rigid body dynamics

Physics-4 is the Physics of the Dynamics of "systems of particles" - including the non-trivial cases involving "rigid bodies" [c.f Ch-2].

It can be described as the union of **Physics-2** and **Physics-3** together with the kinematics needed to cope with the motion of rigid bodies.

$$\text{Physics-4} \equiv \{\mathbb{C}(\mathfrak{Z}_L, \mathfrak{Z}_T) ; E_m, E_p ; Q(\mathfrak{R}^\#) ; \sigma_r \}$$

Historically this Physics was founded by **Huyghens** and developed by **d'Alembert**, the former working largely from the conservation of energy equation and introducing the notion of the **radius of gyration** for a rigid body, whilst the latter set out a **principle** whereby a massive rigid body could be conceived as a collection of particles held together by "apparent" forces between them - these forces maintaining the rigidity via the invariance of the "distance" between them. All these matters were based in the algebra of $\mathfrak{R}^\#$ although later mathematicians, viz., **Hamilton** and **Heaviside**, felt the need to introduce one form or another of a "vector algebra". Perhaps it is true to say that **Huyghens** brought in **Physics-2**, to **Physics-4**, whilst **d'Alembert** anchored it in **Physics-3**.

And this insistence of the "particle" view of the world was no doubt to lead to many complications in the future. In the short term it did, however, bring many dynamical advantages.

[4.2] Radius of Gyration

The application of **d'Alembert's Principle** leads to the result that the overall kinetic energy of a moving rigid body can be expressed as

(K.E. of a particle of total mass, **M**, centred at the C.G.) +

(Rotational energy of the body relative to this C.G.)

This rotational energy was defined by **Huyghens** as ½$I(G)\omega^2$, $I(G)$ being the **moment of inertia** about the axis of rotation through **G**. This **I** was identified by **Huyghens** as Mk^2, **k** being the so-called **radius of gyration** about that axis.

Eddington described these effective "particles" as the **intracule** and the **extracule**.

In 3-dimensions the **I** becomes the so-called **inertia tensor** with components (relative to three perpendicualr axes through **G**) contained in the matrix

$$\begin{bmatrix} A & H & G \\ H & B & F \\ G & F & C \end{bmatrix}$$

[References : Appendix-E [9],[12],[16]].

[4.3] Polhode and Herpolhode cones

When a rigid body is allowed to move freely about a point **O** of itself we can show that the motion can be described as the

rolling of the polhode cone on the herpolhode cone

It is an illustration of the use of a moving frame of reference, viz. one which is fixed in the body, and rotates with it about a point of that body - the point often taken to be the C.G.. Since classical dynamics needs a fixed frame of reference in space, we take this to be a triad, S_0, of mutually perpendicular axes $O(x,y,z)$ and we consider the consequences of this fixed frame being momentarily coincident with a similar triad, **S**, which is moving with the body about the same origin O.

Then suppose that **S** moves with an angular velocity vector $\underline{\omega}$ relative to S_0. The motion of any dynamical vector **F** associated with the two frames will be such that

velocity of **F** relative to S_0 will equal

velocity of **F** relative to **S** + velocity of **S** relative to S_0

Writing the first of these as the usual $d\underline{F}/dt$ and the velocity relative to **S** as $\partial \underline{F}/\partial t$ we obtain
$$d\underline{F}/dt = \partial \underline{F}/\partial t + \underline{\omega} \times \underline{F} \quad (A)$$
since $\underline{\omega} \times \underline{F}$ is the velocity of $\underline{F}$ relative to **S**, $\times$ being the vector product.

If, for example, **S** is spanned by the orthonormal triad **i,j,k**, then the operator $\partial/\partial t$ treats these axes as constant (as far as the differentiation is concerned).

Applying (A) to the angular veclocity vector $\underline{\omega}$ gives the result
$$d\underline{\omega}/dt = \partial\underline{\omega}/\partial t \quad (B)$$
The instantaneous axis of rotation is the axis of $\underline{\omega}$ and this will describe a **cone** in space and another cone in the body itself. The first is called the **herpolhode cone** and the second is the **polhode cone**. But $d\underline{\omega}$ is a displacement in the tangent plane to the herpolhode cone whilst $\partial\underline{\omega}$ is a displacement in the tangent plane to the polhode cone.

Then (B) shows that the two cones touch along a common generator (defined by $\underline{\omega}$) and that sliding does not occur. It follows that the polhode cone rolls on the herpolhode cone.

[4.4] The Lagrange dynamical equations

The comprehensive description of the dynamics of a general system of many bodies is to be found via either the **Lagrange Equations** or the **Hamiltonian Equations**.

These are further discussed in the references : Appendix-E [9],[12],[16],[79],[80],[81] - the most comprehensive of these discussions will be found in **Whittaker's** book, ref [16].

The **Lagrange** equations depend upon identifying the function $L = T - V$ where T = total K.E. of the system, and V = total P.E. of the system. These functions will normally be functions of a set of independent co-ordinates $\{q_r\}$.

If we use Cartesian co-ordinates (x,y,z) and consider the motion of a particle of mass m_i under a force with components (X,Y,Z), subject to **Newton's Laws** of motion, we can write

$$m_i \ddot{x}_i = X_i \quad m_i \ddot{y}_i = Y_i \quad m_i \ddot{z}_i = Z_i$$

In general we can assume that the x,y,z are functions of more general co-ordinates q_r, where $r = 1, 2, \ldots n$. [For example, the q_r can be cylindrical or spherical polar co-ordinates]

Dropping the suffices and multiplying by the factors

$$\partial x/\partial q_r, \; \partial y/\partial q_r, \; \partial z/\partial q_r$$

and summing over all the elements of mass of the system, we get

$$\Sigma \, m \, \{\ddot{x}\partial x/\partial q_r + \text{ etc.}\} = \Sigma \, \{X\partial x/\partial q_r + \text{ etc.}\} \qquad (C)$$

But it is easily shown that, eg, $\ddot{x}\partial x/\partial q_r = \ddot{x}\partial \dot{x}/\partial \dot{q}_r$

which equals $\quad d/dt \, \{\dot{x}\partial \dot{x}/\partial \dot{q}_r\} - \dot{x}\partial \dot{x}/\partial q_r$

so that

$$\ddot{x}\partial x/\partial q_r = d/dt \, \{\partial/\partial \dot{q}_r \, (\tfrac{1}{2}\dot{x}^2)\} - \partial/\partial q_r \, (\tfrac{1}{2}\dot{x}^2) \qquad (D)$$

Since the Kinetic Energy of the whole system is

$$T = \Sigma \, \tfrac{1}{2}m \, \{\dot{x}^2 + \dot{y}^2 + \dot{z}^2\}$$

we can combine (C) and (D) to give

$$d/dt \, [\partial T/\partial \dot{q}_r] - \partial T/\partial q_r = \Sigma \, \{Xdx + Ydy + Zdz\} \qquad (E)$$

The left-hand side of this expression is the Work Done by the external forces and, expressing this in terms of the co-ordinates q_r, we get

$$\Sigma \, \{Xdx + \text{etc}\} = \Sigma \, Q_r$$

giving the **Lagrange Equations** of motion for the system, viz.,

$$d/dt \, [\partial T/\partial \dot{q}_r] - \partial T/\partial q_r = Q_r \quad r = 1, \ldots n \qquad (F)$$

where the Q_r are the components of "force" associated with the displacements dq_r.

These equations are second order differential equations (with respect to time) usually written in the form

$$\frac{d}{dt} \, (\partial L/\partial \dot{q}_r) - \partial L/\partial q_r = 0 \quad \text{for } r = 1 \ldots n. \qquad (G)$$

where the **Lagrange function** $L = T - V$ whenever the applied field of force is a conservative one, with potential funciton $V(q_r)$.

[4.5] Hamilton's dynamical equations

Although the co-ordinates q_r are truly independent it need not follow that the $\dot{q}_r$ are likewise. When the $\dot{q}_r$ are independent we say the dynamical system is **holonomic**, otherwise we say it is **non-holonomic**. [Note: A simple case of a non-holonomic system is provided by a sphere rolling and spinning on a rough surface - because there is a relation between the $\dot{q}_r$].

The **Hamiltonian** equations depend upon identifying the function

$$H = \Sigma_r p_r q_r - L, \quad \text{where } p_r = \partial L / \partial q_r \quad \text{and where}$$

L is regarded as a function $L(q_1, q_2, \ldots q_n, \partial q_1/\partial t, \ldots \partial q_n/\partial t)$

The variables p_r constitute the **generalised momenta** associated with the q_r and the whole set of 2n variables (p_r, q_r) is called the set of **canonically conjugate variables** for the system. We then define a function, called the **Hamiltonian**, H, of the sytem by

$$H = \Sigma_r \, p_r \dot{q}_r - L \qquad (H)$$

where the **Lagrangian** function L is to be expressed in terms of the canonical co-ordinates

$$q_1, q_2, \ldots q_n, p_1, p_2, \ldots p_n$$

by using the definition $\quad p_r = \partial L / \partial \dot{q}_r \quad r = 1, 2, \ldots n$

The definition (H) gives us the important **Hamiltonian Equations**, viz.

$$\dot{q}_r = \partial H / \partial p_r$$

and $\quad \dot{p}_r = - \partial H / \partial q_r \quad r = 1, 2, \ldots n$

When L (and therefor H) does not contain the time t explicitly then the **Hamiltonian** becomes simply the **total energy** of the system, viz.,

$$H = T + V$$

T being the Kinetic Energy and V being the potential energy of the system.

These 2n 1st-order differential equations determine the solution for the motion of the dynamical system.

[Note: For details of these equations and their applications to specific problems the reader is referred to various standard textbooks - for example Appendix-E [16]].

[4.6] Principle of Least Action

The **Lagrange equations** for a dynamical system can also be obtained by asserting the invariance of a certain integral, and by using the techniques of the **Calculus of Variations**.

This integral is $I = \int_C L\,dt$ where **C** is a curve in the dynamical path, starting at time t_0 and ending at time t_1.

The **Principle of Least Ation**, which was introduced by the mathematicians **Maupertuis** (1744), **Euler** (1744) and **Lagrange** (1746), successfully asserts that the "total action" of a system, during the time interval $t_1 - t_2$, viz., $\int_C T\,dt$ is (mathematically) stationary ; in fact it is a minimum, for the actual curve **C** between the two times t and $t + \Delta t$ - as compared with any neighbouring curve between those two times.

This gives prominence to the concept of **action**, which itself is any quantity with the dimensions of **action** - these being [Energy x time], or [momentum x distance], viz. ML^2T^{-1}.

Hamilton's Principle follows the same path and states that for small variations of the curve **C**, betwee the same two time limits, the integral is stationary (being an **integral invariant**).

This means that, in the language of the Calculus of Variations [v. Appendix-E Ref 48],

$$\delta \int_C L\,dt = 0 \quad \delta \text{ denoting any weak variation of the curve } \mathbf{C}.$$

and the **Euler** conditions on this integral immediately give us the **Lagrange equations** of motion.

We can identify this process with that of finding **geodesics** (shortest paths) in a metric space where the metric is an expression ds^2 [v. Appendix-B], for the integral

$\int_C T\,dt$ can also become the integral $\int_C mv\,ds$ or $\int_C p\,dq$

where **p** is the canonical (momentum) variable associated with the position variable **q**.

[4.7] Pfaff's differential form

The work of **Pfaff** in general dynamical systems is discussed in Appendix-E [9],[40]

For a general dynamical system, defined by independent co-ordinates $\{q_r, r = 1 \ldots n\}$ the **Pfaffian** linear form

$$p_1 dq_1 + p_2 dq_2 \ldots p_n dq_n - H\,dt$$

(where the **p**'s and **q**'s have their usual canonical meanings) generates **Hamilton's** equations by way of its **bilinear covariant**, as follows.

Writing $\quad H^* \equiv \Sigma\,(p_r dq_r + 0\,dp_r) - H\,dt \quad$ summed over $r = 1 \ldots n$

and where **H** is the **Hamiltonian** (equal to $(T + V)$ for holonomic systems) then using the definitions of $\quad p_r = \partial H/\partial \dot{q}$ (with $r = 1 \ldots 1$) we obtain the usual **Hamilton's** equations of motion as a consequence of the bilinear covariant of H^*.

This analysis is equivalent to regarding H^* as defined in the **algebraic structure** Λ^n whence the bilinear covariant of the **Pfaffian** linear form H^* is a straightforward generalisation of the vector "curl". That is to say, the mapping which takes the 1-form (in Λ^1) $\Sigma\, X_i dx_i$ summed over i = 1 ... n into the 2-form (in Λ^2)

$$\Sigma\, \{(\partial X_i/\partial x_j - \partial X_j/\partial x_i)\, dx_i \wedge dx_j\} \qquad i,j = 1 \ldots n$$

[4.8] Planetary orbits

We can illustrate this analysis by considering, in **Physics-3**, the planetary orbit under the inverse-square law of force.

[It's using a sledge-hammer to crack a walnut, but the topic is important, v. later]

We use plane polar co-ordinates (r,θ) for the position of the planet (viewed as a particle), relative to its Sun at the origin, O.

These co-ordinates will be the (q_1, q_2) in the **Lagrange/Hamilton** language.

The velocity components are therefore $\dot{r}$, along the radius, and $r\dot{\theta}$ perpendicular to **r**.

The momentum components, for a mass m, are $m\dot{r}$ and $mr^2\dot{\theta}$ - these are the conjugate momenta $\mathbf{p}_1$ and $\mathbf{p}_2$ respectively. We consider momentum per unit mass, so take m = 1.

Assuming the inverse-square law of force we have the potential function $\mathbf{V} = \mu/r$ where $\mu\,(> 0)$ incorporates mass-of-planet, mass-of-Sun, and universal constant of gravitation (**Newton**).

The Kinetic Energy, **T**, is $\frac{1}{2}(\dot{r}^2 + r^2\dot{\theta}^2)$ so we have, in **Hammilton's** theory

$$\mathbf{H} = \mathbf{T} + \mathbf{V} = \tfrac{1}{2}(p_1^2 + p_2^2/q_1^2) + \mu/q_1$$

whilst in **Lagrange's** theory we do not use the $\mathbf{p}_r$ but write

$$\mathbf{L} = \mathbf{T} - \mathbf{V} = \tfrac{1}{2}(\dot{q}_1^2 + q_1^2 \dot{q}_2^2) - \mu/q_1$$

The **Hamiltonian** equations of motion are

$$\dot{q}_s = \partial H/\partial p_s \qquad s = 1,2$$

and $\qquad \dot{p}_s = -\partial H/\partial q_s \quad s = 1,2$

These give $\quad \dot{p}_1 = p_2^2/q_1^3 + \mu/q_1^2 \quad$ and $\quad \dot{p}_2 = 0$

together with $\quad \dot{q}_1 = p_1 \quad$ and $\quad \dot{q}_2 = p_2/q_1^2$

We can integrate these to give

$$p_2 = \text{constant} = q_1^2\, \dot{q}_2 = \text{angular momentum, say h}$$

and $\qquad \ddot{q}_1 = \dot{p}_1 = p_2^2/q_1^3 - \mu/q_1^2$

In the usual notation these give

$$\ddot{r} - r\dot{\theta}^2 = -\mu/r^2$$

together with $\qquad r^2\dot{\theta} = h$

The **Lagrange** equations of motion, being

$$d(\partial L/\partial \dot{q}_s)/dt - \partial L/\partial q_s \qquad s = 1,2$$

give us
$$\ddot{q}_1 - q_1\dot{q}_2^2 = -\mu/q_1^2$$
and
$$d/dt(q_1^2\dot{q}_2) = 0$$

These amount to
$$\ddot{r} - r\dot{\theta}^2 = -\mu/r^2$$
and
$$r^2\dot{\theta} = \text{constant} = \text{angular momentum, say h}$$

as before.

These equations of motion can be solved by changing the variable to $u = 1/r$

whence
$$\dot{r} = -\dot{u}/u^2 = -h\,du/d\theta$$

so that
$$\ddot{r} = -h^2 u^2 d^2u/d\theta^2$$

and the equation of the orbit is a solution of the differential equation
$$d^2u/du^2 + u = \mu/h^2$$

With A, α arbitrary constants this gives us a solution of the form
$$(h^2/\mu)/r = 1 + A\cos(\theta - \alpha)$$

Comparing this with a standard euation of a **conic section**, focus at O (Sun), eccentricity e, and semi-latus rectum ℓ, viz., $\ell/r = 1 + e\cos(\theta - \alpha)$

we obtain an **ellipse** as the orbit, Sun at a focus, and other properties otherwise required by **Kepler** in his previously published **Kepler's laws** - derived by him from a remarkable study of astronomical data provided by the Danish scientist **Tycho Brahe**. [v. Appendix-E [12] [57]].

[4.9] Hamilton's Contact Transformations [v. Appendix-E [9],[82]]

Hamilton was intrigued by the apparent similarity between the trajectories in a dynamical sytem and the paths of optical light rays.

Fermat had already produced his Principle, viz., that the path of a light ray in a medium of refractive index μ iss a consequence of the integral $\int_C \mu(x,y,z)\,ds$ being stationary (in the sense found in the Calculus of Variations), that is to say that
$$\delta \int_C \mu(x,y,z)\,ds = 0.$$

Comparing this with the Principle of Least Action, in which
$$\delta \int_C \{h - \varphi(x,y,z)\}^{1/2}\,ds = 0$$

(where h = the constant energy and φ is the potential function in a conservative system) suggests that the dynamical trajectory behaves like the optical trajectory, provided we can equate the refractive index, μ, with $\{h - \varphi(x,y,z)\}^{1/2}$

Hamilton introduced the idea of a **characteristic function**, $V(x_1,y_1,z_1,x_2,y_2,z_2)$ which was to represent the dynamical "wave-front" as it progresses from the point P_1 to P_2

in a time interval $t_1 \to t_2$.

Following on from **Huyghens' Principle** for the propagation of a wave-front (by the envelope of a succession of wavelets), a **contact transformation** is any function which transforms one wave-front into the "next" one - and it must ensure that the envelope curve remains an envelope curve (must make "contact" with the secondary wavelets).

In a dynamical system this became translated into the following argument.

Suppose a system is defined by the variables $(q_1, q_2, \ldots q_n, p_1, p_2 \ldots p_n)$ and that another set $(Q_1, Q_2, \ldots Q_n, P_1, P_2, \ldots P_n)$ is to correspond to them at some later time (which is either finite or infinitesimally small).

If the equations which express the second set in terms of the first set are such as to make the differential form

$$P_1 dQ_1 + \ldots P_n dQ_n - p_1 dq_1 - p_2 dq_2 \ldots - p_n dq_n$$

into the perfect differential of a function of $(q_1, q_2, \ldots q_n, p_1, p_2, \ldots p_n)$ then this change from the first set to the second set is the **contact transformation** for the dynamical system.

It can be shown [v. Appednix-D [9]] that the conditions for the existence of a contact transformation can be expressd in terms of **Poisson Brackets**, which are defined as follows.

If **u, v** are functions of the canonical dynamical variables

$$(q_1 \ldots q_n, p_1 \ldots p_n)$$

the **Poisson Bracket** is denoted by **[u,v]** and defined by

$$[u,v] = \Sigma_r \{\partial u/\partial q_r \cdot \partial v/\partial p_r - \partial u/\partial p_r \cdot \partial v/\partial q_r\}$$

If $(Q_1 \ldots P_n)$ denote 2n functions of the variables $(q_1, \ldots p_n)$ these constitute a contact transformation whenever

$$[P_i, P_j] = 0 = [Q_i, Q_j] \quad i,j = 1, 2, \ldots n$$

and $\quad [Q_i, P_j] = 0 \quad i \leq j$

whilst $\quad [Q_i, P_i] = 1 \quad$ for $i = 1, 2, \ldots n$

[4.10] **Small Oscillations and Eigenvalues**

Let us first consider a simple dynamical system defined by one co-ordinate **q** and where equilibrium occurs when $\mathbf{q} = \mathbf{q}_0$. Let it be a conservative system, with potential function **V(q)** and kinetic function $\mathbf{T} = \frac{1}{2} f(q) \dot{q}^2$. Then we consider "small oscillations" of the system about the equilibrium position - by this we mean motion where, for small initial values of $|(q - q_0)|$ and of $\dot{q}$, the subsequent values remain bounded and of the same order as the initial values.

Writing $\xi = q - q_0$ then $d\xi/dt = \dot{q}$ and expanding f and V by Taylor's theorem we readily find that, to the second order in ξ and $\dot{\xi}$ the energy integral reduces to the equation viz., $\ddot{\xi} + \{V''(q_0)/f(q_0)\} \xi = 0$

showing that the motion is S.H.M provided $V''(q_0) > 0$.

Thus the position of equilibrium, $q = q_0$, is a **stable** position if $V(q)$ is a **minimum** at $q = q_0$. The motion of small oscillations is then equivalent to those of a simple pendulum with periodic time $2\pi \{f(q_0)/V(q_0)\}^{1/2}$.

This kind of motion is equally observed in a conservative dynamical system possessing **n** independent co-ordinates q_r, $r = 1 \dots n$.

Writing the K.E. as $T = \frac{1}{2} g_{rs} \dot{q}_r \dot{q}_s$ $(r,s = 1 \dots n)$ and $V = V(q_1, q_2, \dots q_n)$ we suppose the equilibrium configuration occurs at $q_r = 0$ for all r. Then for small oscillations we may consider the T and V in the forms

$$T = \tfrac{1}{2} a_{rs} \dot{q}_r \dot{q}_s \quad (1)$$
$$V = b_{rs} q_r q_s \quad (2)$$

where the a's are constants and the b's are independent of the q's.

These expressions are **quadratic forms** and (1) is **positive definite**.

Then we can consider these relations (1) and (2) more succintly expressed by the forms

$$T = \tfrac{1}{2} (\dot{q}.A.\dot{q})$$
$$V = \tfrac{1}{2} (q.B.q)$$

where A and B are matrices of order $n \times n$.

Because T is positive definite it is possible to find a linear transformation matrix Y which transforms both A and B to diagonal form; they become the sums of squares. That is to say we can transform T into (say) $\tfrac{1}{2}\{\dot{\xi}_1^2 + \dot{\xi}_2^2 + \dots + \dot{\xi}_n^2\}$ and V becomes $\tfrac{1}{2}\{\lambda^{-1}\xi_1^2 + \lambda^{-1}\xi_2^2 + \dots + \lambda^{-1}\xi_n^2\}$ the ξ_r being the new co-ordinates and where the λ_r are the values for which

$$\det(A - \lambda B) = 0$$

These λ's are referred to as the **eigenvalues** of the pencil of matrices $A - \lambda B$; the corresponding **eigenvectors** being the vectors $\xi = (\xi_1, \xi_2, \dots \xi_n)$ where

$$(A - \lambda B)\xi = 0$$

When **V** is also positive definite each $\lambda_r > 0$ and the **Lagrangian equations** for the motion result in the S.H.M's, viz., $\ddot{\xi}_r + \lambda_r^{-1}\xi_r = 0$.

The ξ_r are called the **normal co-ordinates** of the system and the motion in which ξ_m varies and ξ_r ($r \neq m$) does not is called the **mth normal mode** of the system.

Chapter-5 /PART-1   Physics-5, Heat and Gas Laws

[5.1] Various Laws

$$\text{Physics-5} \equiv \{\mathfrak{C}(\mathfrak{Z}_L, \mathfrak{Z}_T) ; E_m, E_\theta ; \mathfrak{R}^\# ; \sigma_r \}$$

where E_θ denotes measures of **temperature** on linear and other scales.

The observations on the scales show a loss of "rigidity" and this loss is attributed to **heat** and is taken as a measure of the intuitive "heat". Usually the scales are simply called **thermometers**.

These thermometers define **temperature** as being in a $\Gamma(1\text{-}1)$ with change of geometrical co-ordinates in rigid bodies. [Note : These are explained by introducing the ideas of "coeffients of expansion", etc.]. These latter changes are asumed to be linear - which can only be an approximation to the bilinear form of a $\Gamma(1\text{-}1)$.

The behaviour of gases is defined by measures of **V** (volume), **p** (pressure i.e force/area), and θ (temperature). The early laws (viz. those of **Boyle**, **Charles**) are attempts to find the $\Gamma(1\text{-}1)$ between **p** and **V** (isothermals) or that between **V** and θ (isobaric). But these are contained in

$$\Gamma(1\text{-}1) \equiv pV + ap + bV + c = 0 \qquad (1)$$

where $a, b, c \in \mathfrak{R}^\#$.

The **Boyle/Charles** laws are contained in (1) by considering

$$a = b = 0, c = -R\theta \quad (R \text{ a constant})$$

giving $\qquad pV - R\theta = 0 \qquad (2)$

But (2) is verified only over a limited set of scales, although even this degree of application implies the conditions $\quad a(\theta), b(\theta) \to 0$ as $\theta \to \infty$. $\qquad$ (2a)

Otherwise we can write (1) as

$$(p + b)(V + a) - (ab + R\theta) = 0$$

so that $\quad p + b = [(ab + R\theta)/V]\{1 - a/V + a^2/V^2 - ...\}$

and so for large θ we can obtain

$$p + b = (ab + R\theta)V^{-1} - a(ab + R\theta)V^{-2}$$

we can take (approx) $p = (ab + R\theta)/V$.

Then for a limited range we can consider the cruder approximation $b = k/V^2$ and taking $a = -\delta$, since a,b are of opposite sign. We than see that (1) contains the relation $\qquad (p + k/V^2)(V - \delta) = R\theta \qquad (3)$

the well-known **Van der Waal's** equation of state.

Other equations of state can be derived from (1) by postulating suitable functions for a,b (subject to (2a)). Thus we can see the significance of the following :-

(**Berthelot**) $(p + k/\theta V^2)(V-\delta) = R\theta$

(**Clausius**) $\{p + k/\theta(V + \mu)^2\}(V - \delta) = R\theta$

(**Dieterici**) $p(V - \delta) = R\theta e^{-k/\theta V}$

(**Callendar**) $p(V - \delta) + pc_0[\theta_0/\theta]^n = R\theta$

The merits of these relations are discussed in standard textbooks. [v. Appendix-E [23]]

[5.2] Kinetic Theories

Here we see an attempt to introduce the particle concepts of **Physics-3** into these scales by finding a $\Gamma(1\text{-}1)$ between the element **heat** and the concepts of particle dynamics. Since **Physics-5** has measures which are mapped into $\Re^{\#}$ the concepts must be scalars rather than vectors (in $Q(\Re^{\#})$). The prime scalar quantity which is used is that of **kinetic energy** - in which the word "particle" is replaced by the word **molecule**.

Thus the "kinetic theory" effectively removes "heat" from the generating elements and makes

   heat ≡ kinetic energy of particles in an (abstract) particle picture

But **heat** is restored to the status of a generating element in that part of the **Physics** known as **Thermodynamics** - which does not depend on the above identification. The fact that "heat" is regarded as a "form of energy" (**Mayer**, 1842) does not mean that any appeal to the above identification is made, but merely means that the apparent measures of the element **heat** is mapped into $\Re^{\#}$ as an invariant on the various scales $\mathfrak{C}_a$.

It is an interesting fact that "energy" in thermodynamics is a generating element in this version of **Physics-5** whereas in **Physics-3/Physics-4** it is not. In the latter cases it is <u>derived</u> as a concept from "force", but in thermodynamics there has never been a concept of force to start with.

It was early realised that "heat" is often followed by "motion" and that certain kinds of motion are followed by "heat". Perhaps the kinetic theory (and then **statistical mechanics**) was inevitable.

[5.3] First Law of Thermodynamics

A gas is a special kind of **fluid**, viz., a **compressible fluid** and because of this we find we must deal with **p** (pressure), **V** (volume), and $\mathfrak{I}$ (temperature) in various functional relationships - **Boyle's Law** being the prime example.

But then we must distinguish between situations in which

    (i) pressure is kept constant, and

    (ii) volume is kept constant.

In the first case, a rise in temperature causes an expansion in volume, and this immediately requires that **energy** is changed - either by the expanding gas doing external **work**, or by some external constraining force restricting the expansion.

These situations are accomodated by introducing the notions of

    (i) **internal energy**, dU, of the gas   and

    (ii) **external work**, dW, done by (or on) the gas.

When we specify the quantitiy of **heat/energy**, dQ, involved in any such transaction we can write $\quad\quad dQ = dU + dW \quad\quad$ (4)

this being the First Law of Thermodynamics.

[5.4] Specific heats and Adiabatic expansions

When the pressure is **p** and the change of (local) volume is dV we can obviously write

$$dW = pdV \quad\quad (5)$$

The theory then introduces the notions of **specific heats**, C_p and C_v ; C_p being the specific heat of the gas at **constant pressure** and C_v being that at **constant volume** - **specific heat** being defined as the amount of heat/energy required to raise the temperature of some suitable quantity (eg 1 gm-molecule) of the gas through one degree on an absolute temperature scale. Thus we can write the two cases as

    (i) $C_v d\vartheta = dU \quad$ for the specific heat C_v

and   (ii) $C_p d\vartheta = dU + dW \quad$ for the specific heat C_p

The **Joule, J**, has been taken as an experimental basic unit of energy (eg $1J = 4.185 \times 10^7$ ergs).

[For **Joule's** work v. Appendix-E [22],[23],[29],[30],[61]]

Whilst the "perfect gas" is taken as one which obeys **Boyle's Law**, $pV = R\theta$, actual gases commonly obey a modified law under conditions known as **adiabatic expansion** ; that is to say, when changes take place whereby there is zero (or nearly so) input, or output, of energy/work by the gas.

This leads to an "equation of state" of the form : $\quad pV^\gamma =$ constant where $\gamma =$ the ratio of C_p/C_v . [v. Appendix-E [22],[23],[61]]

Chapter-6 /PART-1 Physics-6

[6.1] This Physics is that of **Electrostatics** and is given by

$$\textbf{Physics-6} \equiv \{\mathbb{C}_\alpha(\mathfrak{Z}_L, \mathfrak{Z}_Q, \mathfrak{Z}_E) ; E_p, E_f, E_v) ; A ; \sigma_r\}$$

where the generating elements are E_p (geometrical points), E_f (line forces), and E_v being the concept of potential energy (derived from **Physics-3**).

the algebra **A** containing projective geometry - whose results have traditonally been expressed in both affine and metric/Euclidean geometry.

The **reflexive measures** associated with the electic charges (assumed to be located at points in the geometry) are comparable with those which define geometrical optics [v. Chapter-01]. What corresponds to the spherical mirror, in that study, is the earthed conducting sphere. The results were first pointed out by **Kelvin** in his theory of **electrostatic images**.

[Note: This is a case where an appeal to topics in projective geometry was used]

Referring to Fig. 6.1 with a point-charge $+\mathbf{p}$ at the point **P**, outside an earthed sphere, centre **O** and radius **r**, let there correspond a charge, say $-\mathbf{q}$, at the internal point **Q**. <u>Stationary Measures</u> of the Base Element $\mathfrak{Z}_Q$, on this Scale will constitute a basic **involution range** {**K,Q,H,P**} with the cross-ratio (**KQHP**) = -1. So **Q** and **P** are **inverse points** with respect to the sphere, giving

$$OQ = r^2/OP$$

We also notice that since Q and P are inverse points the circular section of the sphere (as shown in the Fig.6.1) is defined as the locus of points X which are such that

$$PX/QX = \text{constant} = PH/QH$$

and since we require $V_x = -q/QX + p/PX = 0$ for all such points X, this means that $PH/QH = \mathbf{p/q}$

Therefore, as the notes in Fig. 6.1 indicate, the value of the charge **q** must turn out to be $\mathbf{q = (r/y)p}$

the result which **Kelvin** first demonstrated.

The case of a point-charge outside an earthed conducting plane can be deduced by letting $r \to \infty$.

[6.2] Electrostatic Force Field

This is naturally found from consideration of the homography relating **V** to suitable co-ordinates.

In the field due to a point-charge there will be spherical symmetry, relative to the charge at the centre, and this means that we can write, eg

$$\Gamma(1\text{-}1) \equiv a\mathbf{V}\mathbf{r} + b\mathbf{V} + c\mathbf{r} + d = 0 \qquad (1)$$

where **r** is the radius of the imaginary sphere, and $a,b,c,d \in \Re^{\#}$.

As in other cases this amounts to the **inverse square law of force** since it can be reduced by an origin shift to the form

$$\Gamma(1\text{-}1) \equiv \mathbf{V}\mathbf{r} = \mu \quad \text{a constant.}$$

The **force field** derived from **V** gives the electric force vector **E**. When there is only a single component $\mathbf{E}_r$ in the spherical case this is

$$\mathbf{E}_r = -\partial V/\partial r \quad \text{whence} \quad \mathbf{E}_r = \mu/r^2.$$

Coulomb's Law illustrates this ; in traditional c.g.s. units it is

$$\mathbf{E} = q_1 q_2 / r^2.$$

Gauss' Law is deducible from this (or is often taken as a better foundation for the theory by philosophers who are unhappy with the concept of "action at a distance"), viz.

$$\int_S \underline{\mathbf{E}}.d\mathbf{S} \, \partial \, \text{(total charge within the closed surface S)}$$

[In c.g.s. units the constant of proportionality is 4π]

In the normal 3-space of Euclidean geometry we have the usual form

$$\underline{\mathbf{E}} = \text{grad}V = \nabla V = \partial_x V + \partial_y V + \partial_z V$$

V being a function of (x,y,z).

[The major significance of the electric vector $\underline{\mathbf{E}}$ arises in **Electrodynamics** (v. Ch-8)].

[6.3] Potential barrier

On the other hand, if we take the homography (1) above without shifting the origin, we see that it contains the relation - by writing $r_0 = -b$,

$$V = - (ar + c)/(r - r_0) \qquad (2)$$

and in those fields in which we normally assume **Coulomb's Law**, "a" is very small and "c" = $\pm k$ (the $\pm$ depending on the two kinds of electric charge).

The potential function given by (2) has the graphs shown in Fig. 6.1a. and we can see the two cases of attractive field and repulsive field.

In either of these two cases the graphs show the existence of a **potential barrier** to V measures as the observations move from $r > r_0$ to $r = r_0$. So this r_0 must be regarded as the **least distance** for which the inverse square law holds.

Now if we obtain the electric intensity (field force vector) from (2) we get

$$\mathbf{E} = - dV/dr = + (ar_0 + c)/(r - r_0)^2 \qquad (3)$$

being a repulsive force for $r < r_0$.

It was left to the theory of **wave mechanics** to explain how a nuclear particle might make its escape.

[6.4] Homography in the complex algebra $\mathbb{C}$

By changing the algebra **A** to that of the **complex plane**, $\mathbb{C}$, many problems in electrostatics can be taken as equivalent to simpler ones [v. Appendix-E [13] [15]]. This is because the concepts of **equipotential surfaces** and the orthogonal **E-lines** follow from **Gauss' Law** which gives (in the vector algebra notation) the field equation $\text{div}\,\underline{V} = 4\pi\rho$ (c.g.s units again) ρ being charge density in S.

Together with $\underline{E} = -\nabla V$ we obtain **Poisson's Equation**, $\nabla^2 V = 4\pi\rho$, and the (apparently ubiquitous) **Laplace's Equation**, $\nabla^2 V = 0$.

And the **Cauchy-Riemann Equations**, being a property of all **analytic complex functions**, $f(z)$, viz., with f written as $\varphi + i\psi$ and $z = x + iy$, then

$$\partial\varphi/\partial x = \partial\psi/\partial y \text{ and } \partial\varphi/\partial y = -\partial\psi/\partial x$$

giving the Laplace condition $\nabla^2 \varphi = 0 = \nabla^2 \psi$.

This means that, for problems which can be regarded as being in 2-dimensions (eg by taking a typical cross-section of very long, or infinite, planes/cylinders etc) we can take advantage of the **Cauchy-Riemann** properties of such **regular** complex functions.

In particular we use **conformal transformations** (which preserve the angles at which two curves intersect) so that the "real" and "imaginary" parts of such a function will automatically represent the **equipotential** and **electric force** lines for the problem.

[v. Appendix-E [13], [15], [65] for many worked examples]

[6.5] Conformal mappings

Regarding the conformal transformation as a mapping from the z-plane (in $\mathbb{C}$) to the ζ-plane (where $\zeta = \xi + i\eta$, in $\mathbb{C}$) and writing $\zeta = f(z)$ we can show that, given a twice-differentiable function $V(x,y)$ (eg a "potential function")

then $\qquad \nabla^2 V$ in the z-plane

becomes $\qquad |dV/dz|^2 \nabla^2 V$ in the ζ-plane,

and so it follows that at all points z where $d\xi(z)/dz \neq 0$

$\nabla^2 V = 0$ in the z-plane means also that

$\nabla^2 V = 0$ in the ζ-plane.

Example-1 Given two right circular cylinders of radii **a** and **b**, the one inside the other with their centres d apart : their cross-sections form two co-axial circles with limiting points, say L and L', (v.Fig. 6.2).

Then the transformation $\zeta = (z+c)/(z-c)$ transforms them into two concentric circles since $|\{(z+c)/(z-c)\}| = \lambda$ (the co-axial family)

becomes $|\zeta| = \lambda$ (the concentric family).

As a problem involving such cylinders in an electrostatic field the one in the ζ-plane is easier to solve [v. Ref D-13].

Example-2 The transformation $\zeta = \log z$, so that $\xi + i\eta = \log|z| + i \arg z$, and since we can always write $z = r e^{i\theta}$, gives the situations

(i) if we take η to be the potential function the equipotentials are the planes

θ = constant, and

if we take the special case where $\theta = 0$ and $\theta = \pi$ as the conductors we obtain the field when the two halves of a plane are at different potentials.

The lines of force are then concentric circles (r = constants)

(ii) if we take ξ to be the potential the equipotentials are simply concentric circles and the field is that due to a uniformly charged cylinder.

[6.6] The Schwarz-Christoffel method

This transformation enables us to find a function which maps the interior of any plane polygon (even if it is open to infinity) onto the upper half of the η-plane. The mapping is conformal except at the vertices of the polygon (where angles cannot be preserved).

[Note: Details of the method can be found in Appendix-E [13], [15], [65]].

The method provides the derivative $dz/d\zeta$ as an expression of the form

$$F(\zeta-u_1)^a (\zeta-u_2)^b (\zeta-u_3)^c \ldots (\zeta-u_n)^n$$

where there are n vertices in the polygon.

Fig. 6.3 shows the notation ; vertex z_r mapped onto u_r in the ζ-plane.

The indices are successive values of the form $\alpha_r/\pi - 1$ $r = 1\ldots n$.

The diagram in Fig. 6.4 shows the successive transformations from a section of two parallel planes (one infinite and the other half-infinite) into a straighforward "parallel plate condenser".

It shows how flexible is the method in dealing with strange-looking polygons !

[6.7] Other solutions of Laplace's equation

Solutions of the equation $\nabla^2 V = 0$, can be found by considering functions of the form

$V = R(r)\Theta\varphi$ by introducing "separation constants".

This can be done using various co-ordinate systems, viz.,

        Cartesian (Euclidean) co-ordinates (x,y,z)   (trigonometric/hyperbolic fns)

        Spherical Polar co-ordinates (r,θ,φ) (Various Legendre fns)

        Cylindrical Polar co-ordinates (r,θ,z) (Various Bessel fns)

and quite a few more.

72 MATHEMATICAL PHYSICS

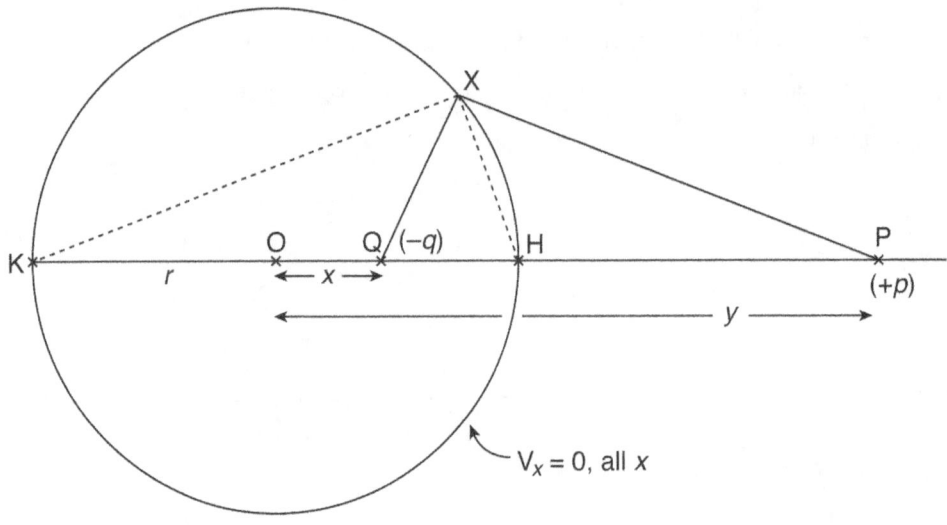

$(KQHP) = -1 \Rightarrow OQ \cdot OP = OH^2$ i.e $xy = r^2$

$V_x = 0 \Rightarrow \dfrac{-q}{XQ} + \dfrac{p}{XP} = Q \Rightarrow \dfrac{q}{p} = \dfrac{QX}{XP} = \dfrac{QH}{HP}$

$\therefore \dfrac{q}{p} = \dfrac{QH}{HP} = \dfrac{r-x}{y-r}$ & $x = \dfrac{r^2}{y} \Rightarrow \dfrac{q}{p} = \dfrac{r}{y}$

Fig 6.1 Kelvin's Result - Image Theory

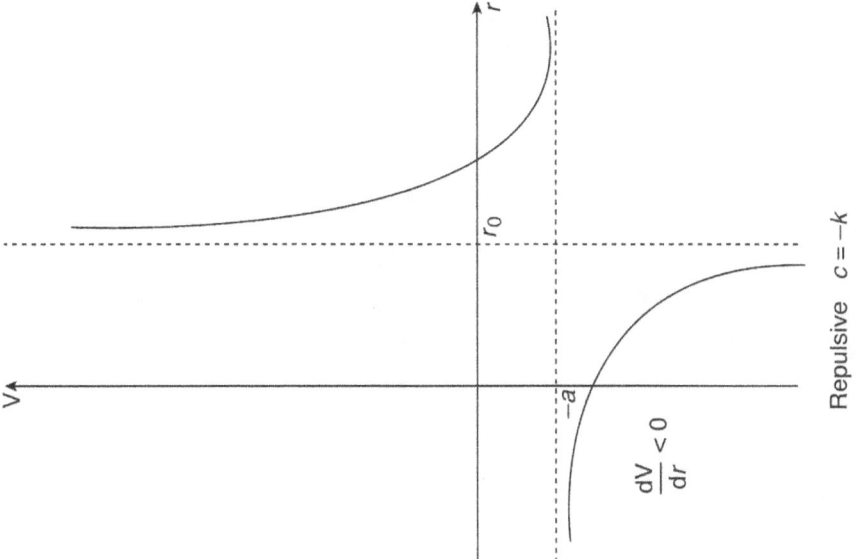

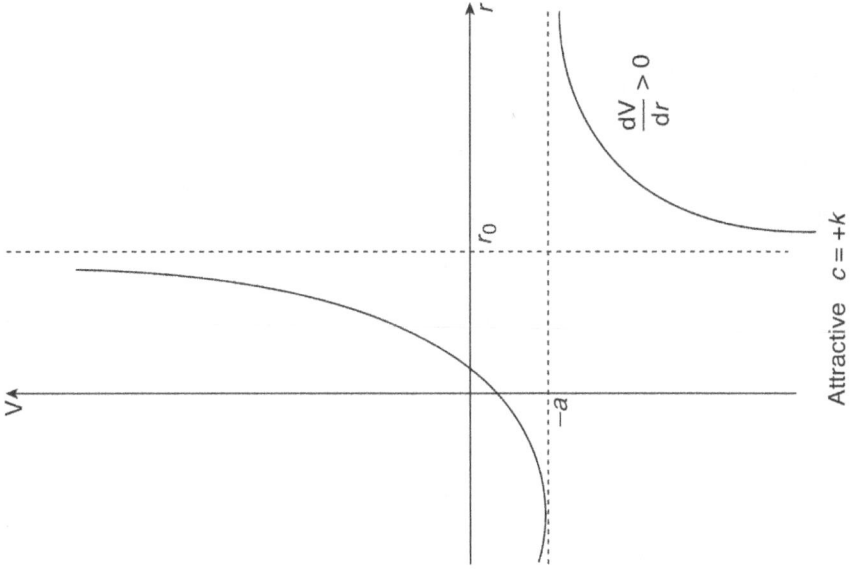

Fig 6.1a

74 MATHEMATICAL PHYSICS

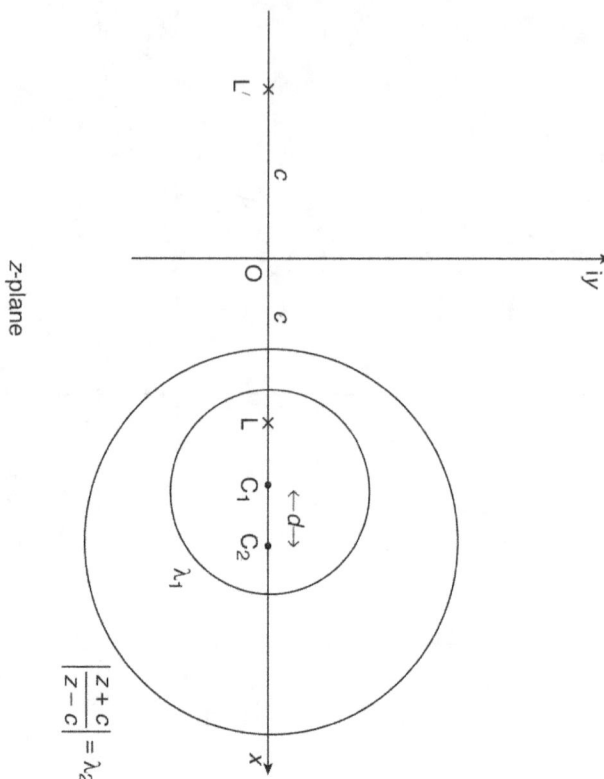

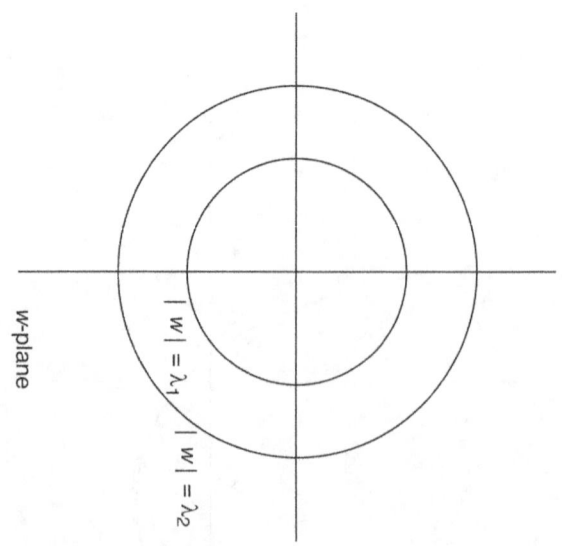

Comformal transformation $w = \dfrac{z+c}{z-c}$; Circle $\left|\dfrac{z+c}{z-c}\right| = \lambda \rightarrow$ Circle $|w| = \lambda$.

Fig 6.2

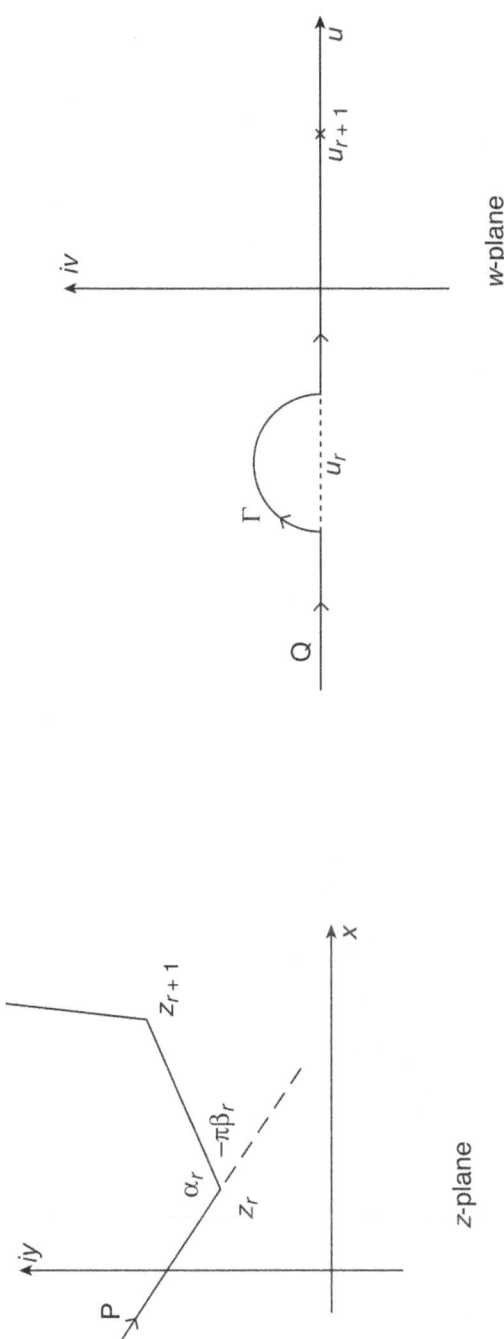

Schwerz-Christoffel transformation: z-plane → w-plane

$$\frac{dz}{dw} = F(w - u_1)^{\beta_1}(w - u_2)^{\beta_2} \ldots (w - u_n)^{\beta_n}; \text{ F a constant, } \beta_r = \frac{\alpha_r - 1}{\pi}$$

As P(z) moves the polygon Q(w) moves along the u-axis, avoiding u_r (image of z_r)

Fig 6.3

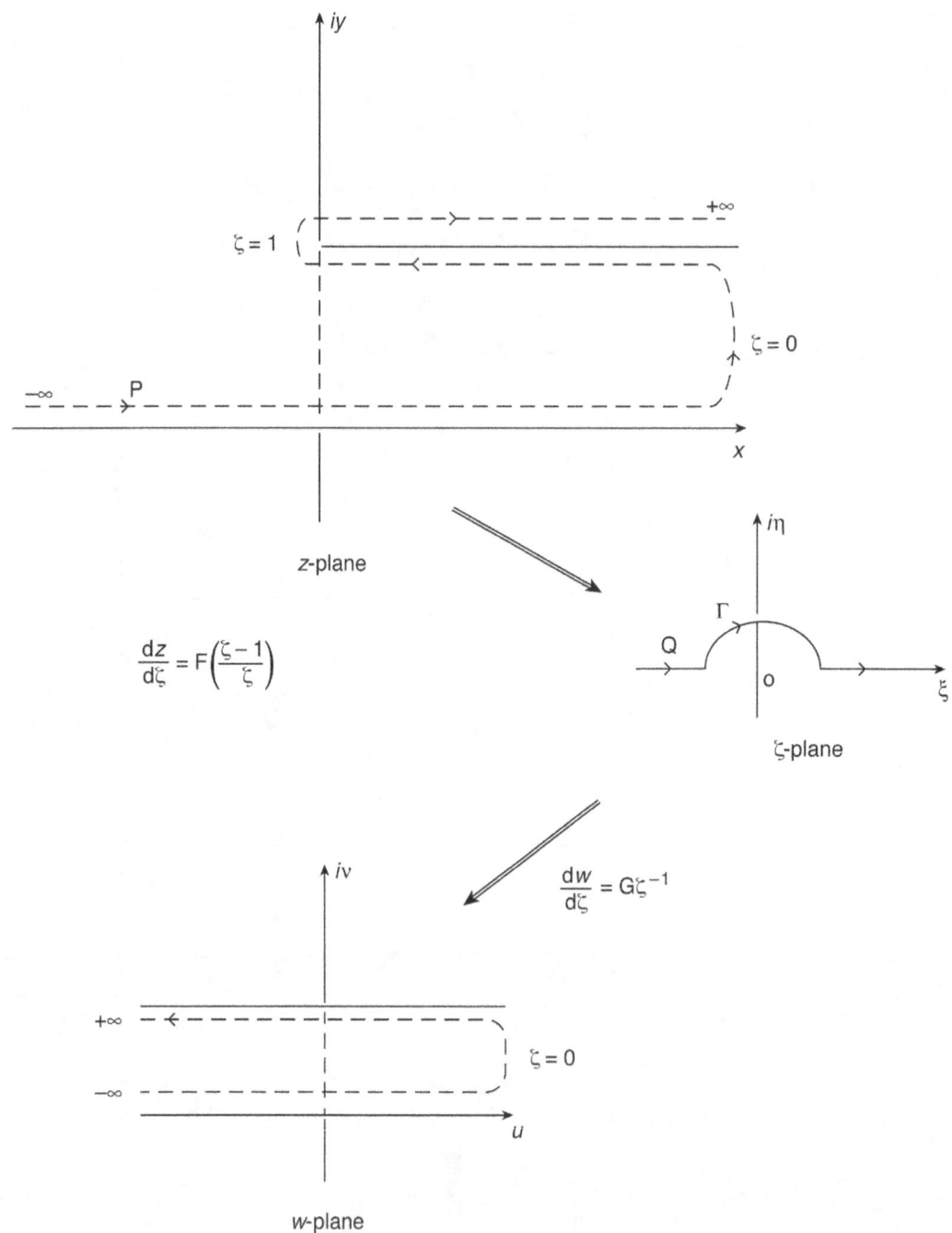

Fig 6.4

Chapter-7 /PART-1   Physics-7 - Magnetostatics

[7.1] Magnetic poles and dipole

This is the Physics of **Magnetostatics** - which was first developed in a manner parallel to that of **Physics-6** (Electrostatics) by the introduction of the idea of a **magnetic pole** (comparable to "electric point charge").

So we have the initial definition as

$$\text{Physics-7} \equiv \text{Physics-4} \cup \{\mathbb{C}_a(\mathfrak{Z}_L,\mathfrak{Z}_H) \,;\, E_h \,;\, A \,;\, \sigma_r\}$$

where the generating element E_h denotes observed magnetic force of attraction and $A \equiv \Re^\#$. This approach allowed the use of the concise homography (in a spherically symmetric case)

(1) $\Gamma(1-1) \equiv \varphi r = $ constant φ being a scalar potential function,

and we have seen how that is equivalent to the "inverse square law of force" between two imaginary "magnetic poles".

The disadvantage of this approach lay in the need to use a limiting process to deal with the actual field of an experimental magnet - which is manifest as the field of a **magnetic dipole**.

[7.2] Magnetic fields in Quaternions $\mathbb{Q}$

This really requires the concept of a **line vector** in space - thus suggesting the need to map the measures of E_h into the quaternion algebra.

In this case we cannot expect to use (1) above, as we did in Chapter-6, since "point" is inadequate to represent a line vector.

Also the generating element E_h must be regarded as **magnetic moment** of the dipole - giving Laws which find their counterpart in the parallel theory of **electric dipoles**. This gives us

(2) $\text{Physics-7} \equiv \text{Physics-4} \cup \{\mathbb{C}_a(\mathfrak{Z}_L,\mathfrak{Z}_H) \,;\, E_h \,;\, Q \,;\, \sigma_r\}$

and the homography must become of the form [v. Fig. 7.1]

(3) $\Gamma(1-1) \equiv (\underline{rr})(\varphi + \underline{A}) + \underline{M} = 0 + \underline{0}$ (the zero quaternion)

because the "potential function" as well as the "double position vector" must be quaternions, and we write (eg) the quaternion potential as "scalar + vector".

Now (3) can be solved for the potential quaternion (since $\mathbb{Q}$ is a division algebra) and we get $\varphi + \underline{A} = -(\underline{rr})^{-1}\underline{M}$.

Since the inverse of a quaterniion q is $q^*/|q|^2$ we use $(\underline{rr})^{-1} = -\underline{r}/r^3$

and obtain $\varphi + \underline{A} = (\underline{r}\,\underline{M})/r^3$

Since in familiar vector notation the quaternion product $\underline{r}\,\underline{M}$ **is equivalent to writing**

$$\underline{r}\,\underline{M} = -(r.\underline{M}) + \underline{r} \wedge \underline{M}$$

the result is the experimentally verified field, viz.,

$$\varphi + \underline{A} = -(\underline{r}.\underline{M})/r^3 + \underline{r} \wedge \underline{M}/r^3$$

[v. Fig. 7.1] The scalar magnetic potential is $\varphi = -(M \cos \theta)/r^2$ whilst the vector magnetic potential is $\underline{A} = (\underline{r} \wedge \underline{M})/r^3$.

[7.3] Magnetic field from the quaternion potential

Parallel to the scalar potential field theories we expect the magnetic force field to be represented by, say, $\underline{H}$ where $\underline{H} = -\nabla$(some potential function).
But now the ∇ operator must be defined in Q. We therefore find the invariant operator introduced by **Hamilton**, viz.,

$$\nabla \equiv \underline{i}\partial_x + \underline{j}\partial_y + \underline{k}\partial_z$$

This gives, in Q,

(1) $\qquad \nabla(\varphi + \underline{A}) = \mathrm{grad}\varphi - \mathrm{div}\underline{A} + \mathrm{curl}\underline{A}$

and since $\underline{H}$ is observed as a vector on the Scales the rhs of (1) must be a vector, giving the well-known equations for the magnetic field

$$\mathrm{div}\underline{A} = 0 \quad \text{and} \quad \underline{H} = -\mathrm{grad}\varphi - \mathrm{curl}\underline{A}$$

We notice also that a **Poisson's equation** comparable to the case in Electrostatics is the same as $\mathrm{div}\underline{A} = 0$ since the "total charge" within an enclosed space must be zero - dipoles consisting of coincident $\pm$ "charges/poles".

[7.4] Magnetic shells $\Phi = \xi\Omega$

The idea of a **magnetic shell** arose from the early studies in magnetostatics.

This is taken to be any magnetised body whose thickness is to be regarded as infinitesimal. If the intensity of magnetisation is, say **I**, and the thickness of the shell is ε, the product $I\varepsilon$ is called the "strength" of the shell. If then dS is any small area of the shell the product "strength" $\times$ dS , viz., $I\varepsilon dS$, is the effective magnetic moment associated with that dS , its axis being along the normal to dS : we denote this moment by ξ.

Any compact finite area S can be regarded as made up of such elements dS and together they will add up to an efective magnet, viz., the magnetic shell in toto.

Now the magnetic potential function Φ at any point P outside such a shell can be found by
$$\Phi_P = \int\int_S \xi \, d\omega$$
where $d\omega$ is the **solid angle** subtended at P by the element dS of the shell.

When the strength of the shell is uniform the function ξ can be taken outside the integral sign, whence the potential function at a point P becomes simply

$$\Phi_P = \xi\, \Omega_P$$

Ω_P being the solid angle subtended at P by the surface **S** of the shell.

This property becomes important when we are dealing with ElectroDynamics [v. Chapter-8].

[7.5] Laplace's equation

Since the magnetic potential function, Φ, also satisfies **Laplace's equation** we can expect the mathematics discussed in Chapter-6 to be applicable - as indeed it is.

We therefore find that magnetostatic problems can be discussed by using the complex algebra, $\mathbb{C}$, and the **analytic** and **regular** functions in that algebra.

We therefore find that the literature referencing magnetostatic fields uses various solutions of **Laplace's** equation, for the magnetic scalar potential.

In particular, this can include expansions in series of the form

$$\Sigma\, \{(r^n + r^{-n})\, P_n\, (\cos \theta)\} \quad \text{when using spherical co-ordinates}$$

Similar series involving various **Bessel** functions can be used when using cylindrical polar co-ordinates (r, θ, z). [v. Appendix-E [65]]

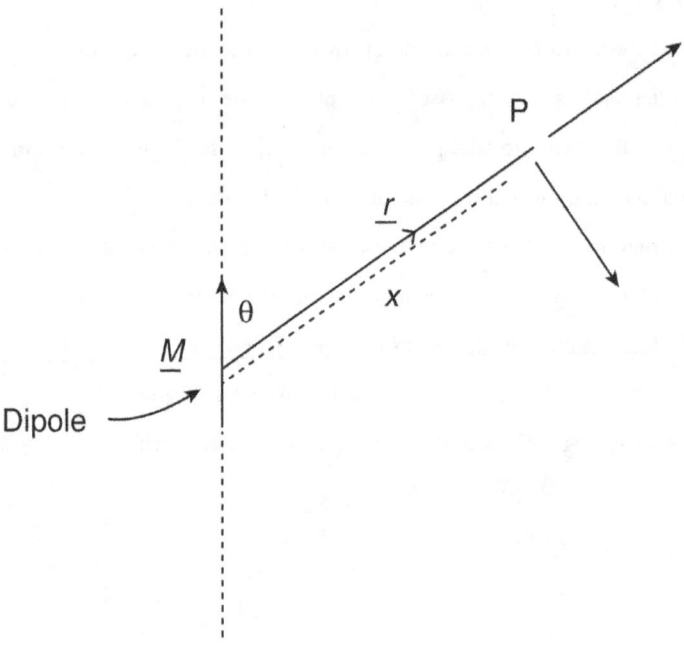

Fig 7.1

Chapter-8 /PART-1   Physics-8 Electrodynamics

[8.1] **Work of Ampere, Oersted, Faraday**

This is **Electromagnetism** - the dynamical extension of **Electrostatics** and **Magnetostatics**.

It can be defined as

$$\textbf{Physics-6} \cup \textbf{Physics-7} \cup \{\mathfrak{C}_a(\mathfrak{Z}_L, \mathfrak{Z}_E, \mathfrak{Z}_H, \mathfrak{Z}_T) \; ; \; \Omega \; ; \; \sigma_r\}$$

Since **q**, charge, is contained in **Physics-6** we notice the simple properties of discrete components via $\Gamma(1\text{-}1)$'s, viz.,

$$q = C^{-1}V \; ; \; \dot{q}\Omega = V \; ; \; \ddot{q}L = V$$

defining **capacity** C, **resistance** Ω, **inductance** L in terms of the electric potential **V**. But the significant relations in this Physics are those between the electric field, **E**, and the magnetic field, **H**.

The basis of these correspondences was laid by the experimental work of **Oersted, Ampere,** and **Faraday** ; the resultant "field theory" being the elegant work of **Maxwell**.

Ampere showed, experimentally, that a coil (closed loop) of wire carrying a current **I** behaves like a magnetic shell with "strength" proportional to the current in the wire. And by choosing suitable units of measurement (the e.m.u) he could write the magnetic potential function as

$$\varphi_P = I\omega_P$$

where ω_P is the solid angle subtended at P by the coil.

This led to **Ampere's circuital theorem** where, still using the concept of a single magnetic pole, it states that :

The energy needed to take a unit magnetic pole once round a current I (e.m.u) is $4\pi I$ ergs

In this system the current **I** is measured in emu via its magnetic effect, and this is not the same as current being measured as "electric charge per second" which is based on electrostatic effects (via **Coulomb's Law**). In this latter case the units will be e.s.u. and the ratio of the emu (measure of current) to the esu (measure of current) is denoted by **c**. This has dimensions of **velocity**, and (surprisingly ?) comes out experimentally as (near enough) 3×10^{10} (the velocity of E-M waves in free space).

An idea of the relative magnitudes of these things follows from a calculation showing that

1 emu of current is equivalent to 10 amperes.

Faraday's work showed that a varying magnetic field, such as could be produced by a coil of wire rotating between the poles of two bar magnets, produces a flow of electric current in the coil. His "magnetic flux" through the area of the coil, say **N**, changing at the rate of dN/dt produces in the wire an electromotive force (e.m.f) given by $-$ dN/dt - being the "back emf" which is the electromagnetic parallel to the "inertia" found in dynamical systems. All these early results were to have profound technological applications - up to the present day.

[8.2] Using the quaternion algebra $\mathbb{Q}$

The important feature of the measures of the field vectors **E** and **H** is expressed in **Euclidean** terms by their being mutually perpendicular. This is in fact a sign that the underlying algebra needs the complex numbers $\mathbb{C}$ for their representation. This in turn suggests that the quaternion algebra should be over $\mathbb{C}$ - which we can write as $\mathbb{Q}$(over $\mathbb{C}$).

We therefore suggest that the algebra should be the Cartesian Product $\mathbb{C} \times \mathbb{Q}$.

Ampere showed that a magnetic field **H** is manifest when an electric charge flows (an electric current) in a line (wire). **Faraday** showed the converse result - that the motion of a magnetic field generates a flow of electic charge - **E** and **H** being always at right angles to each other.

This is all contained in the algebra $\mathbb{C} \times \mathbb{Q}$ by using the "complex quaternion" $e_1 \mathbf{E} + e_2 \mathbf{H}$ for the total field - where $e_1^2 = 1$, $e_2^2 = -1$ - thus we write

$$\mathbf{F} = e_1 \mathbf{E} + e_2 \mathbf{H} \quad \text{as the field quaternion.}$$

If we restrict our Scales to those involving the particle concept the frames of reference will be those of **Physics-3** with generating elements **position** and **time**.

Using the geometry associated with **Physics-3**, viz., P^3 with a Euclidean metrical vocabulary it seems natural to associate **position** and **time** with the basal units of our quaternion algebra $\mathbb{Q}$ (which is required by **Physics-8**).

Thus we may specify a **particle event** by the quaternion

$$ct + ix_1 + jx_2 + kx_3$$

where c has dimensions of velocity. [Compare this with the **Minkowski metric**]. This will also enable us to make use of the **Lorentz-Einstein** formulae derived from the reflexive measures (in involution) of the Base Element $\mathcal{B}_L$.

In the formulation of E-M field equations we have already seen that **E** may be obtained from a scalar potential function **V** whereas **H** may be derived from a vector potential function **A**. If now we extend our quaternion (**Hamiltonian**) operator ∇ to include our 4th variable, (ct), we find that the usual "potential force" relation

Force-vector = - (grad (quaternion potential))

gives us all the important electromagnetic relations derived by **Maxwell**.

To do this we need only introduce the operator

$$\Box \equiv e_1 \nabla + e_2(1/c)\partial/\partial t$$

whence the quaternion field $\underline{F}$ is given by

$$\underline{F} = -\Box(\varphi) \quad \text{with} \quad \varphi = e_1 V - e_2 \underline{A}.$$

Equating the coefficients of e_1 and e_2 gives us the following relations, viz.

$$\text{div}\underline{A} + \frac{1}{c}\frac{\partial V}{\partial t} = 0 \quad (1)$$

$$\underline{E} = -\nabla V - \frac{1}{c}\frac{\partial \underline{A}}{\partial t} \quad (2)$$

and $\quad \underline{H} = \text{curl}\underline{A}. \quad (3)$

from which the **Maxwell** equations in free space follow, viz.

$$\text{curl}\underline{E} = -\frac{1}{c}\frac{\partial \underline{H}}{\partial t}$$

and $\quad \text{curl}\underline{H} = \frac{1}{c}\frac{\partial \underline{E}}{\partial t}$

provided $\quad \text{div}(\text{grad}\underline{A}) = 0.$

In a more general space, with a current density vector **j** and a charge density ρ, we use the **displacement vector, D** and the **magnetic flux vector, B** to give the more general Field equations, viz.,

$$\text{div } \mathbf{D} = 4\pi\rho$$

$$\text{div } \mathbf{B} = 0$$

$$\text{curl } \mathbf{H} = 4\pi\mathbf{j} + (1/c)\partial\mathbf{D}/\partial t$$

$$\text{curl } \mathbf{E} = -(1/c)\partial\mathbf{B}/\partial t$$

and together with these we must take the subsidiary relations [Appendix-E [13]]

$$\mathbf{D} = \kappa\mathbf{E}$$

$$\mathbf{B} = \mu\mathbf{H}$$

$$\mathbf{j} = \sigma\mathbf{E}$$

$$\mathbf{B} = \text{curl } \mathbf{A}$$

$$\mathbf{E} = -(1/c)\partial\mathbf{A}/\partial t - \nabla V$$

$$\text{div } \mathbf{A} = -(\kappa\mu/c)\partial V/\partial t - 4\pi\sigma\mu V$$

[8.3] Invariance under the **Lorentz-Einstein laws.**

The problem is now to discover how the $\Gamma(1-1)$ between the measures $\underline{F}$ (seen by the Scale/Observer $\mathfrak{C}$) and the measures $\underline{F}^*$ (seen by the Scale/Observer $\mathfrak{C}^*$) may be predicted from the relations (1) (2) (3) above.

We will write V_0 for the velocity of $\mathfrak{C}^*$ relative to $\mathfrak{C}$ and, to facilitate the printing, we introduce the operators $\delta_r \equiv \partial/\partial x_r$ for $r = 1,2,3,4$, where $x_4 = ct$.

Then (2) can be written via a matrix operator, viz.,

$$\begin{bmatrix} E_1 \\ E_2 \\ E_3 \\ 0 \end{bmatrix} = \begin{bmatrix} -\delta_4 & 0 & 0 & -\delta_1 \\ 0 & -\delta_4 & 0 & -\delta_2 \\ 0 & 0 & -\delta_4 & -\delta_2 \\ 0 & 0 & 0 & 0 \end{bmatrix} \begin{bmatrix} A_1 \\ A_2 \\ A_3 \\ V \end{bmatrix} \quad (4)$$

and (9) is equivalent to

$$\begin{bmatrix} H_1 \\ H_2 \\ H_3 \end{bmatrix} = \begin{bmatrix} 0 & -\delta_3 & \delta_2 \\ \delta_3 & 0 & -\delta_1 \\ -\delta_2 & \delta_1 & 0 \end{bmatrix} \begin{bmatrix} A_1 \\ A_2 \\ A_3 \end{bmatrix} \quad (5)$$

Since there will be no ambiguity we shall write $e_1 = 1$, $e_2 = i$ whence in $\mathfrak{Q}$ (over $\mathfrak{C}$) we have $\underline{F} = \underline{E} + i\underline{H}$ so that (4) and (5) give

$$\underline{F} = \mathfrak{J} \, W, \quad W \text{ being the quaternion } V + \underline{A}$$

and the matrix $\mathfrak{J}$ being

$$\mathfrak{J} = \begin{bmatrix} -\delta_4 & -i\delta_3 & i\delta_2 & -\delta_1 \\ i\delta_3 & -\delta_4 & -i\delta_1 & -\delta_2 \\ -i\delta_2 & -\delta_1 & -\delta_4 & -\delta_3 \\ 0 & 0 & 0 & 0 \end{bmatrix} \quad (6)$$

To find $\underline{F}^*$ (relative to $\mathfrak{C}^*$) we need the corresponding operator $\mathfrak{J}^*$ such that

$$\underline{F}^* = \mathfrak{J}^* W^* = \mathfrak{J}^* \mathcal{L} W$$

This means that

$$W^* = \mathcal{L} W = \begin{bmatrix} \beta(A_1 - V_0 V/c) \\ A_2 \\ A_3 \\ \beta(V - V_0 A_1/c) \end{bmatrix} \quad (7)$$

Also since $[dx_r] = \mathcal{L}^{-1} [dx^*_r]$ where $r = 1 \ldots 4$

we have the new operators

$$\delta^*_1 = \beta(\delta_1 + V_0 \delta_4/c) \, ; \, \delta_2^* = \delta_2 \, ; \, \delta_3^* = \delta_3 \, ; \text{ and}$$

$$\delta_4^* = \beta(\delta_4 + V_0 \delta_1/c)$$

Using this result and substituting into the corresponding $\mathfrak{J}^*$ we get the following, viz.

$$\underline{F}^* = \beta \, Y \, \underline{F}$$

where $Y = \begin{bmatrix} \beta^{-1} & 0 & 0 \\ 0 & 1 & iV_0/c \\ 0 & -iV_0/c & 1 \end{bmatrix}$ which $= Y_0 + Y_1$

where $Y_0 = \begin{bmatrix} \beta^{-1} & 0 & 0 \\ 0 & 1 & 0 \\ 0 & 0 & 1 \end{bmatrix}$ and $Y_1 = \begin{bmatrix} 0 & 0 & 0 \\ 0 & 0 & 1 \\ 0 & -1 & 0 \end{bmatrix}$

Since $\underline{V}_0$ is parallel to the x_1-axis we can write

$$Y_1 \underline{F} = - \underline{V}_0 \wedge \underline{F}$$

so that $\quad \underline{F}^* = \beta Y_0 \underline{F} - (i\beta/c) \underline{V}_0 \wedge \underline{F}$

It is easy to show that the **dynamical force** $\underline{P}$ (which is defined as $mc\partial_4 (\beta_0 \underline{v})$) transforms by way of

$$\underline{P}^* = \beta Y_0 \underline{P}$$

and this enables us to deduce the **Lorentz Force** on a charged particle in the field $\underline{F}$ as follows.

If the particle carries a charge of e units and moves with a velocity V_0 along the x_1-axis (with respect to the Scale $\mathfrak{C}^*$) the force in $\mathfrak{C}^*$ is simply

$$\underline{P}^* = e\underline{E}^*$$

Hence the force measured by the Scale $\mathfrak{C}$ will be

$$\underline{P} = \beta^{-1} Y_0^{-1} \underline{P}^*$$

which $\quad = \beta^{-1} Y_0^{-1}$ (Real part of $e\underline{F}^*$)

which $\quad = e\{Y_0^{-1} Y_0 \underline{E} - (\underline{V}_0/c) Y_0^{-1} Y_1 \underline{H}\}$

and this gives

$$\underline{P} = e\underline{E} + (e/c)\underline{V}_0 \wedge \underline{H}$$

Which is the **Lorentz Force** exerted by the field $\underline{E} + i\underline{H}$ on a charged particle (charge e).

[8.4] **Electromagnetic waves**

The equations in [8.3] immediately give

$$\text{curl } H = 4\pi\sigma E + (\kappa/c)\partial E/\partial t$$

and since $\quad$ curl curl $\mathbf{F} = \text{grad div } \mathbf{F} - \nabla^2 \mathbf{F}$

this gives us $\quad \nabla^2 H = (4\pi\sigma\mu/c)\partial H/\partial t + (\kappa\mu/c^2)\partial^2 H/\partial t^2$

Simialrly we can deduce the identical equation for the vector **E**.

In a non-conducting medium $\sigma = 0$ and so we get, eg,

$$\nabla^2 E = (\kappa\mu/c^2) \partial^2 E/\partial t^2$$

with the equivalent relation for the magnetic field vector **H**.

This shows that the E-M Field can be propagated as a **wave motion** with phase velocity

vel = $c/(\kappa\mu)$ - which in free space is simply c, identifying it with the velocity of light.

[8.5] Plane E-M waves

It can be shown (Appendix-E [13] [16] [22] [65]) that the rate of flow of E-M energy across unit area is given by the **Poynting** vector $\Pi = (c/4\pi)(\mathbf{E} \times \mathbf{H})$, using the vector product notation.

Now an E-M plane wave travelling in an isotropic medium (with $\sigma = 0$) can be represented by vectors of the form

$$\mathbf{E} = \mathbf{E}_0 \, f(\mathbf{n}.\mathbf{r} - vt) \quad \text{and} \quad \mathbf{H} = \mathbf{H}_0 \, g(\mathbf{n}.\mathbf{r} - vt)$$

and the **Fourier** component of any such wave-form will look like

$$\cos \text{ (or sin) } (\mathbf{n}.\mathbf{r} - vt)$$

and **n** will define the normal to the wave-front - that is to the direction of the energy flow. Since this must be the direction of the **Poynting's** vector we deduce that in such a plane wave the electric and magnetic vectors **E** and **H** must be

(a) perpendicular to each other, and

(b) perpendicular to the direction of propogation of the wave-form.

Chapter-9 /PART-1   Physics-9 : Fluid Mechanics

[9.1] Compressible and incompressible Fluids

Physics-9 is **HydroMechanics**, including **Hydrostatics** and **HydroDynamics**.

The Physics deals with measures on **fluids**, and these are usually characterised as either **compressible** or **incompressible**.

The former is the study of **gases** ; (see the earlier Chapter-5), so here we consider statical and dynamical properties of incompressible fluids in general. Naturally it is water which comes readily to mind, but other more viscous fluids are important in the world of modern technology.

[9.2] Hydrostatics

In dealing with **Hydrostatics** the fluid is at rest in equilibrium under given external forces and constrained within specified boundaries.

The **base elements** are those required by the experimental measures - chiefly $\mathcal{B}_L$; $\mathcal{B}_T$ is not involved.

The **generating elements** are $\mathcal{B}_P$ (pressure = force per unit area) , $\mathcal{B}_V$ (volume), and $\mathcal{B}_\rho$ (density = mass per unit volume).

The common force field under which the fluid is at rest is that of **gravity** (assumed constant) but other mechanically contrived forces can also arise.

The treatment of fluids, either in statical equilibrium or in dynamical motion, does not depend upon any attempt to interpret the fluid as a collection of particles. It therefore depends heavily on ideas of a **field theory** in a **continuum** of material.

For an incompressible fluid the density ρ can therefore be taken as constant - since the liquid cannot be deformed under pressure. And in cases where the external field is that of the constant gravity, **g**, we find a simple $\Gamma(1\text{-}1)$ between pressure, **p**, and depth, **z**, viz.,

$$\mathbf{p} = Kz + H \qquad \text{K being a constant and H being the atmospheric pressure}$$

exerted at the surface of the liquid.

Using dimensional analysis we have $[H] = [p] = [\text{Force}][\text{Area}]^{-1} = MLT^{-2}.L^{-2}$ so that $[K] = ML^{-1}T^{-2}L^{-1}$ and since $[g] = LT^{-2}$ we see that $[K/g] = ML^{-3}$ and these are the dimensions of the density ρ (which = mass/volume).

We therefore have the simple $\Gamma(1\text{-}1) \equiv \mathbf{p} = \rho g z$.

[9.3] The principle of Archimedes

This ancient discovery explains the forces acting on a solid body which is wholly or partially immersed in an incompressible static fluid.

It is usually stated along the lines of :

> The resultant thrust on the body is a force (of buoyancy) acting upwards and equal to the weight of fluid displaced by the body - the thrust acting in a vertical line which passes through the centre of gravity of the displaced fluid (centre of buoyancy)

The centre of gravity will be at a depth, $\overline{z}$, given by integration over the the volume, v, which has been displaced, i.e by $V\overline{z} = \int_v z dV$ ρ being constant.

By hindsight it is not difficult to argue this from the first principles of Statics [Appendix-E 51].

It leads to an understanding of the **stability** of solid bodies floating in a fluid (eg ships).

[9.4] Buoyancy, Metacentre, and Stability

In the equilibrium condition the solid body must be such that its weight, **W** is balanced by the **Archimedian** upthrust, acting at the **centre of buoyancy H**, which itself must lie in a vertical line with the centre of gravity, **G**, of the solid.

In the event of a small rotational displacement of the solid (v. Fig 9.1) there is induced a torque (couple) - either increasing the displacement (unstable) or reducing it (stable).

This arises because $H \rightarrow H^*$ and the vertical through H^* meets the original line GH in, say **M**. This **M** is called the **metacentre** and its position - above or below **G** - determines the turning effect of the resulting couple, providing stability or instability.

These situations correspond to the cases

      (i) **G** lies above **M** (unstable)   and

      (ii) **G** lies below **M** (stable)

If it so happens that $G \equiv M$ then the small displacement has no effect.

[9.5] Equation of continuity in Fluid flow

The study of fluids in motion is usually based on an approach by **Euler**, viz., to consider some co-ordinate-based point **(a,b,c)** in the space encompassing the flow and thereby to examine generating elements, such as **density velocity pressure**, associated with the fluid. Regardless of whether the fluid is compressible or incompressible the idea of conservation of matter leads to a general **equation of continuity**.

This expresses the idea that throughout any small element of volume centred on **(a,b,c)** (and excludng the presence of any **sinks** or **sources** therein) the rate of change of the mass of fluid in this volume must equal the difference between the rates of change of the inflow and outflow of the fluid crossing the boundary of the volume.

If $dv = dxdydz$ is such a volume and ρ is the density this mass will be the product ρv and its rate of change will be $(\partial \rho/\partial t)\, dxdydz$.

This analysis seems to fall naturally into the realm of an **exterior algebra** so that, writing the products as $\wedge$-products, the element of volume becomes $dx \wedge dy \wedge dz$, the element of surface becomes $dx \wedge dy$ (and similar) and the flow across an element of surface becomes $\rho\underline{v}$ - the vector with components $dx \wedge dy, dy \wedge dz, dz \wedge dx$, where $\underline{v}$ is the velocity vector at the chosen point $(\mathbf{a,b,c})$.

Throughout a given region Ω of the space the conservation of matter then results in the integral relation

$$\int_\Omega (\partial\rho/\partial t)\, dx \wedge dy \wedge z \;=\; -\int_\omega (\rho\underline{v}.dS)$$

S being the bounding surface of the volume Ω and ω being the total surface of Ω.

Gauss' theorem then gives us with (in the language of vector algebra)

$$\int_\omega \rho\underline{v}.dS \;=\; \int_\Omega \mathrm{div}\,(\rho\underline{v})\, dx \wedge dy \wedge dz$$

Since this must be true for volumes throughout Ω we deduce the equation of continuity at all such points as $(\mathbf{a,b,c})$, viz.,

$$\partial\rho/\partial t + \mathrm{div}\,(\rho\underline{v}) = 0 \qquad (A)$$

When the fluid is homogeneous and incompressible ρ is constant and (A) reduces to

$$\mathrm{div}\,\underline{v} = 0 \qquad (B)$$

[Note : refer to Chapter-17 for a derivation of (A) in a localised projective space of **poins**]

[9.6] Stream Lines and velocity potential

A stream line, or line of flow, is a curve such that, at any moment in time, the tangent to thc curve gives the direction of motion of the fluid.

If the velocity, $\mathbf{q}$, at a point (x,y,z) has components (u,v,w) then the equations of the stream lines will be $\quad dx/u = dy/v = dz/w \qquad (C)$

In the case of steady motion the components (u,v,w) will not be functions of the time $\mathbf{t}$, albeit functions of position (x,y,z).

These **stream lines will be cut at right angles** by the surfaces given by the differential equation $\quad udx + vdy + wdz = 0 \qquad (D)$

and the condition for the existence of such surfaces is that (D) admits of a solution of the form $\varphi(x,y,z) = C$, C being a constant $\in \Re^{\#}$.

The vectorial condition is that, writing **q** as the vector (**u,v,w**),

$$\mathbf{q} \cdot \text{curl } \mathbf{q} = 0 \qquad (E)$$

When the expression (D) is an exact differential, viz. $-d\varphi$, so that

$$\mathbf{q} = -\text{grad } \varphi$$

then φ is called the **velocity potential** function.

This is a statement in the exterior algebra Λ since the exterior derivative, δ^1, of the 1-form in (D) is curl **q** and the problem is to find a 0-form (ie φ) such that

$$\delta^1(\delta^0) = 0 \qquad \text{(because in vector algebra curl grad} = \underline{0})$$

When curl **q** = 0, so each component vanishes, the motion of the fluid is called **irrotational**; otherwise it is naturally called **rotational**.

The exterior derivative, δ^2, operating on a 2-form in Λ, is the well-known **divergence**,

div (2-form) → (3-form) - and div curl (1-form) = 0 (pseudo scalar in Λ^3).

[Note : Exterior algebra is discussed in Appendix-A]

[9.7] Bernoulli's equation

This is applicable to any fluid which is

(i) in steady flow, and

(ii) whether or not a **velocity potential** function exists.

At a specified point in space, where the fluid is flowing, we consider

(a) the change in the total energy per unit mass, dE, of an infinitesimal element of volume, and

(b) the pressure change, dp, occurring at that point

Under (a) we take the value as ρdE, ρ being the **density** at the point and under (b) we take dp, and we notice that from considerations of dimensions

$$[\rho dE] = ML^{-3}L^2T^{-2} \quad \text{whilst} \quad [dp] = MLT^{-2}L^{-2}$$

each of which has dimension $ML^{-1}T^{-2}$.

Such measures therefore qualify as **poins** in our localised projective space [v. Chapter-17] and therefore ρdE and dp, can be regarded as Generating Elements on the observational Scales, and when these measures are in a $\Gamma(1-1)$ and in a Harmonic Range we can anticipate the discussion in Chapter-17 and contemplate the Cross-Ratio ($\rho dE \; 0 \; dp \; \infty$) If this is a localised harmonic range, at a point, we have

$$(\rho d\mathbf{E}\ 0\ d\mathbf{p}\ \infty) = -1$$

and this reduces to $(1/\rho)d\mathbf{p} + d\mathbf{E} = 0$

Whence we get the usual form of **Bernoulli's** equation, viz.,

$$\int (1/\rho)d\mathbf{p} + \mathbf{E} = \text{constant}$$

where **E** will be ½(vel)² + **V** ; **V** being the potential of any applied field, and the integration being taken along a **stream line**.

[9.8] Laplace's equation

There are many situations where the behaviour of the fluid can be easily studied in 2-dimensions, so that with incompressible irrotational fluids the equation of continuity reduces to $\quad \nabla^2 \varphi = 0 \quad$ (The 2-dimensional **Laplace's** equation)
As we have seen in Chapter-6 this means that we have all the advantages of using the complex algebra $\mathbb{C}$ for analysis of the motion, the stream lines and the velocity potential curves forming orthogonal families - and where the presence of **sources** or **sinks** appear as **singularities** in the potential functions.

[Examples of these techniques can be found in Appendix-E [25]]

[9.9] Comparison with Electrostatics

There are many other comparisons with the analysis of electromagnetism - in particular the use of **Kelvin's** method of images, which brings out once more the significance of the homography/involution relations.

For example, we find that in the problem of a fluid **source** near to a boundary (plane), the lines of flow are found by considering its image in the boundary (as a mirror). This results from the involution range $\quad (-a\ 0\ +a\ \infty) = -1\quad$ so that the flow requires an equal source at the position -a (v. Fig. 6.1). The boundary reflects the source by another source (not a sink) in the symmetrical image position.

[9.10] Allowing for Viscosity

The property of **viscosity** of a fluid is defined by the $\Gamma(1\text{-}1)$

$$\tau\ dy - \mu\ dx = 0$$

where μ is the coefficient of viscosity and τ is the **shear stress** between adjacent layers of fluid. The **kinetic viscosity** is taken as $\nu = \mu/\rho \quad \rho$ being the density function.

Chapter-10 /PART-1   Theories of Wave Motion

[10.1] The basic equation, d'Alembert's solution

So far we have studied the impact of a $\Gamma(1\text{-}1)$ in a (projective) space of measures. But now we shall consider how the introduction of a **wave theory** has enabled us to cope with a $\Gamma(1\text{-}n)$, where $n \geq 1$.

It follows from our earlier discussion of various field theories that the field equations give rise to what is known as the **equation of wave motion**, viz. any equation of the form

$$\nabla^2 \xi - c^{-2} \partial^2 \xi / \partial t^2 = 0 \qquad (1)$$

where **t** denotes a measure of time, c is the wave velocity (not necessarily the velocity of light !), and ξ is a relevant variable on appropriate Scales.

Assuming only that ξ is a function of **x** and **t** we can write new variables

$$p = x - ct \quad \text{and} \quad q = x + ct$$

whence (1) gives us
$$\partial^2 \xi / \partial p \partial q = 0$$

with a general solution (after **d'Alembert**), viz.,

$$\xi = f(p) + g(q)$$

f and **g** being arbitrary (but suitably differentiable) functions.

$f(x-ct)$ represents a wave-form travelling along the $+x$-axis whilst

$g(x+ct)$ represents a wave-form travelling along the $-x$-axis.

For example, **f** is such that $f([x+x_0]-[ct+ct_0]) = f(x-ct)$ whenever $x_0 = ct_0$. That is to say the wave-form is reproduced along the x-axis with velocity $c\ (= x_0/t_0)$.

Since the **wavlength**, λ, and **frequency**, ν, are related to the velocity c by the equation $\qquad \lambda \nu = c$

we can always think of the argument in the functions **f** and **g** as being of the form

$$(x/\lambda \pm \nu t).$$

In a metric 3-space we replace p and q by $\mathbf{r.n} \pm ct$, where **n** is the normal to a plane and which is called the "wave-front".

Such solutions are examples of **progressive waves** (v. seq).

[10.2] Waves on a stretched string

Waves on a finite stretched string, such as in musical instruments, give us examples of what are called **staionary waves**, or **standing waves**.

This is because the string imposes **boundary conditions** on the solutions - such as, that the string being tied at end-points and lying along the x-axis, must mean that the displacement **y** (perpendicular to the x-axis) is constrined to vanish ,for all **t** ; the end-points being $x=0$ and $x=\ell$.

A **Fourier analysis** of the plucked string profile results in solutions of the typical form

$$y = \Sigma_r \{A_r \sin(r\pi x/\ell) \cos(r\pi ct/\ell)\} \quad r = 1 \ldots \infty$$

the A_r being the amplitude of the **rth normal mode** of the oscillations. Each normal mode corresponds to an equation of S.H.M - and we have seen a particular example of this in the description of small oscillations in a dynamical system [v.Ch-4].

This illustrates the role of **J** (the integers) in the set of measures, for the normal modes form a **countable set** of observations - being in a $\Gamma(1-1)$ with the natural numbers.

It is in fact one of the first hints of the **notion of quantisation** on the Scales - a notion which was not to be elevated into a principle until the illusive hunt for the continuum of real numbers, $\Re^{\#}$, in Physics measures was to be brought up against the brick wall of the reality of Scales [v. earlier chapters].

[The reader will find a comprehensive account of this work in Appendix-E [49]]

[10.3] Sound Waves - Pitch and Timbre

Assuming that sound waves commonly occur in compressible fluids, and since the changes of **pressure** and **density** take place over very small time intervals, the **Boyle's Law** relation between "pressure" and "volume" (or **density**) must be replaced by that appropriate to **adiabatic changes** (that is to say, without any changes in internal energy).

This means that we must use the relation $\quad p = kv^\gamma \quad$ k a constant

where γ is the ratio of the specific heats $\quad C_p/C_v \quad$ [v. Ch-5].

The sound waves are **longitudinal waves** in the medium, and obey the familiar equation of wave motion, viz., $\quad \partial^2\xi/\partial x^2 = v^{-2}\partial^2\xi/\partial t^2$

"v" being the wave velocity, and where ξ = the longitudinal displacement of a particle of the medium, and where $v^2 = dp/d\rho$; ρ being the density and **p** being the pressure at the location under consideration.

In a musical context a **Fourier** analysis of the sound wave-form will produce a range of frequencies from

(i) the fundamental - which defines the **pitch** of the note

(ii) a range of **harmonics**, with higher frequencies, - which defines the **timbre**.

It is this distribution of waves, with different frequencies v_0, v_1, v_2 etc., which characterises the sounds of different musical instruments ; v_0 is the fundamental (or pitch) and the higher frequencies v_r determine the instrumental timbre.

Fortunately the human ear only has a sensitivity to a limited range of frequencies - so the **Fourier analysis** does not need to $\to \infty$.

[10.4] Surface waves

These can occur on liquids - particularly on water surfaces - and can be discussed as follows.

Bernoulli's equation gives us

$$\tfrac{1}{2} u^2 + V - \partial \varphi / \partial t = \text{constant}$$

where **u** is the velocity of the fluid at a point, φ is the velocity potential (given by $\mathbf{u} - \nabla\varphi$) and V is the potential function of the applied conservative field of force.

Tidal Waves occur where the wavelength is much greater than the depth of the water, otherwise they are called Surface Waves. [v. Appendix-E [22] [25]

[10.5] Huyghen's Principle

The introduction of the **Wave Theory of Light** into problems found in geometrical optics led to a new understanding of that subject.

The early pioneering work was done by **Fresnel** and **Hughens** [v. Appendix-E Ref 10,52] and forms the foundation of what is often called **physical optics**. **Huyghens** first proposed what is now known as **Huyghens' Principle**. It supposes that the **wave-front** is propagated through space via the following process, viz.,

> Each point on the wave-front acts as a source of a secondary wave (or wavelet) and the resultant wave-front, at a later time Δt, is the **envelope** of all such wavelets.

This is illustrated in Fig. 10.1, and a thorough mathematical discussion of the Principle can be found in Appendix-E [54].

The first triumph was the explanation of the **law of refraction** [v. Chapter-1] and from there we find that experiments by **Young** (the **Young's Slits** experiment) led to the explanation of **interference phenomena** in that subject.

Thus, if we take a simple mathematical description of two waves, in 1-dimension and being $2\pi\delta/\lambda$ out of phase, these being superimposed (via some Scale arrangement), viz.,

$$\varphi_1 = A \sin 2\pi(x/\lambda - vt)$$

$$\varphi_2 = A \sin 2\pi([x+\delta]/\lambda - vt)$$

then the resultant effect, at **x** & **t**, will be respresented by the sum of these,

giving $\quad \varphi_1 + \varphi_2 = 2A \cos \pi\delta/\lambda \sin 2\pi([x+\delta/2]/\lambda - nt)$.

So the amplitude of the effect varies as δ between 0 (when $\delta = \lambda/2, 3\lambda/2, ...$)
and 2A (when $\delta = 0, \lambda, 2\lambda, ...$).

In terms of light these will be correspondingly **dark** and **bright** regions - the well-known **interference fringes**.

[10.6] Interfereence and Diffraction

The fascinating field of the phenomenon of **diffraction** also follows from the application of this **Wave Theory** of light. The early history of its development by **Fresnel** and **Hughens** can be found in references [Appendix-E [10], [29], [49], [52], [54]].

[Note : This pheneomenon again illustrates how the **Wave Theory** can accomodate the notion of a $\Gamma(1-n)$ correspondence, where $n \in J^+$. and also that such a $\Gamma(1-n)$ is part of the expression, in Physics measures, of so-called **quantisation**. [v. Chapter-14] **Quantisation** is not something **imposed** on certain measures. It is something **inherent** in the whole process of taking **any measures** on a Scale in any **Physics**].

[10.7] Electro-magnetic waves

An important development in the significance of wave motion was the deduction made from the **Maxwell field equations** that the **electromagnetic field, E-M** can also be transmitted as a wave-form.

From the field equations discussed in Ch-8 we may easily deduce [Appendix-E [13]] that, in free spac the electric field vector $\underline{E}$ and the magnetic field vector $\underline{H}$ both satisfy the basic equation of wave motion (1). [v. Appendix-E [13] [34] [49]]

In any medium other than free space the vectors $\underline{E}$ and $\underline{H}$ must be replaced by the electric **displacement vector** $\underline{D}(= \kappa\underline{E})$ and the magnetic **induction vector** $\underline{B}$.

In each case the field is propagated with a velocity $c/\sqrt{(\kappa\mu)}$.
The identification of this "c" with the velocity of light led to the understanding of the position of light in the spectrum of E-M wavelengths.

It will be shown in Part-2, and is well known, that the light signal can be identified with the field generated by $E + iH$ (using $\mathbb{C}$) and we shall find a significant role for it as a Base element $\mathcal{B}_L$ in an extended version of the **Lorentz/Einstein** equations of **Special Relativity**.

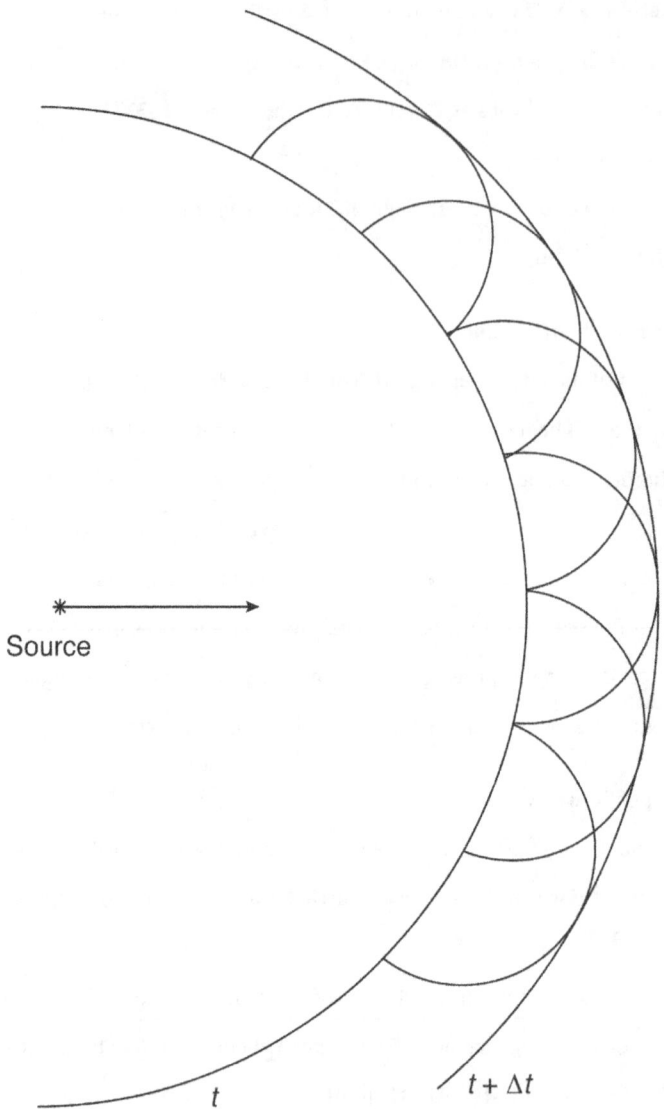

HUGHENS' Construction for propation of a spherical wave

Fig 10.1

Chapter-11 /PART-1 Invariants

[11.1] Algebraic invariants

We have seen that the experimental basis of any Theoretical Physics lies in the search for **invariants** among a suitably selected set of **Scales**, $\mathfrak{C}_\alpha$. And this means that the search for **algebraic invariants** in any of the algebras we use becomes of paramount importance. No doubt the work of **Einstein**, in his use of the ideas of expressions which are invariant under linear transformations of co-ordinate **frames of reference** has brought this concept centre-stage in modern thinking.

We shall now look more closely at this whole question - chiefly for the benefit of readers who are particularly interested in the pure mathematics of it.

[11.2] Quadratic differential forms

We have seen in **Physics-4** that the **first Pfaff's sytem** of differential equations, which are derived from a certain **linear form**, amount to the canonical dynamical equations of **Hamilton**. We have also seen that when restricting ourselves to a **metrical geometry**, such as the metrics via **Pythagoras, Lorentz, Minkowski, Einstein**, etc., we are searching for invariants associated with a **quadratic differential form**, viz. ds^2, - which itself must be an invariant.

In the latter case all the measures were supposed to be subject to the invariance of the metric
$$ds^2 = g_{\lambda\mu} \, dx^\lambda \, dx^\mu \qquad (1)$$
and this invariance to be under all point transformations in the space defined by the co-ordinates $\{x^i\}$ where $i = 1,2,3,4$ and $x^4 = ct$.

The theory put forward by **Weyl** was an attempt to produce field equations independently of this metric invariance [v. Ch-12].

He did this by considering, together with (1), a **linear affine space** of the form $\varphi_i \, dx^i \quad (i = 1,2,3,4)$
and was thus able to produce both the **Maxwell** (E-M) field equations as well as the **Einstein** gravitational field equations. The functions φ_i were to be interpreted as the **(E-M)** field potential components, viz., $V + \mathbf{A}$ [v. Ch-8].

[11.3] Projective invariants w.r.t a groundform

These examples show that the Field Theories have been obtained by reference to both **linear** and **quadratic** forms - the field equations thereby derived giving us relations which are to hold at **points** in the geometry $\mathbf{P}^n$.

It is helpful therefore to consider the essential algebraic properties of what are called **Projective Invariants**.

Suppose we are given a transformation (a Linear Mapping), say,

$$\Im : \underline{x} \to \underline{x}^*$$

from the vector space $\{\underline{x}\}$ to $\{\underline{x}^*\}$ together with a **polynomial** $I(a,b,c, ...)$ of the **coefficients of given groundforms** $f(\underline{x})$, $g(\underline{x})$ ... then

$I(a,b,c, ...)$ is a **relative projective invariant** of the groundforms, under $\Im$, if

$$I(a^*,b^*,c^*,...) = |M|^w I(a,b,c,...) \qquad (2)$$

where $|M|$ = the determinant of the matrix (of) $\Im$; w being an integer, called the **weight** of the invariant. [v. Appendix-E [17]].

$\Im$ is to be homogeneous in the components of $\underline{x}$, this being appropriate to the use of homogeneous co-ordinates in the projective space. Thus, for example, in a vector space of two dimensions (viz. the underlying vector space for a 1-dimensional projective space) we have $\underline{x} = (x_1, x_2)$ and

$$\Im \equiv \left\{ \begin{array}{l} x_1^* = a_{11} x_1 + a_{12} x_2 \\ x_2^* = a_{21} x_1 + a_{22} x_2 \end{array} \right\} \qquad (3)$$

where $(a_{11}a_{12} - a_{12}a_{21}) \neq 0$.

We notice that if we introduce the inhomogeneous co-ordinates $x = x_1/x_2$, $x^* = x_1^*/x_2^*$ (3) is the same as the $\Gamma(1-1)$ already discussed at some length, viz.,

$$x^* = (ax + b)/(cx + d).$$

[11.4] Absolute invariants

When $w = 0$ we call $I(a,b,c,...)$ an **absolute invariant** of the groundforms f,g ; these polynomials being algebraic expressions in the components of $\underline{x}$.

Example-1 With a single groundform $f(\underline{x}) = ax_1^2 + 2hx_1x_2 + bx_2^2$ ($f(\underline{x})$ being a **binary quadratic**) the **discriminant** $(ab - h^2)$ is a relative invariant of weight 2.

Example-2 With two binary quadratics as groundforms, viz.,

$$f(\underline{x}) = ax_1^2 + 2hx_1x_2 + bx_2^2$$

and $\quad g(\underline{x}) = Ax_1^2 + 2Hx_1x_2 + Bx_2^2$

we have an invariant of weight = 2 in the expression   aB + bA - 2hH.

We notice that the vanishing of the invariant in Ex-1 is the condition for the two roots of $f(\underline{x}) = 0$ to coincide - an intrinsic property which a linear transformation cannot change.

Similarly we notice that in Ex-2 the vanishing of the invariant is the condition that the roots of $f(\underline{x}) = 0$ and the roots of $g(\underline{x}) = 0$ should form a **harmonic range**; again a property which is preserved under all $\Gamma(1\text{-}1)$'s.

[11.5] Covariants, contravariants, mixed concomitants

In addition to the idea of an absolute invariant there are also expressions which are known as **covariants, contravariants** and **mixed concomitants**.

covariants are expressions $H(a,b,c, \ldots \underline{x})$, satisfying (2) above, but which contain the variables $\underline{x}$ (which are of course transformed by the $\Im$).

contravariants are expressions $K(a,b,c, \ldots \underline{u})$, satisfying (2) above, but which contain the contragredient variables $\underline{u}$; these latter are transformed by the transpose of the inverse mapping, viz., $\Im^{-1*}$. Variables such as $\underline{u}$ and $\underline{x}$ are those related by a

scalar product (eg $u_1 x_1 + u_2 x_2 + \ldots u_n x_n = 0$). We naturally speak of such variables $\underline{u}$ and $\underline{x}$ as vectors in a **dual space**. Thus, in plane geometry,

line co-ordinates and **point co-ordinates** are in a dual space - the equation of a line/point being of the form $\ell x + my + nz = 0$; ℓ, m, n being the line co-ordinates whilst x, y, z are the point co-ordinates.

As one might expect, **mixed concomitants** contain any mixture of variables whatever.

It may also be noticed that all the relative projective invariants of weight w can be made into absolute projective invariants by restricting the mappings $\Im$ to a subgroup $\Im_+$, defined by

$\Im_+ \in \{\Im\}$ with the property $|M_+| = +1$.

In this case $\Im_+$ is called an **orthogonal transformation** and in a space with **Euclidean metric** it preserves length (interval). Thus in **Physics-4** $\Im_+$ is equivalent to **rotation** of rigid bodies. Similarly if we take the subset (not a subgroup) $\{\Im_-\}$ defined by $|M_-|$ we obtain mappings under which all relative invariants of even weight become absolute invariants.

In **Physics-4** the mapping $\Im_-$ corresponds to a rotation and a reflection (mirror image).

Both of these mappings are contained in the subgroup for which $|M|^2 = +1$.

[11.6] Differential operators, Aronhold operator

An extension of these ideas is the concept of an **invariant process**, commonly involving a **differential operator**.

Thus if we have a function $I(a_i)$ of the **coefficients**, a_i, of a form $f(\underline{x})$, and if we can apply an operator R to this to give us the property

$$R(a^*_i) \, I(a^*_i) = |M|^w \, R(a_i) \, I(a_i)$$

then R is an **invariant operator**, having generated the invariant $R \, I$.
It actually generates new invariants out of old invariants.

An example of this is the **Aronhold operator** which generates from an invariant of one form a simultaneous invariant of two forms. If the first form is the binary n-ic

$$(a_0, \ldots a_n \!\!\int x_1, x_2)^n \qquad \text{[v. Appendix-E [17]]}$$

and the second is the binary n-ic $(b_0, \ldots b_n \!\!\int x_1, x_2)^n$ the Aronhold operator is

$$R \equiv \Sigma_r \, b_r \partial/\partial a_r \qquad \text{summing over } r = 1 \ldots n.$$

Example-3 In the case of two quadratics we notice that

$$(A\partial/\partial a + B\partial/\partial b + C\partial/\partial c)(ab - h^2) = Ab + Ba - 2hH$$

illustrating the above process. [v. above examples]

[11.7] Cogredient and contragredient Tensors

A particular case of the **Aronhold** operator is the **polarizing process** whereby polar forms may be generated from (eg) the binary n-ic $(a_0, a_1, \ldots a_n \!\!\int x_1, x_2)$ by the operator

$$R \equiv y_1 \, \partial/\partial x_1 + y_2 \, \partial/\partial x_2$$

For example, the first polar of the quadratic $(a_0, 2a_1, a_2 \!\!\int x_1, x_2)$ is the expression

$$(a_0 x_1 y_1 + a_1 x_1 y_2 + a_1 x_2 y_1 + a_2 x_2 y_2) \qquad (4)$$

which is a bilinear form in the two sets (x_1, x_2) and (y_1, y_2).

The expression (4) is a special case of the general **multilinear form** in variables $\underline{x}, \underline{y}, \underline{z}, \ldots$ (where eg $\underline{x}$ denotes $x_1, x_2, \ldots x_n$)

viz. $\quad f = \Sigma a_{ijk\ldots} x_i y_j z_k \ldots \qquad (5)$

the summation being over each subscript in turn (from 1 to n).

In general we may suppose that the variables $x, y, z \ldots$ will include variables $u, v, w \ldots$ which transform contragrediently to the $x, y, z \ldots$ (via the transpose reciprocal $\Im^{-1*}$). To keep the distinction in mind we may denote the coefficients of f by $a_{ijk\ldots}^{pqr\ldots}$; the lower suffices corresponding to the cogredient variables and the upper to the contragredient variables (in the dual space).

When we write our multilinear form as (eg)

$$F = \Sigma A_{ijk}{}^{pqr} x_i y_j z_k u^p v^q w^r \quad (6)$$

the quantities $A_{ijk}{}^{pqr}$ are called **tensor components** ; the set of these is called a **tensor** of orders (3,3), being of order 3 in both the cogredient and contragrediant variables.

In what is commonly called the **tensor calculus** a symbol of this kind is said to be **covariant** of order 3 and **contravariant** of order 3. But we should notice that in the theory of **algebraic projective invariants** these words are used to mean different types of **concomitants**.

Thus, for instance, let us consider the **work function**

$$\delta W = Q_1 dq_1 + Q_2 dq_2 + Q_3 dq_3$$

for force components Q_r and co-ordinate changes dq_r.

In the language of the tensor calculus we say :

(a) δW is an invariant **scalar** ; and this means that the change of variables (from one basis to another) induce by operators $\{\Im\}$ leave δW unaltered.

(b) The dq_r form a **contravariant vector** (normally writtern as dq^r),

(c) The Q_r form a **covariant vector**.

But in the language of **algebraic invariants** we would speak of these "vectors" as particular kinds of **concomitants**.

[11.8] Differential forms in General Relativity theory

The aim of the tensor calculus is to obtain invariants from **differential forms**, the idea being that we can thereby include in our mappings $\{\Im\}$ more general functional transformations. So if we consider a mapping defined by a more general function **f**, say, viz.,

$$\Phi : \underline{x} \to \underline{x}^* = f(\underline{x}) \quad (7)$$

(which means $x_i^* = f_i(x_1, x_2 ... x_n)$, with $i=1...n$)

Then (7) is equivalent to a **linear transformation** in the space of the differentials, viz.,

$$\Phi \text{ inducing } \quad \Im : d\underline{x} \to d\underline{x}^* \quad (8)$$

via $\quad dx_i^* = \Sigma_i (\partial f/\partial x_i) dx_i) \quad$ for $i = 1 ... n$.

It was established by **Christoffel** that the general problem of the tensor calculus (finding concomitants) is identical with the problem of finding projective invariants. From our point of view we expect to find inportant examples of **field theories** in the applications of the tensor calculus.

If our Physics measures are to be mapped into a space which is dual to the one spanned by the differentials dx_r then the measures will be represented by variables which transform contragrediently to this vector base. If we are also, for example, taking these measures in a geometry such as that of **Physics-3** with a metric (**Riemannian**) given by the invariant

$$ds^2 = g_{ij}dx_i dx_j \qquad (9)$$

then the invariants we are looking for in our field theory will be projective invariants of differential groundforms **f, g** ... to which are adjoined the form (9). These will be **projective concomitants** of the forms **f, g** ... adjoined to (9).

We shall see (v.Ch-12) that such operators, R, acting in tensor fields forms the basis of **Einstein's** work in his gravitational theory.

In general, if R is an invariant operator and Θ is a mixed concomitant, then

$$\mathrm{R}\Theta, \mathrm{R}^2\Theta, \mathrm{R}^3\Theta \ldots$$ will be mixed concomitants of various kinds.

Then the kind of field equations any theory is likely to find most welcome will be those of the form

$$\mathrm{R}^k \Theta = \mathbf{I} \qquad (10)$$

where R^k <u>has reduced</u> Θ <u>to an invariant</u> **I**.

This is because we are most likely to **observe** an invariant among our Scale measures.

<u>Example-3</u> Relation (10) is well illustrated in classical field theories by way of

(i) $\nabla^2 V = 0$; $\nabla^2 V = -4\pi\rho$

[$\mathrm{R} \equiv \nabla$, $\Theta = \mathbf{V}$, k=2, ρ is invariant charge density]

(ii) $\nabla^2 V = 0$; $\nabla^2 V = 4\pi\gamma m$

[$\mathrm{R} \equiv \nabla$, $\Theta = \mathbf{V}$, k=2, ρ is invariant mass, **m**]

(iii) $\Box^2 \varphi = 0$;

[$\mathrm{R} \equiv \nabla$, $\Theta = \mathbf{V}$, k=2, Zero is invariant; all in 4-space]

These examples are related to certain metric quadrics in **Euclidean space**, the last one being associated with the groundform

$$c^2 dt^2 - dx^2 - dy^2 - dz^2 \qquad (11)$$

This metric (11) is said to be defined in the **space-time continuum** since the theoretical physics is assumed to require algebra of the reals, $\mathfrak{R}^\#$. This supports the significance of the **Lorentz-Einstein** transformations $\mathscr{L}$, in that we can always consider the invariant operator as being the product $\mathscr{L}(\mathrm{R})$ and **I** as $\mathscr{L}\mathbf{I}$ on any Scale.

Since the coefficients of the groundform in (11) are constants, in $\mathfrak{R}^\#$, the invariants under $\mathcal{L}$ occur at **every point** of the 4-space. This is not necessarily so when the groundform is the **Einstein-Riemannian** ds^2, since the coefficients, g_{ij}, are functions of the four co-ordinates x_i. In this case we expect to find invariants which are **locally** significant - a characteristic of field theories which brings out the essential difference between the **Special** and the **General** relativity theories.

Example-4 Suppose **V** is a scalar invariant (eg electrostatic potential) with the differential form
$$dV = \Sigma_r(\partial V/\partial u_r \, du_r) \qquad r = 1,2,3$$
and the u_r are position co-ordinates. Let the metric be the general orthogonal (**Euclidean**) one, so that
$$ds^2 = \Sigma_r(h_1^2 du_r^2)$$
Then the electric intensity (field vector)
$$\underline{E} = (E_1, E_2, E_3)$$
is mapped onto the coefficients in the linear form
$$- \Sigma_r(h_r^{-1} \partial V/\partial u_r \, ds_r) \qquad r = 1,2,3$$
which coefficients are invariant under mappings which preserve ds^2 and dV.

Example-5 The mappings of Ex-4, which preserve ds^2 also ensure that the expression
$$(h_1 h_2 h_3)^{-1} \{\partial(h_2 h_3 F_1)/du_1 + \ldots \text{similar}\}$$
denoted by div $\underline{F}$ is an invariant. Thus if div $\underline{F} = 0$ on one Scale then it is zero on all Scales which are related by transformations $\{\vartheta\}$. In electrostatics the field equation
$$\text{div } \underline{E} = 4\pi\rho$$
relates one invariant ρ to a concomitant $\underline{E}$.

Example-6 Another invariant operator is that denoted by **curl** - which generates one vector from another (eg $\underline{H} = \text{curl } \underline{A}$). The components of curl $\underline{F}$ are obtained, in a space with metric
$$ds^2 = dx^2 + dy^2 + dz^2$$
as the coefficients of the **Pfaffian bilinear covariant** of the linear form
$$F_1 dx + F_2 dy + F dz$$
thus giving us four components for the curl, viz., F_{ij}.
If the vector $\underline{F}$ becomes a 4-vector the components of curl $\underline{F}$ are obtained from the bilinear covariant of the differential form
$$F_1 dx_1 + \ldots + F_4 dx_4$$
giving six independent components, F_{rs}.

Example-7 In Newtonian dynamics "mass" and "force" are mapped onto the coefficients of dv and dt in the linear form

$$mdv - Fdt$$

i.e. onto the **dual space** of the vector space spanned by {dv, dt}, and the mappings which the **Galilean-Newtonian** theories allow are those of **affine geometry**, viz.,

$$\Im_G : \begin{Bmatrix} v \rightarrow v^* = v + V_0 & \text{and} \\ t \rightarrow t^* = t + \alpha \end{Bmatrix} \quad V_0 \text{ and } \alpha \text{ being constants.}$$

Example-8 We have already seen (**Physics-4**) that the **first Pfaff's system** of the differential form

$$p_1 dq_1 + p_2 dq_2 + \ldots + p_n dq_n + 0 dp_1 + \ldots + 0 dp_n - H dt$$

form the **Hamiltonian** field equations for a system defined in this $(2n+1)$ space. The **Hamiltonian**, H is the invariant of the system - the q_r are the independent co-ordinates of the system and the p_r are the generalised components of **momentum**.

The measures mapped onto the dual space are identified as those of **momentum** and **energy**. The question of a metric quadric does not arise ; the permitted mappings are those which preserve the bilinear covariant of the above **linear differential form**. These mappings are traditionally called **contact transformations** [v. Appendix-E [9]].

Example-9 **Maxwell's equations** are field equations in **Physics-9**, under the invariant operator $\square$. In this case the transformations are those which preserve the metrical quadric

$$dx^2 + dy^2 + dz^2 - c^2 dt^2 = 0$$

Hence the mappings include those of the form $\mathscr{L}$ - and so the **Maxwell** field equations are **relativistically invariant**.

Chapter-12 /PART-1 Einstein's General Relativity Theory

[12.1] Einstein's use of the Riemannian metric

The general theory of relativity, [v. Refs Appendix-E 6,28], is concerned with finding **field equations** in **Physics-3** under the general set of mappings $\{\Im\}$ which are to be subject to retaining the invariance of the differential form (the **Riemannian** metric)

$$ds^2 = g_{ij}dx_i dx_j \qquad (1)$$

and where we adopt the summation convention that a repeated index is to be summed over 1,2,3,4 ; also the fourth variable x_4 is to be ct.

Since **Einstein** produced the generalised law of gravity by considering invariants as functions of the coefficient g_{ij} a great deal of effort has been committed to somehow including in this process the equally important, and invariant, **Maxwell equations** for the **E-M** field. This was known as the search for a **unified field theory**.

In the light of the discussions in this thesis this seems to be a fruitless pursuit - because if, in fact, the **E-M** field and the **Gravitational** field are truly **Base Elements** in the appropriate Physics, then they can hardly be combined into a single Base Element.

In this respect it is proposed that, after **Einstein** the major contribution to the question has been provided by **Weyl**.

[12.2] The Riemann/Christoffel tensor

The theory has always been expressed in terms of the tensor calculus [v. Ch-11] in which g_{ij} is a second order covariant tensor (ds^2 being invariant) and dx^i a contravariant vector (tensor of order 1).

We have seen in Chapter-3 that in **Physics-3** force is not a generating element and therefore the **gravitational force** of **Newton's** universe is really a defined quantity in terms of the generating elements "mass" and "acceleration". Of course, in **Newton's** dynamics mass-measures are always invariant whereas in Special Relativity it is subject to the transformation $\mathscr{L}$.

The point of this is that it should be possible to derive field equations for **Physics-3** in which the "law of gravity" (**Newton's** well-known "inverse-square law") is deducible by invariants associated with the invariance of (1). [v. our discussion of this via a simple homography, in Ch-3].

This was achieved by **Einstein**, and others, by using the the groundform (1) and by searching for suitable concomitants among the abstract measures mapped onto the coefficients

g_{ij} ; other groundforms may also be introduced to help in this search.

Using the language of the tensor calculus we must be aware of the fact that the differentiation of a vector (tensor of order 1) is only another tensor if we interpret this operation as **covariant differentiation**. That is to say, if the vector be denoted by X_μ ($\mu = 1,2,3,4$) the covariant derivative is $X_{\mu,\nu}$ (possessing 16 components) where

$$X_{\mu,\nu} = \partial X_\mu/\partial x^\nu - \Gamma_{\mu\nu}{}^\alpha \quad (2)$$

where the comma indicates the covariant differentiation, and where

$$\Gamma_{\mu\nu}{}^\alpha = \tfrac{1}{2} g^{\alpha\lambda} \{\partial g_{\mu\lambda}/\partial x^\nu + \partial g_{\nu\lambda}/\partial x^\mu - \partial g_{\mu\nu}/\partial x^\lambda\} \quad (3)$$

where the summation convention means the summing over λ.

This is called a **Riemann-Christoffel symbol of the second kind** ; the corresponding **symbol of the first kind** being denoted by $\Gamma_{\mu\nu\alpha}$ and defined by

$$\Gamma_{\mu\nu\alpha} = \tfrac{1}{2} \{\partial g_{\mu\alpha}/\partial x^\nu + \partial g_{\nu\alpha}/\partial x^\mu - \partial g_{\mu\nu}/\partial x^\alpha\} \quad (4)$$

Here th g^{ij} are related to the g_{ij} by the relation

$$(g^{ij}) = (g_{ji})^{-1}$$

so that g^{ij} is the minor of g_{ij} divided by $\det(g_{ij})$; since $g_{ij} = g_{ji}$ each matrix in the above is symmetric.

We should also notice that

$$\Gamma_{\mu\nu}{}^\alpha = \Gamma_{\mu\nu\lambda} g^{\alpha\lambda}$$

and that the symbols Γ do not form a tensor.

Successive covariant differentiation of a vector X_μ gives the relation

$$X_{\mu,\sigma\nu} - X_{\mu,\sigma\nu} = X_\lambda B^\lambda{}_{\mu\nu\sigma} \quad (5)$$

and $B^\lambda{}_{\mu\nu\sigma}$ is a tensor of the 4th order - and known as the **Riemann-Christoffel tensor**.

We complete this display of "heavy artillery" by listing the expression for this tensor **B**, viz.

$$B^\lambda{}_{\mu\nu\sigma} = \Gamma^\alpha{}_{\mu\sigma} \Gamma^\lambda{}_{\alpha\nu} - \Gamma^\alpha{}_{\mu\nu} \Gamma^\lambda{}_{\alpha\sigma} +$$
$$\partial \Gamma^\lambda{}_{\mu\sigma}/\partial x^\nu - \partial \Gamma^\lambda{}_{\mu\nu}/\partial x^\sigma \quad (6)$$

The vanishing of all the components of **B** at all points of the space-time ensures that successive covariant differentiations is independent of the order.

In the context of the differential geometry of surfaces in 3-space (where $i,j = 1 .. 2$) the components of **B** are associated with the **curvature** of the surface (which is a 2-space). Thus the non-vanishing of the components of the **B** of (6) is often referred to as a measure of the curvature of space-time - its deviation from "flatness" (in which a dynamical particle would presumably move in a monotonous manner, unaffected by "force" or "gravitation").

[12.3] The Schwarzchild metric

To obtain a 2nd-order differential operator for the gravitational field (based on the invariance of (1) above) we must look for some tensor based on **B** - and we cannot simply equate it to zero else we get only zero gravitation.

Einstein proposed that the **contracted Riemann-Christoffel tensor** (obtained by writing $\lambda = \sigma$, and then summing) should express the desired field equation, viz.,

$$\Gamma_{\mu\nu} = \mathbf{B}^{\lambda}{}_{\mu\nu\lambda} = 0 \qquad (7)$$

Thus $\Gamma_{\mu\nu} = \Gamma_{\mu\sigma}{}^{\alpha}\Gamma_{\alpha\nu}{}^{\sigma} - \Gamma_{\mu\nu}{}^{\alpha}\Gamma_{\alpha\sigma}{}^{\sigma} +$

$$\partial(\Gamma_{\mu\sigma}{}^{\sigma})/\partial x^{\nu} - \partial(\Gamma_{\mu\nu}{}^{\sigma})/\partial x^{\alpha}$$

Then (7) can be solved for the g_{ij} [v. Appendix-E [6],[28],[47]] and assuming they are independent of x_4 (the time variable), and expressing the result in spherical polar co-ordinates, the metric becomes

$$ds^2 = -(dr)^2/k - r^2 d\theta^2 - r^2 \sin^2\theta \, d\varphi^2 + k dt^2 \qquad (8)$$

with $k = 1 - 2m/r$, m being a constant. This is the **Schwarzchild metric**.

The motion of a particle is now the path of a **geodesic** in this 4-space ; that is to say, it moves along a path which is the shortest distance (this being measured by (8)).

These geodesics give the planetary orbits of the **Newtonian Theory**, but with additional correction terms which help to account for the well-established rotation of the elliptical axes (the advance of the perihelion).

The gravitational force on a particle (of mass **m**) follows via the potential function

$$V(r) = -\kappa M/r - \kappa M h^2/r^3$$

h being the constant angular momemtum per unit mass, **M** the mass of the particle at the origin and κ being the **universal gravitational constant**.

This gives us the **gravitational force**, in a spherically symmetrical free space, as

$$F(r) = -\kappa M/r^2 - 3\kappa M h^2/r^4$$

which is a modification of **Newton's** formula - and should be compared with the result we obtained in Chapter-3 (**Physics-3**). [v. Appendix-B [28] for a full discussion].

In the latter case, where the algebra used was that of the quaternions, $Q(\Re^{\#})$, we noticed that the field also included the phenomenon of **spin** (associated with the mass-particle). **Einstein's** theory does not include any reference to this intinsic spin of the particle - yet all the planets possess an instrinsic spin about their axes.

To obtain additional correction terms **Einstein** later tried the equation

$$\Gamma_{\mu\nu} = \lambda\, g_{\mu\nu} \quad (\lambda \text{ being very small})$$

but we shall not pursue this here.

[12.4] Forsyth's derivation of invariants

It is important for us to notice that, in our algebraic theory of concomitants, the quadratic differential form ds^2 is an invariant and that the **Riemann-Christoffel tensor** (and its contracted form $\Gamma_{\mu\nu}$) is a mixed concomitant - and that the only concomitants associated with the form (1) above, and which do not contain derivatives of higher order than the second, can be expressed as functions of $g_{\mu\nu}$ and $B_{\mu\nu\sigma}{}^{\lambda}$.

The purely algebraic foundations (as opposed to that of the tensor calculus) of the above gravitational theory was neatly expressed in a paper by **Forsyth** [v. Appendix-E [83]].

In this paper the quadratic form, such as (1), was written without differentials by taking variables (X_1, X_2, X_3, X_4) in place of $(dx_1, \ldots$ etc). In looking for mixed concomitants the dual variables (U_1, U_2, U_3, U_4) were introduced via the

linear form $\quad \Sigma_r U_r X_r = \Sigma_r U^*_r X^*_r$

together with the line-variables P_r, $r = 1 \ldots 6$ - defined in terms of the $X_r X^*_s$ by the usual formulae, being the six independent components in the anti-symmetrical matrix

$$(A_{rs}) \quad \text{where} \quad A_{rs} = X_r X^*_s - X_s X^*_r$$

The set of permissible transformations were regarded as the **Lie Group** of continuous transformations,

$$X^*_r = X_r - \varepsilon\, \xi_r(X) \qquad (9)$$

where ξ_r ($r = 1,2,3,4$) are arbitrary continuous functions, and ε is infinitesimally small.

In general a mixed concomitant will now be a function such as

$$\varphi\,(g_{rs}, X, P, U)$$

and such that, under (9),

$$\varphi^* = \Omega \varphi \qquad (10)$$

where $\quad \Omega = $ the determinant of the transformations

$\qquad\qquad\; = $ the Jacobian, to the first order in ε,

$\qquad\qquad\; = 1 - \varepsilon(\xi_1 + \xi_2 + \xi_3 + \xi_4)$

and where $\varphi^* \equiv \varphi(g^*_{rs}, X^*, P^*, U^*) \qquad (11)$

Of course, the g_{rs} will transform contragrediently to the X_r and cogrediently to the U_r.

By substituting from (9) into (10) and (11) the relation (10) gives a set of equations obtained by equating coefficients of the powers of ε to zero, and by using the fact that the ξ_r are arbitrary (but continuous) independent functions.

The result is a set of 136 zero relations giving φ in terms of $(g^*_{rs}X_i^* \ldots g_{rs}X_i)$.

This set splits up into three groups of equations containing, respectively, 80, 40, 16 differential equations of the 3rd, 2nd, and 1st orders.

The group containing 80 equations is a complete **Jacobian** linear system and possesses 20 independent integrals which are related to the 20 independent components of the **Riemann-Christoffel** symbols. The remaining groups are also complete **Jacobian** systems each of which possesses 30 independent integrals.

Out of these Herculean efforts **Forsyth** obtained the **Riemannian curvature** as an **invariant** (this being the invariant of $g^{\mu\nu}G_{\mu\nu}$) and he integrated the differential equations to obtain the **Schwarzchild mertric** (8).

In obtaining the metric (8) it was not necessary to make the **Einstein** assumption that the $g_{\mu\nu}$ are independent of time (x_4), but only that as $r \to \infty$ the metric approaches the form $\quad -dr^2 - r^2 d\theta^2 - r^2\sin^2 d\varphi^2 + c^2 dt^2$

[12.5] Weyl's attempt at a unified field theory

Following **Einstein's** work on his General Theory there began a search for the so-called **unified field theory**, the aim of which was to incorporate the **Maxwell** (**E-M**) field equations into the metrical space which **Einstein** had used.

We have already pointed out that, subject to the notions of **Base Elements**, this is likely to be a confusing and pointless exercise. But this has not, of course, discouraged scientists from pursuing that particular road to fame.

An early attempt was made in the work of **Weyl**, and he realised that the so-called **gravitational potentials**, $g_{\mu\nu}$, must be enhanced in some way.

His method was to introduce, in addition to the $g_{\mu\nu}$, a **linear form** $\varphi_i dx^i$ into the discussion. This was to be invariant under suitable **gauge transformations** so as to produce the **Maxwell** field equations, beside the gravitational field otherwise derived from the metric $\quad ds^2 = g_{\mu\nu} dx^{\mu\nu}$.

A fuller discussion of this will be found in the references - Appendix-E, [6],[28],[47].

Chapter-13 /PART-1   Products of Algebras

[13.1] Outer product of algebras A and B

We consider dealing with the combination of two (distinct or identical) algebras, and we do this by forming the **outer product** (sometimes called the **direct product**) of two algebras, say **A** and **B**.

If **A** has a finite basis $\{\underline{e}_1, \underline{e}_2, \ldots \underline{e}_n\}$ and **B** has a basis $\{\underline{f}_1, \underline{f}_2, \ldots \underline{f}_n\}$ the outer product $\mathbf{A} \times \mathbf{B}$ has the finite basis $\{\underline{g}_{ij}\}$ where

$$\underline{g}_{ij} = \underline{e}_i \underline{f}_j \qquad (i,j = 1 \ldots n) \qquad (1)$$

The $\underline{g}_{ij}$ are exhibited in the matrix array formed by the rows $\{\underline{e}_i\}$ and the columns $\{\underline{f}_j\}$.

If we take the particular case of $n = 3$ and suppose that

$$a_1 \underline{e}_1 + a_2 \underline{e}_2 + a_3 \underline{e}_3$$

is an element of **A** whilst

$$b_1 \underline{f}_1 + b_2 \underline{f}_2 + b_3 \underline{f}_3$$

is an element of **B** then these give rise to the element with the n^2 components

$$a_i b_j \, \underline{e}_i \, \underline{f}_j \qquad i,j = 1 \ldots n \qquad (2)$$

of the outer product of the two algebras, viz., $\mathbf{A} \times \mathbf{B}$.

The basis of $\mathbf{A} \times \mathbf{B}$ clearly contains 9 elements (n^2 in general) ; a typical member of the algebra, such as (2), might often be abbreviated to the matrix form $(a_i b_j)$ provided the basis is always understood.

The possibilities of a linear change of basis, for either **A** or **B** or both, can then always be regarded as a linear transformation in the algebra $\mathbf{A} \times \mathbf{B}$.

But this situation has already been considered in our discussion of **tensors** (v. Appendix-A) - the $a_i b_j$ (or c_{ij}) being the coefficients in a **multilinear form**.

The question of **covariance** or **contravariance** is then the question of **cogredience** or **contragredience** between, say, the c_{ij} and the basis $\underline{e}_i \underline{f}_j$.

[13.2] Matrix representation - Trace as an invariant

We are here only interested in the significance of this algebraic discussion in a Physics.

Since there is no reason why the two algebras **A** and **B** should not be identical we can consider the outer product $\mathbf{A} \times \mathbf{A}$, whence (1) becomes

$$\underline{g}_{ij} = \underline{e}_i \, \underline{e}_j \qquad (3)$$

Considering the dominance of the algebras $\Re^{\#}$, $\mathbb{C}$, $\mathbb{Q}$ (being division algebras) we look at algebras with bases

$$\{1\} \text{ in } \Re^{\#}$$

$$\{1, i\} \text{ in } \mathbb{C} \quad \text{and}$$

$$\{1, i, j, k\} \text{ in } \mathbb{Q}$$

By way of illustration, we recall that in **Physics-2** the relevant algebra is $\mathbb{Q}$ since this carries measures of quaternions (vectors) of the form $v_1 i + v_2 j + v_3 k$, the v_i being reals.

Then what sort of measures can we expect to be represented in $\mathbb{Q} \times \mathbb{Q}$?

Such an element will be of the form

$$\underline{v}\,\underline{v} = (v_1 i \ldots) \times (v_1 i \ldots)$$

which written out in full is

$$\begin{array}{l} v_1^2 ii + v_1 v_2 ij + v_1 v_3 + \\ v_2 v_1 ji + v_2^2 jj + v_2 v_3 jk + \\ v_3 v_1 ki + v_3 v_2 kj + v_3^2 kk \end{array} \quad (3)$$

Of course a general element in $\mathbb{Q} \times \mathbb{Q}$ will contain $4^2 = 16$ members ; (3) contains only 9 because we have restricted the example to "vectors".

From the point of view of Physics, in which we are looking for possible **invariants** for measures on different scales $\mathfrak{C}$, a significant invariant of (3) lies in the **spur** (or **trace**) of the inherent matrix.

This **spur** is the sum of the leading **diagonal elements**, viz.,

$$\text{spur }(v_i v_j) = v_1^2 + v_2^2 + v_3^2 \quad (4)$$

It follows that in **Physics-2** , among the Scales $\{\mathfrak{C}_\alpha\}$ in which the velocity vectors $\underline{v}$ are linearly related, we shall find **energy measures** as **invariants** of the outer products $\underline{v}\,\underline{v}$, viz., $\quad$ spur ($\frac{1}{2}$ m $\underline{v}\,\underline{v}$)

And when dealing with the **angular energy** of a rigid body we take the counterpart of (3) and require the invariant $\quad$ spur $\{\frac{1}{2}$ M $\underline{\omega}\,\underline{\omega}\}$

where $\underline{\omega}$ is the angular velocity vector and both it and **M** are in $\mathbb{Q}$.

By taking **M** as a symmetrical matrix, viz.

$$M = \begin{bmatrix} M_{11} & M_{12} & M_{13} & M_{14} \\ M_{12} & A & -H & -G \\ M_{13} & -H & B & -F \\ M_{14} & -G & -F & C \end{bmatrix}$$

and $\underline{\omega}$ as

$$\Omega = \begin{bmatrix} 0 & \omega_1 & \omega_2 & \omega_3 \end{bmatrix}$$

$$\begin{bmatrix} 0 & 0 & 0 & 0 \\ 0 & 0 & 0 & 0 \\ 0 & 0 & 0 & 0 \end{bmatrix}$$

then the **spur** of the product $\tfrac{1}{2} \mathbf{M} \Omega^T \Omega$ is the **angular energy** expression viz.,

spur $(\tfrac{1}{2} \underline{M} \, \underline{\omega} \, \underline{\omega}) = A\omega_1^2 + B\omega_2^2 + C\omega_3^2$

$\qquad\qquad\qquad - 2F\omega_2\omega_3 - 2G\omega_3\omega_1 - 2H\omega_1\omega_2$

[v. eg Refs D-[12], [16]]

The constants A, B, ... are the **moments** and **products** of **inertia** of the body relative to standard **Euclidean** axes.

[13.3] The mapping $\pi : \mathcal{Q}^2 \to \mathcal{Q}$

So far we have not suggested that any attempt should be made to relate (eg) ij to ji, and this is because the question is algebraically irrelevant - but it may be relevant in any Physics whose measures are in $\mathcal{Q}^2$.

In $\mathcal{Q}^2$ ij and ji are distinct basal units although in $\mathcal{Q}$ ij = - ji. Hence if we form an element of $\mathcal{Q}^2$ and then decide to write ij = -ji (and similar) we are effectively projecting the generating element into $\mathcal{Q}$ - via a map $\pi : \mathcal{Q}^2 \to \mathcal{Q}$ - and this might then allow us to use it in conjunction with other elements which are otherwise always to be found in $\mathcal{Q}$.

Thus, in $\mathcal{Q}^2$, the element whose components are of the form (basal units understood)

$$\begin{bmatrix} 0 & 0 & 0 & 0 \\ 0 & F_{11} & F_{12} & F_{13} \\ 0 & F_{21} & F_{22} & F_{23} \\ 0 & F_{31} & F_{32} & F_{33} \end{bmatrix}$$

will project into the element of $\mathcal{Q}$, viz.

$\qquad -(F_{11} + F_{22} + F_{33}) + (F_{23} - F_{32})i + (F_{31} - F_{13})j + (F_{12} - F_{21})k \quad (5)$

or $\qquad$ scalar + vector $\qquad$ (in $\mathcal{Q}$)

and if $\underline{a}$ and $\underline{b}$ are two quaternions this process of projection will result in the the quaternion $\underline{a}\,\underline{b}$ in $\mathcal{Q}$, viz., (using vector algebra notation)

$$-(\underline{a}.\underline{b}) + \underline{a} \wedge \underline{b}.$$

[13.4] The Electromagnetic field - Poynting vector

We illustrate these ideas by obtaining certain results which are well known in Electromagnetism, **Physics-8**.

We have already appealed to the algebra $\mathbb{C} \times \mathcal{Q}$, and this enabled us to obtain results using the elements $\underline{E} + i\underline{H}$ in that algebra - $\underline{E}$ and $\underline{H}$ being quaternions in $\mathcal{Q}$.

We now extend the algebra to $\mathbb{C} \times \mathbb{Q}^2$ and consider the element of this algebra,

viz. $\quad\quad\quad\quad (\underline{\mathbf{E}} + i\underline{\mathbf{H}})(\underline{\mathbf{E}} + i\underline{\mathbf{H}}) \quad\quad\quad (6)$

where the base units of $\mathbb{C}$ must not be confused with those of $\mathbb{Q}$.

This element of $\mathbb{C} \times \mathbb{Q}^2$ will have a matrix representation, say,

$$\{Y_{rs}\} \quad ; r,s = 1,2,3,4$$

equal to the product of

$$\begin{bmatrix} 0 & E_1+iH_1 & E_2+iH_2 & E_3+iH_3 \\ E_1+iH_1 & 0 & 0 & 0 \\ E_2+iH_2 & 0 & 0 & 0 \\ E_3+iH_3 & 0 & 0 & 0 \end{bmatrix}$$

with the matrix

$$\begin{bmatrix} 0 & E_1-iH_1 & E_2-iH_2 & E_3-iH_3 \\ E_1-iH_1 & 0 & 0 & 0 \\ E_2-iH_2 & 0 & 0 & 0 \\ E_3-iH_3 & 0 & 0 & 0 \end{bmatrix}$$

and this gives $\quad \{Y_{rs}\} = \{Y_{rs}^+\} + i\{Y_{rs}^-\} \quad\quad (7)$

where $\{Y^+\}$ is a symmetric matrix, not containing i

and $\{Y^-\}$ is an antisymmetric matrix, not containing i.

If we now project this element back into $\mathbb{C} \times \mathbb{Q}$, via the mapping π, we obtain

$$\text{spur } \{Y_{rs}\} = -(\underline{\mathbf{E}}^2 + \underline{\mathbf{H}}^2) = -W, \text{ say} \quad (8a)$$

and $\quad \text{vector } \{Y_{rs}\} = i\,\underline{\mathbf{E}} \wedge \underline{\mathbf{H}} = i\,\underline{\Pi}, \text{ say} \quad (8b)$

It follows that the element (9) in $\mathbb{C} \times \mathbb{Q}^2$ corresponds to a set of measures mapped into $\mathbb{C} \times \mathbb{Q}$, viz.,

$$-W + i\underline{\Pi} \quad\quad (8c)$$

Of these W is the **electromagnetic energy in free space** and

$\underline{\Pi}$ is the **Poynting** vector.

[13.5] E-M field ; Quaternions over $\mathbb{C}$

If we treat (8c) as a generalised quaternion potential for the field in free space we can derive the quaternion **force** therefrom by using the operator $\square$ (v. supra), viz.,

$$\square \equiv \nabla + (i/c)\partial/\partial t$$

and we then have the force in $\mathbb{C} \times \mathbb{Q}$ as

$$\mathbf{X} + i\mathbf{Y} = -\square(-W + i\underline{\Pi})$$

and this $\quad = \nabla W + i \text{ div } \underline{\Pi} - i \text{ curl } \underline{\Pi} + (i/c)\partial W/\partial t + (1/c)\partial\underline{\Pi}/\partial t$

with $\mathbf{X}$ and $\mathbf{Y}$ in $\mathbb{Q}$.

But this field is self-propagating ; there are no measurable quantities $\mathbf{X}$ and $\mathbf{Y}$.

Hence $\mathbf{X} = \mathbf{Y} = 0 + \underline{0}$ in $\mathbb{Q}$. We therefore get the equations

$$-\nabla W = (1/c)\partial \underline{\Pi}/\partial t \qquad (9)$$

$$-(1/c)\partial W/\partial t = \text{div } \underline{\Pi} \qquad (10)$$

and

$$\underline{0} = \text{curl } \underline{\Pi} \qquad (11)$$

Equation (10) expresses the fact that $\underline{\Pi}$ is an **energy flow** vector.

Equations (9) and (11) both say the same thing, viz., that $\underline{\Pi}$ can be expressed as

grad $\varphi(t)$ where $\varphi(t)$ is a scalar function of $\underline{r}$ and t.

Hence $\underline{\Pi}$ may be regarded as a vector which acts along the normal to a surface

$\varphi(t) = \varphi(\underline{r})$ - which surface may be supposed to represent a general **wave-front**.

This means that the propagation of E-M waves may be contained in the algebraic element $\{20_{rs}\}$ in $\mathbb{C} \times \mathbb{Q} \times \mathbb{Q}$.

It follows that (9) and (10) give

$$(1/c^2) = (1/c) \text{ div } \partial \underline{\Pi}/\partial t = \text{div grad } W$$

or

$$\nabla^2 W = (1/c^2) \partial^2 W/\partial t^2 \qquad (12)$$

being the ubiquitous equation of **wave motion** (for the E-M field).

The **Poynting** vector, which is perpendicular to both $\underline{E}$ and $\underline{H}$, gives us the flow of energy normal to the plane wave-front.

[13.6] E-M field energy in $\mathbb{C} \times \mathbb{Q} \times \mathbb{Q}$

The above discussion means that we can state the following :-

The propagation of E-M energy in space-time is observed as wave-motion by measures mapped into $\pm 2 \times \mathbb{Q}$ and this wave motion is the projection, from $\mathbb{C} \times \mathbb{Q} \times \mathbb{Q}$, of the

energy law $\frac{1}{2}(\underline{E} + i\underline{H}) \text{ I } (\underline{E} - i\underline{H}) = \text{constant} \qquad (13)$

The equations in $\mathbb{C} \times \mathbb{Q}$ are obtained from (13) by

$$\Box \{(\underline{E} + i\underline{H}) \text{ I } (\underline{E} - i\underline{H})\} = 0 \qquad (14)$$

The I in the above being the **idemfactor** in $\mathbb{Q} \times \mathbb{Q}$; thus the concept of the **conservation of energy** finds its application when the formal expression for energy in **Physics-8** is embedded in the algebra $\mathbb{C} \times \mathbb{Q} \times \mathbb{Q}$.

[13.7] Homographies in Clifford algebra

An homography $\Gamma(1\text{-}1)$ expressed in a **Clifford** algebra, $\mathbb{G} \times \mathbb{G}$ (over $\mathfrak{R}^{\#}$), occurs by considering its typical homogeneous form :

$$\Gamma(1\text{-}1) \equiv axy + bxs + cyt + dst = 0$$

and taking a basis for a **Clifford** algebra as {i j}, so that projecting an element into
$\Re^{\#}$ is achieved by writing   ii = 1 = jj [we shall see that ij = -ji does not arise]
that is to say we take the projection π by

$$\pi : \text{Aii} + \text{Bjj} \to A + B \quad \in \Re^{\#}.$$

The $\Gamma(1\text{-}1)$ can be described by using the two "vectors" ($\in \mathfrak{G}$) viz., (xi + tj) and (yi + sj) together with the "matrix", $\in \mathfrak{G} \times \mathfrak{G}$, viz.,

$$M = \begin{bmatrix} \text{aii} & \text{bij} \\ \text{cji} & \text{djj} \end{bmatrix}$$

whence the projection π sends the statement

$$(xi + tj) \, M \, (yi + sj)^T = 0 \quad (\in \Re^{\#})$$

into   (axy + bxs) ii + (cyt + dst) jj = 0

and thence to the form of $\Gamma(1\text{-}1)$.

[13.8] Eddington's Sedenion algreba   [Ref Appendix-E [5]]

We have already seen, [v. Appendix-A], how we can build a **wedge-space**, $\Lambda(V_n)$, over a vector space with a basis of n elements.

If we apply this to the quaternion basis {**1,i,j,k**} we obtain a wedge-space $\Lambda\mathcal{Q}(\Re^{\#})$ where

($\mathcal{Q}_0$) : $\Lambda^0\mathcal{Q}$ being the field $\Re^{\#}$ or $\mathcal{C}$ (scalars) ; dimension = 1
    1 base unit "1"

($\mathcal{Q}_1$) : $\Lambda^1\mathcal{Q}$ being 1-forms (algebra $\mathcal{Q}$) ; dimension = 4
    4 base units E_α

($\mathcal{Q}_2$) : $\Lambda^2\mathcal{Q}$ being 2-forms ; dimension = 6
    6 base units $E_{\alpha\beta}$

($\mathcal{Q}_3$) : $\Lambda^3\mathcal{Q}$ being 3-forms ; dimension = 4
    4 base units $E_{\alpha\beta\gamma}$

($\mathcal{Q}_4$) : $\Lambda^4\mathcal{Q}$ being 0-forms (pseudo-scalars) ; dimension = 1
    1 base unit $E_{\alpha\beta\gamma\delta}$

The **exterior algebra**, $\Lambda \cdot \mathcal{Q}$, is now the direct sum of the wedge powers ; and because $2^4 = 16 = 4^2$ the algebra is equivalent (isomorphic) to the direct algebra $\mathcal{Q} \times \mathcal{Q}$.

Thus if we map physics measures into $\mathcal{Q} \times \mathcal{Q}$ we may equally well map them into $\Lambda \mathcal{Q}$.

This idea was pursued by **Eddington** [v. Ref above]- who called it **Sedenion algebra**

[In honour of the Professorial Chair he occupied in Cambridge University]

and which he thought of as a representation of the outer product $\mathfrak{G} \times \mathfrak{G}$.

He was largely concerned with developing an algebra which could be used to bridge

Quantum and Relativity Theories, and this involved regarding **Physics-3** as a **Physics-4** - and not the other way round.

It was because of this latent aspect of his work that he referred to it all as **Spin Theory**.

The idea is, in general, that measures in **Physics-4** require two vector representations

(i) a linear momentum vector, components p_1, p_2, p_3, p_4 defined in the 4-dimensional space-time (invariant under the **Lorentz** operator $\mathcal{L}$), and

(ii) an angular momentum vector, components $p_{12}, p_{13}, p_{14}, p_{23}, p_{24}, p_{34}$.

Also **Physics-8** requires a general 4-vector $\{\varphi, \underline{A}\}$ invariant under $\mathcal{L}$ - and also two scalars viz., the magnetic and electric charge (also invariant).

It follows that the exterior algebra $\Lambda\mathcal{Q}$ has just the right dimensions to host the 16-component physics measures found in **Physics-4** $\cup$ **Physics-8** viz.,

"charge" $(\mathcal{Q}_0 \cup \mathcal{Q}_4) + p_r(\mathcal{Q}_1) + p_{rs}(\mathcal{Q}_2) + (\varphi, \underline{A})(\mathcal{Q}_3)$

In fact **Eddington** built up his Sedenion Algebra by multiplying each base unit by the complex symbol **i** ($i^2 = -1$).

It is worth pointing out the special nature of **Eddington's** notation - since he represented the base unit by symbols E_{rs}, with two suffixes, whereas some of them should have been in $\mathcal{Q}_3$ or $\mathcal{Q}_4$.

Since he was primarily thinking of the Sedenion algebra as a means of mapping measures into **Physics-4** $\cup$ **Physics-8**, and thereby keeping the vectors of dimensions (1,4,6,4,1) separate, he used a pattern of suffixes which reminded him of these vector spaces.

The 4-vector $\{p_r\}$, $r = 1..4$, for linear momentum (invariant under $\mathcal{L}$) was renamed $\{p_{r5}\}$ and the 4-vector for **Physics-8**, viz. $(V,\underline{A})$, was denoted by $\{p_{0r}\}$ whilst the remaining 6-vector was left as $\{p_{12}, p_{13}, p_{14}, p_{23}, p_{24}, p_{34}\}$. These 14 components in the algebra now complete a matrix representation (4 x 4) apart from the two scalars. **Eddington** denoted one of these by p_{05} and the other by p_{16} - but the p_{16} is not "p-one-six" but "p-sixteen" (the last one!). This set of suffixes was carried over into the base unit E_{rs}. [v. Fig. 13.1 for the layout].

The p_{16} can be placed anywhere down the diagonal of zeros.

If the base units of $\mathcal{Q}$ are denoted [v. Appendix-A §(A4)] by $\{\hat{e}_0(=1), \hat{e}_1, \hat{e}_2, \hat{e}_3\}$

then **Eddington's** base units for $\Lambda\mathcal{Q}$ are :-

$E_{01} = i(\hat{e}_0 \hat{e}_1)$ $E_{02} = -i(\hat{e}_0 \hat{e}_3)$ $E_{03} = -i(\hat{e}_2 \hat{e}_2)$

$E_{04} = -i(\hat{e}_3 \hat{e}_2)$ $E_{05} = -i(\hat{e}_1 \hat{e}_2)$

$E_{12} = i(\hat{e}_0 \hat{e}_2)$ $E_{13} = i(\hat{e}_2 \hat{e}_3)$ $E_{14} = i(\hat{e}_3 \hat{e}_3)$

$E_{23} = -i(\hat{e}_2 \hat{e}_1)$ $E_{24} = -i(\hat{e}_3 \hat{e}_1)$ $E_{34} = -i(\hat{e}_1 \hat{e}_0)$

$E_{15} = i(\hat{e}_1 \hat{e}_3)$ $E_{25} = -i(\hat{e}_1 \hat{e}_1)$ $E_{35} = i(\hat{e}_3 \hat{e}_0)$

$E_{45} = -i(\hat{e}_2 \hat{e}_1)$ $E_{16} = i(\hat{e}_0 \hat{e}_0)$

Then if each of the base units of Q is given its 2 x 2 matrix representations each of the E_{rs} has a 4 x 4 matrix representation, and the commutative properties of the $\hat{e}_r$ produce their counterpart in properties among the E_{ij}.

Unfortunately this notation disguised the dimensionalities of the vectors involved - which dimensionalities are clearly exhibited by using the wedge-space $\Lambda\mathcal{E}$.

[v. Ref Appendix-E [5] for Eddington's full discussion - where he extends his E-numbers to a 256-space (!) of EF-numbers.
These ideas were not well received by post-Eddington physicists, and although he felt the need for a new algebra for Physics he did not extend his thinking to the idea that a new geometry might be a better bet.]

118 MATHEMATICAL PHYSICS

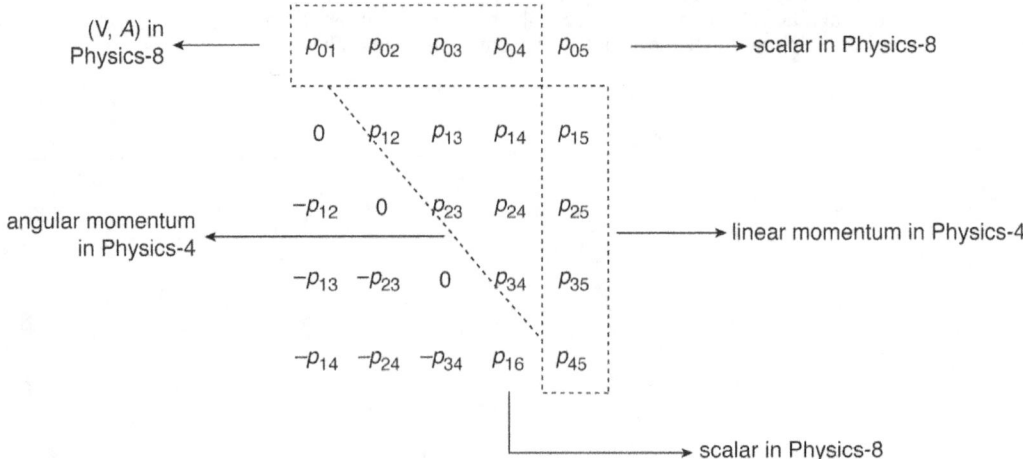

Fig 13.1

Chapter-14 /PART-1 Quantum Matters

[14.1] Quantisation - various

This really amounts to the fact that the Physicist's acknowledgement that he cannot measure anything diectly into the irrational numbers - but only into the **rationals**, $\Re$. This is because the **rational numbers** are **countable** - that is to say, they are in a $\Gamma(1\text{-}1)$ correspondence with the **integers**, **J**. More precisely, there is an **onto mapping**

$$\rho : \mathbf{J} \times \mathbf{J}^+ \rightarrow \Re$$

in that $\rho\ (p,q) \rightarrow p/q$ (and $q \neq 0$).

Fig. 14.1 illustrates how the counting can be envisaged for $\mathbf{J}^+ \times \mathbf{J}^+$ - by following the arrows.

Measures of the generating elements $\{\mathbf{E}\}$ will therefor naturally come in countable "packets".

For example, electric charges are multiples of the $\pm$ (electronic charge) ; lengths are ultimately multiples of wavelength ; sounds are multiples of frequency ; countable normal modes of oscillation ; countable eigenvalues of a matrix operator ; countable components of a vector (or tensor) with respect to a countable set of axes ; countable interference or diffraction maxima ; countable **action** via **Planck**'s constant **h** ; countable **energy levels** via **h**ν.

[14.2] Pythagoras' black sheep

The Pythagoreans were Physicists firstly and Mathematicians secondly, because they knew that all measurements were ratios of integers. But since they thought that Physics and Mathematics are identical they did not look kindly on the real mathematician in their midst - one **Hippasus of Metapontum**, who proved that (eg) $\sqrt{2}$ is not rational, and reputedly threw him overboard [metaphorically, we hope]. Perhaps the belief is still not dead ?

[Ref. Appendix-E [14] - Kline]

[14.3] Interference patterns, using $\Re^\#$ and $\mathbb{C}$

Young's slits experiment is a demonstration of how the measures of light naturally relate one algebra, $\Re^\#$, to another, $\mathbb{C}$.

Although Physicists have the need to attribute experimental measures to real numbers, in $\Re^\#$ this does not necessarily mean that they should be so exclusively described - for other algebras can contain $\mathfrak{R}^{\#}$ as a subalgebra (eg the complex algebra $\mathbb{C}$).

Suppose, for example, we have a Scale $\mathfrak{C}_\alpha$ on which we "see" the element $\mathbf{E}_p$ as a measure in $\mathfrak{R}^{\#}$, and that we also have a Scale $\mathfrak{C}_\beta$ which exhibits the measures of $\mathbf{E}_p$ in the complex algebra $\mathbb{C}$.

Contemplating these measures consecutively on $\mathfrak{C}_\alpha$ and $\mathfrak{C}_\beta$ we obtain the scheme shown in Fig. 14.1a, and this may be contained in the Scale $\mathfrak{C}_\gamma \equiv \mathfrak{C}_\beta\{\mathfrak{C}_\alpha\}$. We then need to find a way of restricting a number, like $X + iY$, into a real number - and to "see" this result in an experimental arangement. **Young's** experiment did just that.

Writing $X + iY = a\, e^{i\theta}$ we see that the measures of $\mathbf{E}_p$ will be real whenever

$$\theta = n\pi \quad \text{where } n = 0, 1, 2, \ldots \quad \text{(a countable set)}$$

and that is when the "values" become $a(-1)^n$.

$$\mathbf{E}_p\, (X + iY = a\, e^{i\theta})$$
$$\downarrow$$
$$\downarrow$$

	...	+a	-a	+a	-a	+a	-a	+a	-a	+a	...
n =		-4	-3	-2	-1	0	+1	+2	+3	+4	

[This is illustrated in Fig. 14.1b.]

This situation is essentially that of a **quantisation** of the discrete observable states, as opposed to that of continuous states.

On $\mathfrak{C}_\alpha$ the $\mathbf{E}_p$ have continuous values ; on $\mathfrak{C}_\beta$ they have continuous but complex values ; on the final Scale $\mathfrak{C}_\gamma$ they have discrete but real values.

Thus we would map the final measures into the **integral domain**, J, via the sequence

$$x(\text{on } \mathfrak{C}_\alpha) \to (X + iY) \text{ on } \mathfrak{C}_\beta \to J\, (\text{on } \mathfrak{C}_\gamma)$$

and this would be accomplished via a mapping ζ where

$$\zeta : x\, (\text{on } \mathfrak{C}_\alpha) \to a\, e^{i\theta}\, (\text{on } \mathfrak{C}_\beta)$$

Also since we are considering observations of the Base Element $\mathfrak{Z}_L$ themselves the measures will be **reflexive** and so, on $\mathfrak{C}_\gamma$, the images will occur within the involution

$$(P\ I\ Q\ J) = -1$$

I,J being the double points - but with the proviso that (forced on $\mathfrak{C}_\gamma$) the points I,J must have co-ordinates $\pm i$ - then this amounts to identifying the existence states with a variable λ given by

$$(\lambda\ i\ 0\ -i) = -1$$

or with a variable θ (**Laguerre's** theorem) where $\quad e^{2i\theta} = -1$

This again gives $\pm\,\theta = (2n-1)\pi/2$ - which is a set of discrete states.

[Note : This is equivalent to taking the **projective metric** on $\mathfrak{C}_\gamma$,

viz. $1/2i \ln (P\,I\,Q\,J)$ which is $\theta \pm 2n\pi$: v. Appendix-B]

So this "length" on $\mathfrak{C}_\gamma$ proceeds by way of **quantum jumps**.

It follows [v. Fig. 14.1c] that the metric properties are governed by a **discrete set of metric conics** - points on each of which may be "seen" - and an observable metric (absolute) conic will possess the **isotropic lines** as tangent lines. Thus the absolute conics will form a **confocal family** with foci at the points A and B (the slits) being a family of hyperbolae. Furthermore since the reference points I,J are on $\mathfrak{C}_\gamma$ this line acts as an absolute line, so that all the conics of this family cut this line. These conics alternate with another family of "dark" conics (also hyperbolae).

In metrical language the "bright" absolute conics are given by the discrete family of hyperbolae,

viz., those defined by $\qquad AP - BP = \pm\, nk$

where $n = 0, 2, 4, \ldots$ and k is a Scale constant

whilst the "dark" absolute conics are given by the relation $\quad AP - BP = \pm\, mk$

where $m = 1, 3, 5, \ldots$. In each case the foci are the points A and B.

We notice also that the mapping ζ which occurs in the above analysis is traditionally the mapping which is the basis for a successful **wave theory** (of light).

[14.4] An experimental model producing quantisation

A model to demonstrate quantum states can be provided by the following simple experimental arrangement.

Take a Scale $\mathfrak{C}_\alpha$ which consists of a conducting circular plate with an electric potential applied between the centre and the outer rim, and take with it a non-conducting rod AB with electrical contacts at the extremeties A and B.

The Scale $\mathfrak{C}_\alpha$ is designed to measure the **length** L^α as follows :

    (a) anchor the rod AB with A at the centre of $\mathfrak{C}_\alpha$ in such a way that it may lie flat on the disc and free to rotate about that end.

    (b) the rod is arbitrarily spun about its end A and, with B making contact with the plate, the potential difference is measured by a voltmeter

    (c) we agree to say that the "length" of AB is observed to be

$$L^\alpha = V_B - V_A$$

where $V_B > V_A$.

We shall find that, wherever the rod AB is to be found, its "length" will be a constant.

Now consider another Scale $\mathfrak{C}_\beta$ consisting of a plane grid structure of parallel conducting wires between two electrical terminals $\ominus$ and $\oplus$ [v. Fig. 14.1d].

The rod is pinned at the end A on one of the conducting wires and observations are made by spinning it arbitrarily about that end. It is also assumed that the grid is always wide enough to ensure that the end B remains within its confines.

When the rod comes to rest the voltages V_A and V_B are observed and the "length" is again defined as $$L^\beta = V_B - V_A$$ whenever $V_B \neq 0$.

It is now clear that we can observe any of the following :

(d) L^β cannot be observed because B does not make contact with any one wire

(e) when L^β can be observed its value will be quantised - because V_B can only take values which are multiples of the potential difference h

(f) L^β can take both +ve and -ve values

(g) If C is another point on the grid, and within B's range then
since $V_B - V_A = (V_B - V_C) + (V_C - V_A)$
and it follows that $L^{\beta,AB} = L^{\beta,AC} + L^{\beta,CB}$

The condition (d) means that $L^{\beta,AB}$ obeys

  (i) the triangle law of addition   and

  (ii) the **Ritz Combination Rule**.

We also notice that $\mathfrak{C}_\beta$ selects only trivial cases of triplets which obey the theorem of **Pythagoras**, viz. those for which $V_B - V_C = 0$.

The important difference between L^α and L^β lies in the fact that

  (iii) on $\mathfrak{C}_\alpha$ the rod AB possesses a unique measure, constant everywhere

  (iv) on $\mathfrak{C}_\beta$ the rod possesses two countable (discrete) sets of values, viz., a quantum number ,n, which anchors the end A on any wire, and another quantum number, say m, which can take any of $2p+1$ values - for with A anchored the end B can contact an odd number of wires, p being specific to the rod.

If now we assume that these quantum states are equally likely then the probability of a measure giving (say) the rth value will be $1/(2p + 1)$, and the **expectation value** of this rth. measure will be $(2p + 1)^{-1} \times$ (value of this rth measure)

The actual values these measures give will clearly be

$$0, \pm h, \pm 2h, \pm 3h, \ldots \pm ph$$

giving the expectation value for the rth measure as $rh/(2p + 1)$ and the Scale $\mathfrak{C}_\alpha$ will correspond to the case of $h \to 0$.

We can also expand on the differences bewteen the two Scales as follows.

Since there is a simple $\Gamma(1\text{-}1)$ between L^α and its measures on $\mathfrak{C}_\alpha$ we can say that this E_p (i.e L^α) behaves like a **particle** - so it has no internal structure, being entirely an **external** particle, characterised by quantum numbers $n = m = 0$.

So as far as $\mathfrak{C}_\beta$ is concerned this E_p cannot be a particle. It must have some internal structure, and this must be determined by the quantum numbers associated with it. Thus $\mathfrak{C}_\beta$ has **discovered** the internal structure of E_p (i.e. L^{AB}).

We can give a name (or names) to this internal structure of E_p. For example, we can say that E_p is a **vector** quantity in a 2-dimensional space and that its measures are the **projections in an assigned direction** viz. the direction of the maximum value L_0. But the projections may only be taken from quantum specified positions ; the permissible angles θ are thereby quantised.

On the other hand we can say that since the $(2p + 1)$ values of E_p are independent (in the sense that the probabilities refer to independent (**Bernoullian**) events) then we should construct an **abstract vector space** of $(2p + 1)$ dimensions and represent the E_p in this. That is to say, map the E_p measures into a $V_{(2p+1)}$.

If we denote the basis of this space by the unit vectors $\{\hat{e}_r\}$ then the E_p should be associated with an **N**-vector (given $N = 2p+1$) via

$$\mathbf{x} = x_1\hat{e}_1 + x_2\hat{e}_2 + \ldots + x_N\hat{e}_N \qquad (1)$$

In (1) the x_r should actually be probabilties, since these are the things which are truly independent, and so there should be a sense in which the **total probability** should be 1. Furthermore we can regard the statement (1) as telling us the "state of the system" and which lays out the possible values of its measures.

If we impose the usual **Euclidean** metric on V_N we can arrange the independence feature via the **orthogonality condition**, viz., $\hat{e}_r \cdot \hat{e}_s = \delta_{rs}$ (the **Kronecker** delta).

This ensures that the **modulus** of such a vector $\mathbf{x}$ is

$$\|\mathbf{x}\| = |x_1|^2 + |x_2|^2 + \ldots |x_N|^2 = 1$$

We have now associated our measure, E_p on $\mathfrak{C}_\beta$, with a **normalised vector** in the vector space V_N - with a **Euclidean** metric imposed.

?-Question :- Are these differences intrinsic properties of the generating element E_p, which we are calling length, or are they properties of the Scales $\mathfrak{C}_\alpha$ and $\mathfrak{C}_\beta$?

?-**Answer** : Without the scales $\mathfrak{C}_\alpha$ there can be no measures of the $\mathbf{E}_p$, and without the generating elements $\mathbf{E}_p$ there are no measures on the scales $\mathfrak{C}_\alpha$ - which suggests that the whole operation is only an expression of the union of the two.

[14.5] Planck's quantisation of Action

The classical statistical analysis of **Maqxwell-Boltzmann** required each degree of freedom of each member of (say) a gas to contribute an amount ½kT to the average energy of the gas ; k being Boltzmann's constant and T the absolute temperature. For black body radiation, such as is found in a temperature enclosure, the degrees of freedom become the number of normal modes of the wave-form. The energy of the radiant heat for all those frequencies lying between ν and $\nu + d\nu$ gives the function

$$E(\nu, \nu + d\nu) = 8\pi\nu^2 c^{-3} \cdot \tfrac{1}{2}kT \, d\nu$$

which is known as **Rayleigh's Law** .

Unfortunately it is contradicted by experiment. [Ref Appendix-E [23]]

It was **Planck** who found the correct formula by the postulate :-

The Harmonic Oscillator exhibits only distinct (countable) energy levels

and he suggested that the energy of an oscillator should be of the form

$$\mathbf{E}_\nu = rh\nu \quad \text{where } r = 0, 1, 2, 3, \ldots$$

that is to say, $r \in \mathbf{J}^+ \cup \{0\}$.

The quantity **h**, known as **Planck's constant**, has a value of 6.628×10^{-27} erg-sec and its dimensions are those of **action** [energy x time, or linear momentum x distance].

The replacement for **Rayleigh's Law** thus became

$$E(\nu, \nu + d\nu) = 8\pi\nu^2 c^{-3} \cdot h\nu \cdot F(\nu) \, d\nu$$

where $F(\nu) = \{e^{-h\nu/kT} - 1\}^{-1}$ - and is in satisfactory agreement with experiment.

[14.6] Bohr's theory of the hyrogen atom

Planck's proposal began the search for other consequences of this quantisation of action and also of energy levels, and the first triumph was provided by **Bohr's** theory of a model of the simple hydrogen atom.

Earlier scattering experiments by **Rutherford** in the Cavendish Laboratory in Cambridge, had confirmed the idea of the simple atom as electrons in orbit about a positive proton.

This idea was still firmly ensconced in **Physics-3**, that of particle dynamics, and was clearly influenced by the classical successes in understanding **planetary motion**.

Bohr introduced **Planck's** suggestion into the **action** of a simple picture of an electron circulating around a proton [v. Fig. 14.2].

Taking the canonical (**Hamiltonian**) co-ordinates, **p** = angular momentum & **q** = angle, he proposed that the action associated with a whole orbit will be $\int_c p\,dq$ and that this will be "quantised", à la **Planck**, thus giving

$$\mathbf{p} = nh/2\pi \quad \text{where} \quad \mathbf{p} = ma^2\dot{\mathbf{q}} \quad m = \text{mass of electron}$$

the resulting possible radii of the circular orbits being given by

$$a_n = n^2h^2/(4\pi^2 me^2)$$

The smallest of these (the so-called ground-state) will be given by $n = 1$ and

$$a_0 = 0.528 \times 10^{-2} \text{ cm} \quad (\text{the } \mathbf{Bohr} \text{ radius of the atom})$$

The corresponding **energy levels** of the electrons-in-orbit become

$$E_n = -2\pi^2 me^4/n^2h^2. \qquad (1)$$

[14.7] Spectral series

Bohr's analysis had immediate results in the experimental data in **spectral analysis**.

He deduced that the frequencies of spectral lines in absorption and emission spectra are to be given by

$$E_n - E_m = h\nu_{nm} \qquad (2)$$

where ν_{nm} is the frequency of the radiation given out when the electron "jumps" from orbit n to orbit m ($m > n$) [v. Fig. 14.3].

More generally, if m, p, n are positive integers and $m > p > n$ then (2) gives us

$$E_n - E_p + E_p - E_m = E_n - E_m$$

which means that $\quad \nu_{np} + \nu_{pm} = \nu_{nm}$

a relation discovered by **Ritz** (in 1905) and called the **Ritz Combination Rule**.

It also follows from (1) and (2) that

$$\nu_{nm} = 2\pi^2 me^4 h^{-3} \{1/m^2 - 1/n^2\} \qquad (3)$$

Now the **Balmer Series** was a well-known sequence in the hydrogen spectrum and the frequencies were known to be given by

$$\nu_n = R\{1/2^2 - 1/n^2\} \quad \text{with } m = 2 \text{ and } n = 3, 4, \ldots$$

The constant **R** (the **Rydberg** constant) has an estimated value of 1.09677×10^5 cm^{-1} and the **Bohr** formula above gave a value for **R**, viz., $2\pi^2 me^4 h^{-3}$, and this evaluates to 1.0955×10^5 - regarded as a convincing argument for the new theory.

Other series of spectral lines, eg

Lyman (m = 1 ; in the ultra-violet)

Paschen (m = 3 ; in the infra-red)

Brackett (m = 4 ; in the infra-red)

also followed.

[14.8] Particle or Wave ?

There are a few naive questions to be answered by theories, such as that of **Bohr**, such as :

(1) **Physics-3** is a traditional particle-oriented approach, whereas it had become necessary to introduce field theories (eg the **wave theory**) in other Scale measures,

(2) "action at a distance" theories had always been difficult to accept ; for example, what occupies the apparent "spaces" between the proton and the electron ; is there not a need for some "thing" in the space to allow transmission of efffect, eg how can **Newton**'s gravitational force get from one object to another ?

(3) is the **Bohr** atom merely a **symbolic model** or is it **really real**?

These questions cannot be tackled by scientists who have not been able to distinguish the mathematical language from the physical Scale observations. For many scientists there is still the question "Is it a particle or is it a wave ?".

A mathematician can only answer "one or both, or maybe neither" - but the language we use with which to discuss the observational measures is neutral.

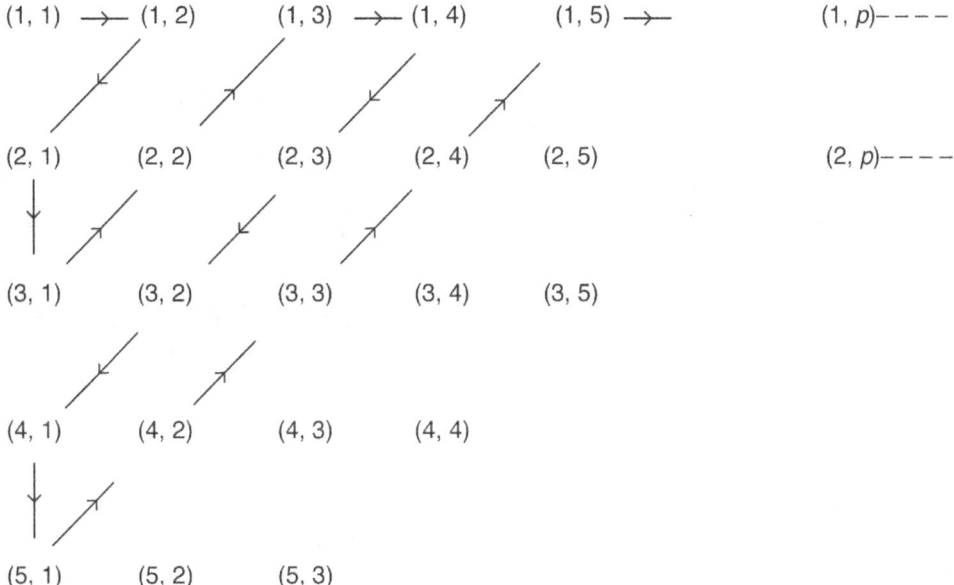

Countable Rational Numbers $(q, p) \in J^+ \times J^+$

$(q, p) \longleftrightarrow \frac{p}{q}, p, q \in J^+ ; \frac{p}{q} \in R$

Fig 14.1

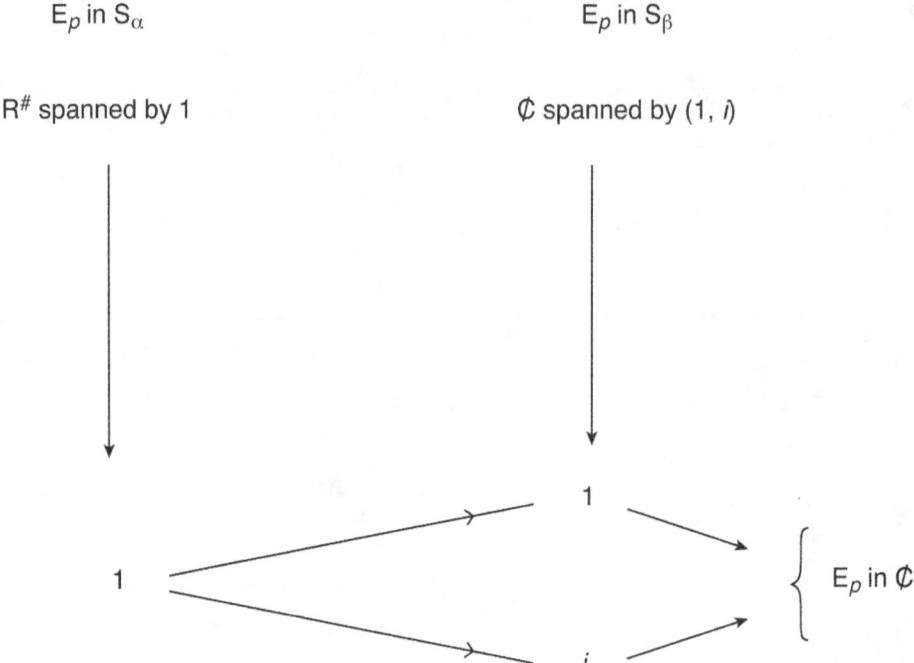

Fig 14.1a

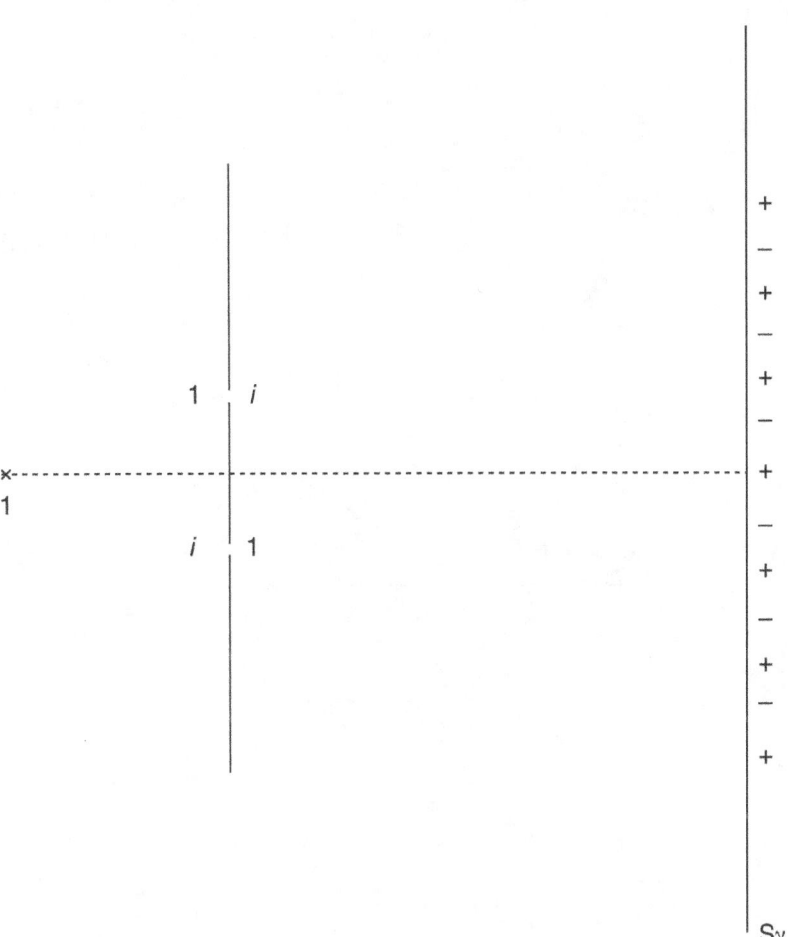

Fig 14.1b

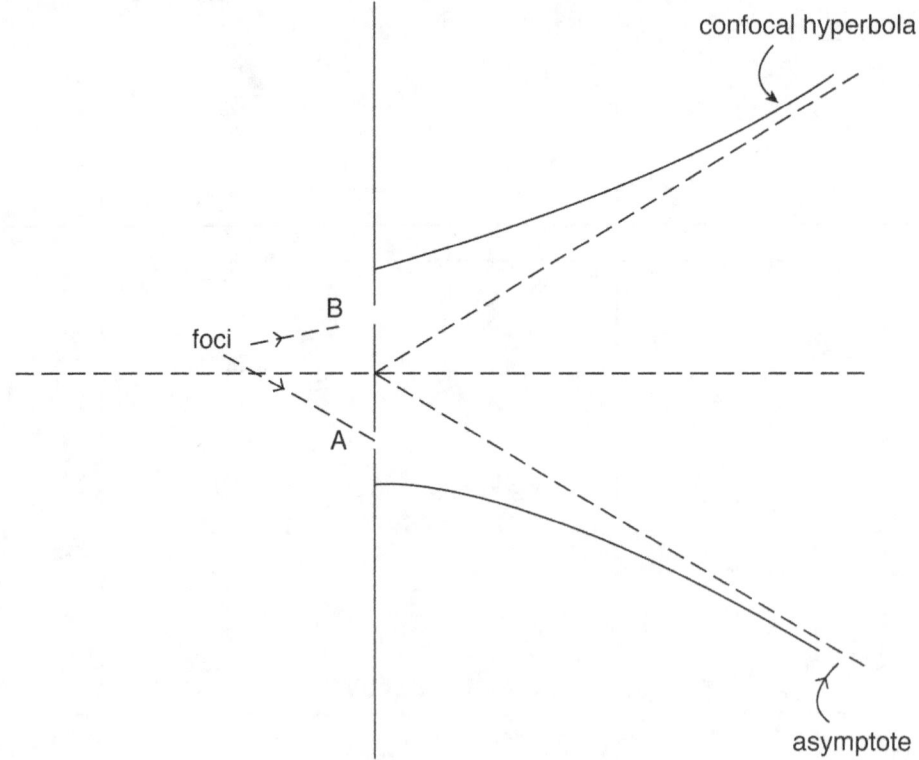

Fig 14.1c

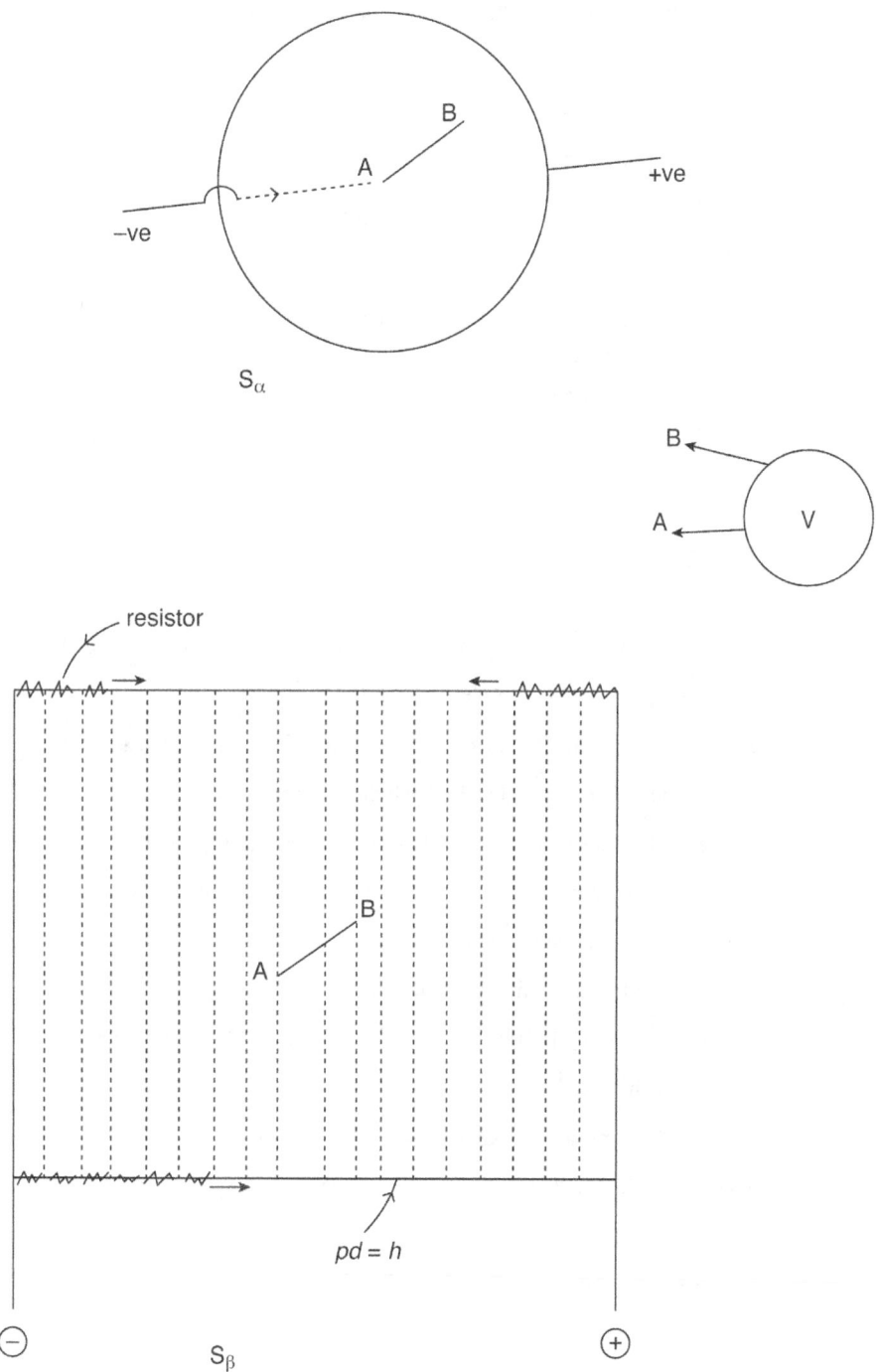

Fig 14.1d

Bohr's theory of the Hydrogen atom (1913)

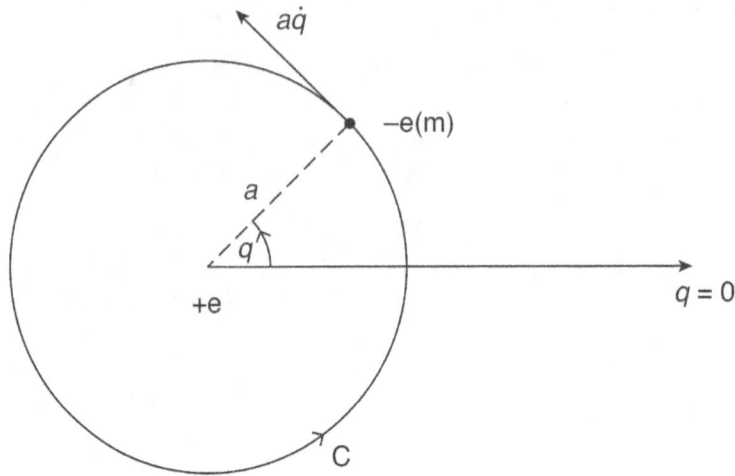

Neils Bohr introduced the idea of **Planck**'s constant (h) into a theory of the H-atom in the classical theory of motion-in-a-circle in the following way.

The angular momentum, p, is the usual quantity $m \times a\dot{q} \times a$ which is $ma^2\dot{q}$ and this is a constant of the motion. Since both h and p have dimensions of Action he proposed that for the orbit around the circle this should be $\int_0^{2\pi} p\,dq$ - which equals $2\pi p$ - and that this should be (à la **Planck**) nh, where n is any positive integer; so $ma^2\dot{q} = nh/2\pi$. Assuming the inverse-square law of attraction beteen the electron and the proton nucleus the circular motion classically requires that $ma\dot{q}^2 = e^2/a^2$.

So then we have $\quad ma \times [nh/2\pi ma^2]^2 = e^2/a^2 \quad$ and solving for a (now being a_n) we get $\qquad a_n = n^2h^2/4\pi^2 me^2$.

When we put n=1 we get what is known as the **Bohr** radius.

And the energy E (which is K.E + P.E) becomes quantised as

$$E_n = -2\pi^2 me^4/n^2 h^2$$

being the energy levels of the hydrogen atom.

Fig 14.2

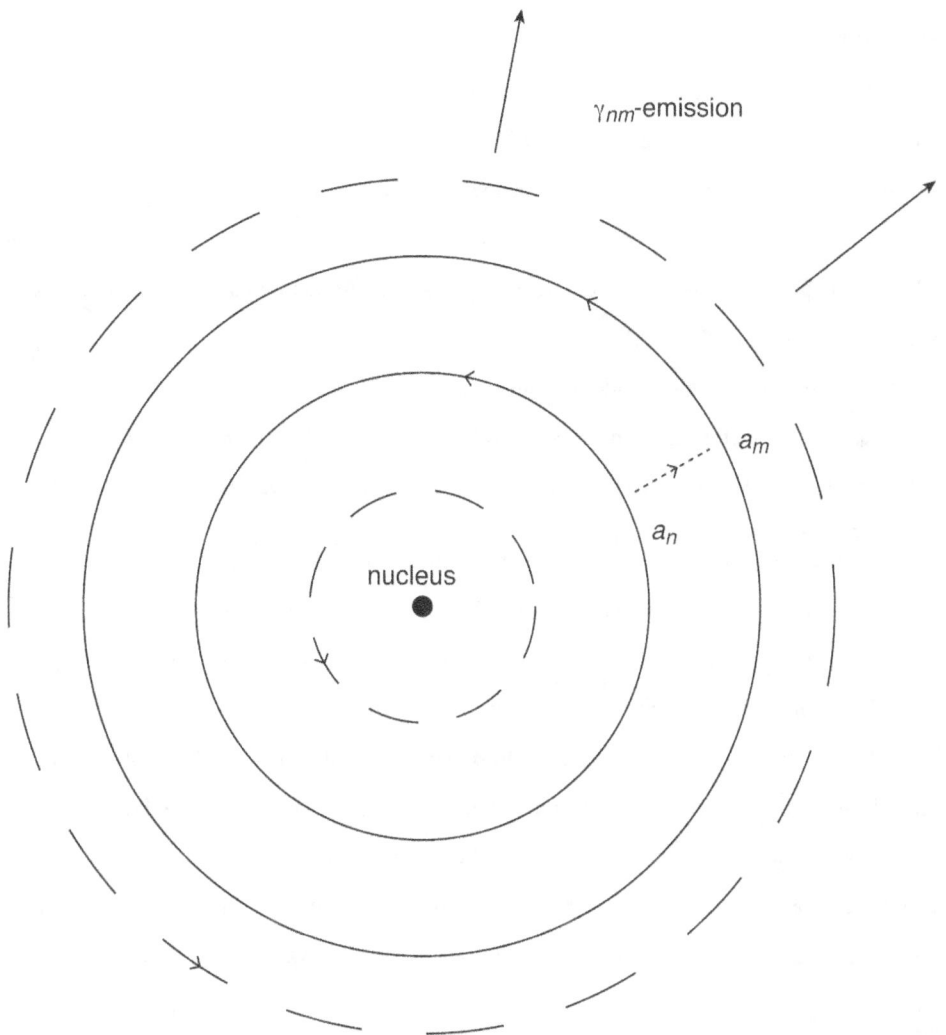

Generating Spectral sequence

Frequencies γ_{nm} given by $E_n - E_m = h\gamma_{nm}$

Fig 14.3

Chapter-15 /PART-1 Heisenberg's Matrix Mechanics

[15.1] Hilbert Space

The success of **Bohr**'s atomic structure immediately led to an extension of the approach into more general dynamical systems - now that the classical **Newtonian** dynamics was shown to be inadequate for a description of the postulated atomic structure.

Using some of the accepted countable observations, mentioned in [14.2], **Heisenberg** produced what he called **matrix mechanics**, basing his ideas on the assumption that (eg) **energy levels**

must be   (i) real numbers, $\Re^{\#}$

and also   (ii) countable eigenvalues of some matrix.

This required that such matrices must be **Hermitian** matrices - since these, when reduced to a diagonal form by suitable transformations, always have real eigenvalues.

[Note : A **Hermitian** matrix (named after the French mathematician **Hermite**) has complex components and off the diagonal each a_{rs} is the complex conjugate of a_{sr}.]

He therefore proposed that (eg) momentum **p** should be regarded as a matrix **P** and that the co-ordinate **q** should be regarded as a matrix **Q** (these being **Hermitian**). The energy **E** would become a diagonal matrix with energy levels E_n.

All of these matrices would need to be defined over an **infinitely countable** set of axes $\{\hat{e}_n\}$, and this meant that the vectors and matrices must be defined in a **Hilbert Space** [v. Appendix-B]. Furthermore these matrices would naturally be **linear operators** on the vectors of the space - which themselves would define the **state** of the system. So the result of an observation of some dynamical generating element would be tountamount to evaluating an **eigenvalue** for that element, and associating with it an **eigenvector**.

[15.2] Born, Jordan, Dirac; Poisson brackets

The relationship of these revolutionary ideas to classical dynamics was to be governed by **Bohr**'s **correspondence principle** which stated, quite reasonably, that **h** must be introduced into the "new" mechanics in such a way that when we let $h \to 0$ the dynamical laws would revert to the classical case.

Now for any pair of canonical co-ordinates (**q**,**p**) in classical dynamics we know that $pq - qp = 0$ and since in the matrix algebra we know that $PQ - QP \neq O$ it was necessary

to try $PQ - QP = h\,C$ where **C** is to be some constant matrix.

It was **Born** and **Jordan** who first pointed out that this constant matrix must be taken as $(h/2\pi i)\,\mathbf{I}$, **I** being the **unit matrix** in the algebra.

Dirac proposed a connection with classical dynamics by considering the so-called **Poisson Brackets**. [v. Chapter-4 and Appendix-E [66], [73]]

{Reminder: If **u**, **v** are functions of the canonical dynamical variables
$$(q_1 \ldots q_n, p_1 \ldots p_n)$$
the **Poisson Bracket** is denoted by [u,v] and defined by

$$[u,v] = \Sigma_r(\partial u/\partial q_r \cdot \partial v/\partial p_r - \partial u/\partial p_r \cdot \partial v/\partial q_r)$$

Then if f is some function of the canonical variables (which might represent some dynamical property of the system) and if **H** is the **Hamiltonian** of the system, we get

$$df/dt = [f, H]$$

special cases being

$$\dot{q}_r = [q_r, H] \quad \text{and} \quad \dot{p}_r = [p_r, H] \}$$

Heisenberg therefore proposed that the appropriate equations of motion for the new mechanics should be along the same lines, but with **matrix operators** taking the place of the classical variables (q_r, p_r).

He also pointed out that the measures, of say **H**, being real numbers should be found as the eigenvalues of the appropriate operators - and that these will be real whenever the matrix operators are **Hermitian**.

[15.3] PQ - QP = ih/2π

Using these ideas the "position" operator **Q** and the "momentum" operator **P** can be written in the form

$$Q = Q_+ + iQ_- \quad \text{and} \quad P = P_+ + iP_-$$

where eg Q_+ is a real symmetric matrix whilst Q_- is a real antisymmetric matrix ; these matrices operating in the **Hilbert Space** which we can denote by V_∞, a vector space with basis $\mathbb{C}$.

It follows that the matrix product

$$QP = \{Q_+P_+ - Q_-P_-\} + i\{Q_+P_- + Q_-P_+\}$$

and $\quad PQ = \{P_+Q_+ - P_-Q_-\} + i\{P_-Q_+ + P_+Q_-\}$

and since eg Q_+ is real symmetric and Q_- is real antisymmetric we have

$$Q_+P_+ = P_+Q_+ \quad \text{and} \quad P_-Q_+ = -Q_+P_-$$

Therefore $\quad QP - PQ = i\{2Q_+P_- + 2Q_-P_+\}$

which means that $\quad QP - PQ = i \times$(some real matrix}.

Now if we contemplate a $\Gamma(1\text{-}1)$ between pairs $(\mathbf{Q},\mathbf{P})$ in this space $\mathbf{V}_\infty$ we require complex numbers a,b,c such that

$$QP + aQ + bP + cI = 0 \qquad (4)$$

But since the algebra over $\mathbf{V}_\infty$ is non-commutative there will also be numbers a",b",c" together with

$$PQ + a"Q + b"P + c"I = 0 \qquad (5)$$

(4) and (5) now give the "Poisson Bracket" as

$$(QP - PQ) + a^!Q + b^!P + c^!I = 0 \qquad (6)$$

where $a^!,b^!,c^!$ are in $\mathcal{C}$.

Now the relation (6) can only be open to observables if $(\mathbf{QP - PQ})$ **is Hermitian but if it is written as** $A_+ + iA_-$ **the** A_- is not neccessarily antisymmetric. And since $(QP - PQ)$ **cannot be related to either Q or P the only possibility is that it is related to the identity matrix I. We therefoe must have**

$$(QP - PQ) = ikI \qquad (7)$$

k being a real number, and a constant.

It is also the same as $(\mathbf{PQ - QP} = -ikI \qquad (7a)$

The search for a suitable value for the real number k leads to the following argument.

Using (7) and (7a) and repeatedly multipying, on the left and the right, by powers of $\mathbf{P}$ we get, eg,

$$P^2Q - PQP = -ikP \text{ and } PQP - QP^2 = -ikP \text{ giving } P^2Q - QP^2 = -2ik$$

and generally $\quad P^nQ - QP^n = n(k/i)P^{n-1} \qquad (8)$

The **Born/Jordan** requirement amounts to taking the value of k as $h/2\pi$, or $\hbar$, so (8) will finally become

$$P^nQ - QP^n = nh/2\pi i \qquad (8a)$$

This will be discussed further in the next chapter.

[15.4] The Harmonic oscillator in matrix mechanics

We can use (7) to study the **Harmonic Oscillator** from the point of view of this **matrix mechanics**. Since the analysis gives us the acceptable quantised energy levels with $k = h/2\pi$ we shall adopt that form for the relation (7), viz.

$$(PQ - QP) = (h/2\pi i)I \qquad (8)$$

Since the classical SHM is governed by the equation $\ddot{q} = -\omega^2 q$ the potential function is $V = \omega^2 q^2/2$ and the **Hamiltonian** is therefore

$$H = \tfrac{1}{2}p^2 + \tfrac{1}{2}\omega^2 q^2$$

p being the momentum and $\omega = 2\pi\nu$ (ν being the frequency of oscillation).

In matrix mechanics we replace these variables with matrix operators **P** and **Q** and write

$$\mathbf{H}_{op} = \tfrac{1}{2}\mathbf{P}^2 + \tfrac{1}{2}\omega^2\mathbf{Q}^2 \qquad (9)$$

and suppose that $\mathbf{H}_{op}$ is the diagonal matrix with elements

$$E_{rs} = 0 \text{ when } r \neq s, \text{ and } E_{rr} \text{ is a real number, } r,s \in \mathbf{J}^+$$

We also require the matrices **P** and **Q** to be Hermitian.

Now write $\quad \mathbf{A} = \mathbf{P} + i\omega\mathbf{Q}$ and $\mathbf{B} = \mathbf{P} - i\omega\mathbf{Q}$

so that $\quad \mathbf{P} = \tfrac{1}{2}(\mathbf{A} + \mathbf{B})$ and $i\omega\mathbf{Q} = \tfrac{1}{2}(\mathbf{A} - \mathbf{B})$

and (8) becomes $\quad \tfrac{1}{2}(\mathbf{BA} - \mathbf{AB}) = h\nu\mathbf{I}$

whilst (9) is $\quad \tfrac{1}{2}(\mathbf{BA} + \mathbf{AB}) = 2\mathbf{H}_{op}$

These imply $\quad \mathbf{BA} = 2\mathbf{H}_{op} + h\nu\mathbf{I}$

and $\quad \mathbf{AB} = 2\mathbf{H}_{op} - h\nu\mathbf{I}$

Thence $\quad h\nu\mathbf{A} = \mathbf{H}_{op}\mathbf{A} - \mathbf{A}\mathbf{H}_{op} \qquad (10)$

If we write $\mathbf{A} = \{a_{nm}\}$ and $\mathbf{H}_{op} = \{E_n\delta_{nm}\}$ and equate corresponding matrix elements on either side of (10) we get

$$h\nu a_{nm} = (E_n - E_m)\, a_{nm}$$

so that $\quad a_{nm}\{h\nu - (E_n - E_m)\} = 0 \quad$ for all n,m

This implies that $a_{nm} = 0$ except when $E_n - E_m = h\nu$.

Similarly we can show that $b_{nm} = 0$ except when $E_n - E_m = -h\nu$

These results suggest that we might write

$$E_n = \varepsilon + nh\nu \text{ for all n } (\varepsilon \text{ being a constant})$$

Whence $\quad a_{nm} = 0$ except when $n - m = 1$, i.e. $a_{n,n-1} \neq 0$

and $\quad b_{nm} = 0$ except when $m - n = 1$, i.e. $b_{n,n+1} \neq 0$

Since **P** and **Q** are Hermitian this means that $b_{n,n+1} = a^*_{n+1,n}$

and writing $b_{n,n+1} = r_n e^{i\theta_n}$ with $a_{n+1,n} = r_n e^{-i\theta_n}$ we can consider the case given by $n = 1$ and deduce, eg,

$$b_{12}a_{21} = r_1^2 = 4(\varepsilon + h\nu)$$

and $\quad b_{12}a_{21} = r_1^2 = 2h\nu$

consequently $\quad 2\varepsilon = -h\nu$.

The consequent energy levels for the oscillator now become

$$E_n = (n - \tfrac{1}{2}) h\nu \qquad n = 1, 2, \ldots$$

or, equivalently

$$E_r = (r + \tfrac{1}{2}) h\nu \qquad r = 0, 1, \ldots$$

This solution is clearly that required by **Planck's Postulate**, differing only in the **zero point energy**, viz., $\tfrac{1}{2} h\nu$.

The essential points involved are

    (i)   The energy matrix must be diagonal

    (ii) The "co-ordinate" matrices must be Hermitian

    (iii) The commutation relation (equation (8)) must hold.

We shall see that although this approach might seem to deal with the quantised variables (a countable spectrum) it posed a new problem for those looking for an **uncountable (continuous) spectrum** - since this is where modern physicists still believe the measures should be found.

In terms of the **Heisenberg** matrix mechanics this required that the co-ordinate axes of reference, for the vectors and matrices, should now constitute an **uncountable continuous set**. This would then provide us with the possibility of a **continuous range of eigenvalues**, to be found in $\Re^{\#}$, still making the **Hilbert Space** inadequate for the Physicists' analysis.

Chapter-16 /PART-1   Quantisation in a Continuum

[16.1] An uncountable numbers of axes

Field theories, with their need for a **continuous manifold** and the
algebra of the real numbers $\Re^{\#}$, have always involved the idea of **operators**
acting on a suitable function (often called a "wave") and at a point.
These operators are based on the Calculus and, for example, move the study from
a curve/surface (of points) to the tangent/envelope space (of lines) - providing
for the dynamics of movement (velocity, acceleration).

And in order for the Physicists to have theories based on the real continuum, $\Re^{\#}$,
it seemed to be necessary to extend the **Hilbert Space** used in **matrix mechanics** so that
not only should the basis of unit vectors $\{\hat{e}_r\}$ form a countable (but infinite) set but
that they should now be replaced by an **uncountable infinite** set. In other words there
should be as many (and as closely packed) as the real numbers themselves.

This signals mathematical difficulties with manipulating the matrices which are supposed
to represent the operators **Q**, **P**, etc.. For one thing it needs to replace a "state vector"
in **Hilbert Space**, based on an enumerable set of axes $\{\hat{e}_n\}$, by an infinite uncountable
set, say $\hat{e}_x$, - indexed by the real numbers $x \in \Re^{\#}$. At the least it will require "sums"
to be replaced by "integrals", and matrix operators to be replaced by a new breed of operator
viz. linear operators : **Q**, **P**, **H**, etc. which will have the same names and significance,
as in **Hamiltonian** dynamics, but will be defined in a new abstract space.

[16.2] Orthodox derivation of the momentum operator

Before we come on to **Dirac's** notation in this context we can outline an analysis
which can be viewed as relying on the intuitive concepts behind **Heisenberg's** matrix mechanics.

We consider the case of a continuous spectrum for **Q** by assuming them to be of the form
$n\varepsilon$ and, when it seems convenient, we let $\varepsilon \to 0$ in a controlled way.

Following on from equation (8a) in Chapter-14 (above) we contemplate (in a heuristic sense)
a particular **unitary transformation** on the **Hilbert Space**, V_∞, viz.,

$$U(\lambda) = I + (i\lambda/\hbar)P + (1/2!)(i\lambda/\hbar)^2 P^2 + \ldots$$
$$= \exp(i\lambda/\hbar)P \qquad (1)$$

with elements (in $\mathbb{C}$) $[\hat{u}_{n,m}]$. This is a unitary matrix (remember **P** is a matrix) which
is one which peserves "length", or $\|\ \|$, and which may be used to change the basis of V_∞

since it does not alter the interpretation of a state vector as one carrying the probabilties of the measures. Also since $\mathbf{P}^n$ occurs for all n = 0, 1, 2, 3, ... the operator can affect the whole of **Hilbert Space**.

[Note: A unitary matrix is one whose elements $u_{r,s}$ are the complex conjugates of $u_{s,r}$; that is to say, across the main diagonal the elements are complex conjugates - so the diagonal elements are real numbers]

It follows that a unitary matrix/linear-operator can only "rotate the axes" (however many there are).

Using (1) and equation (8a) from Chapter-15 we see that

$$\mathbf{UQ} - \mathbf{QU} = \lambda \mathbf{U}$$

and so
$$\mathbf{U}(\lambda)\mathbf{Q} = (\mathbf{Q} + \lambda \mathbf{I})\mathbf{U}$$

and so $\mathbf{U}$ affects the eigenvalues of $\mathbf{Q}$ according to the relation

$$\hat{u}_{n,m}(\lambda) \times (m\varepsilon) = (n\varepsilon + \lambda)\hat{u}_{n,m}(\lambda)$$

since we are taking the eigenvalues of $\mathbf{Q}$ to be $n\varepsilon$ and those of the rhs to be $m\varepsilon$.

Taking $\lambda = \varepsilon$, since λ is arbitrary we get

$$\hat{u}_{n,m}(\varepsilon)(m - n - 1) = 0$$

Hence $\hat{u}_{n,m} = 0$ unless $m = n+1$, so $\hat{u}_{n,n+1} \neq 0$.

Similarly, writing $\lambda = -\varepsilon$ we have $\hat{u}_{n+1,n} \neq 0$.

But since $\mathbf{U}(-\lambda)\mathbf{U}(\lambda) = \mathbf{I}$ we must have

$$\hat{u}_{n,n+1}(\varepsilon) \times \hat{u}_{n+1,n}(-\varepsilon) = 1$$

and since $\hat{u}_{n,n+1}$ is the complex conjugate of $\hat{u}_{n+1,n}$ it follows that

$$\hat{u}_{n,n+1}(\varepsilon) = e^{i\theta_n} \quad \text{where } \theta_n \in \Re^{\#}.$$

We can now transform the orthonormal basis $\{\hat{u}_n\}$ to, say $\{\hat{e}_n\}$ by way of

$$\hat{e}_n = e^{-ia_n} \quad \text{where } (a_{n+1} - a_n = \theta_n)$$

Then the "matrix" $\mathbf{U}(\varepsilon)$ is a matrix with all zeros except for those off the main diagonal, viz. those for which col = row + 1, when each is unity, 1.

Thinking of the typical "probability vector" in $\mathbf{V}_\infty$ as the expression

$$\underline{\mathbf{X}} = \Sigma_r x_n \hat{e}_n \quad r = 1, 2, \ldots n, \ldots$$

we see that the transformation $\mathbf{U}$ affects it by way of

$$\mathbf{U}(\varepsilon) \underline{\mathbf{X}} = \Sigma_r x_{n+1} \hat{e}_n$$

showing that the transformation $\mathbf{U}$ rotates the axis $\hat{e}_n$ into the axis $\hat{e}_{n-1}$, since it now associates the component x_{n+1} with $\hat{e}_n$ - and that is the same as associating the component x_n with $\hat{e}_{n-1}$.

It now follows that $\quad \mathbf{U}(k\varepsilon) \underline{\mathbf{X}} = \Sigma_r x_{n+k} \hat{e}_n \quad$ (2)

where k is a positive integer.

Proceeding to a limit in which we let $\varepsilon \to 0$ in such a way that $k\varepsilon \,(= a)$ remains finite (a non-trivial requirement) we can say that (2) is equivalent, in the **function space** which is obtained by this extension to **Hilbert Space**, to

$$U(a)\, \psi(q) \equiv \psi(q + a)$$

We can rewrite this as

$$\exp\{(i\hbar/a)P\}\, \psi(q) \equiv \psi(q + a)$$

and expanding the rhs by **Taylor's** theorem we can identify the operator **P** via

$$(i/\hbar)^n/n!\, P^n\, \psi(q) = (1/n!)\, d^n/dq^n\, (\psi(q)$$

that is to say   **P** is the operator $(\hbar/i)\, \partial/\partial q$ (3)

[16.3] Dirac's Space - bra and ket vectors

The task of extending the concept of **Hilbert Space** to one with a continuous range of "axes" was advanced by **Dirac**, so that **Hilbert Space** became **Dirac Space** by the following analysis. [v. Appendix-E [66], [73]].

He introduced an abstract symbolism involving a generalised **linear operator** and a generalised "vector".

Let the "state" of the system (which is to be the counterpart of a classical **Hamiltonian** dynamical system) be represented by the new "vector" called a **ket** and written as

$$|\,P\rangle.$$

[The word "vector" is in quotes because its components cannot be separately distinguished]

There is also to be a "dual vector" called a **bra** and written as

$$\langle Q\,|$$

and the "scalar product" of two such will be written as

$$\mathbf{bra(c)ket} \leftrightarrow \langle Q\,|\,P\rangle$$

this being a number ($\in \mathbb{C}$) - so the **norm** of a **ket/bra** will be $\|\langle P\,|\,P\rangle\|$.

Since the summation cannot be indexed in the usual way with vectors it would seem inevitable that this "scalar product" will have to be expressed as an **integral** over a suitable range. Also a **ket** will be something defined on a range of "axes" which is indistinguishable from a range of $\mathfrak{R}^{\#}$. The "coefficients" of these "vectors" (**bra** or **ket**) are also assumed to be found in the complex field $\mathbb{C}$. The **ket**, $|\,P\rangle$, will therefore be a **function** defined in $\mathfrak{R}^{\#}$. So although this scheme is an extension of the **Hilbert Space** assumed in the **Heisenberg** matrix mechanics it does not exclude the possibility of (eg) a **ket** being a vector in that space - that is to say, having a countable (discrete) set of axes. Indeed **Dirac** makes use of this possibility in some of the applications to experimental situations.

The action of a **linear operator**, α, on a **ket** produces another **ket** via the notation

$$|F\rangle = \alpha |A\rangle$$

and its induced effect on a **bra** is defined via

$$\{\langle B\alpha\} |A\rangle = \langle B|\{\alpha|A\rangle\}$$

Dirac discusses various algebraic properties which these entities must possess - in particular the equation for the existence of **eigenkets** and **eigenbras** as solutions of the equation

(eg) $$\alpha|P\rangle = \lambda|P\rangle$$

λ being an eigenvalue and $|P\rangle$ being an associated **eigenket**.

Dirac required that an **observable** should satisfy the following conditions :-

(a) it should be represented in his scheme as a **linear operator**, eg α,

(b) such an α operates upon a **ket** (and dually upon a **bra**) which itself is to represent the **state** of the system,

(c) the possible **values** of the observable are to be **real numbers** (in $\Re^\#$)

(d) these values (and only these) must be the **eigenvalues** derived from α.

This means that although countable (quantised) eigenvalues can be accommodated so also can uncountable (continuous) ones. No doubt he had in mind that (eg) an **electron** can exist in a **bound state** in an atom (quantised states) but also that it can become **free** and move around in a (supposedly) continuous field of positions.

[16.4] Eigenkets in a continuum

When the eigenvalues for say, **q**, form a range of real numbers a **ket** $|P\rangle$, as the counterpart of a classical vector, will be indexed by, say $q \in \Re^\#$, and written

with components $$\psi_q \hat{e}_q$$

and over the range $R = [-\infty, +\infty]$ the normalisation of this vector will be given by the integral

$$\int_R \langle \psi(q) | \psi \rangle \, dq = 1$$

where the **bra** and **ket** are complex conjugates. Such normalisation is required if the componets of the "vector" are to be interpreted as probabilities - adding up to 1.

In general **Dirac** argues that when the eigenvalues consist of both a range of reals as well as a set of discrete (countable) values, then a **ket** will be expressible as

$$|P\rangle = \int |q\rangle \, dq + \Sigma_r |q'_r\rangle.$$

From this foundation **Dirac** was able to justify the **Heisenberg** and the **Schrödinger** equations of motion for a generalised (and quantisable) dynamical systems.

[v. Appendix-E [66], [73] for a relevant analytical discussion]

For example, the **Schrödinger** representation is obtained by noting that, for the calculus of the reals, the "position" **ket**, $|q\rangle$, becomes the **wave function** $|\psi(q)\rangle$ defined over the range $(-\infty, +\infty)$, and the observable **momentum** operator becomes $-i\hbar\partial/\partial q$, - where it is usual to write the symbol $\hbar$ for the combined quantity $h/2\pi i$.

[16.5] de Broglie's postulate

As **Einstein** in 1905 had explained the photo-electric effect by regarding light as equivalent to "particles" of energy, so in 1924 **L. de Broglie** (Appendix-E [74]) assigned a "wave-length" to a moving particle. His theoretical suggestion was that a particle of mass m moving with a velocity v will exhibit wave-like properties with a wave-length of

$$\lambda = h/mv$$

where h is **Planck's** constant.

Clearly it is only likely to apply to particles of very small mass - such as electrons ?

Direct experimental verification was obtained by **Davisson** and **Gerner** in 1927, who bombarded nickel crystals with a beam of electrons.

In 1928 **G.P.Thomson** passed a beam of electrons through a thin metal foil. The random arrangement of the crystal surfaces in the foil gave symmetrical diffraction ring spacing in the emergent beam. Measurement of the ring spacing gave numerical confirmation of **de Broglie's** formula.

This gave **Schrödinger** a foundation for a "wave equation" for the behaviour of an electron in its orbit in the H-atom, by using the standard differential equation of wave motion and incorporating **de Broglie's** result.

Letting the wave disturbance be described by a function $\Psi(q,t)$ - **q** representing any of the system's position co-ordinates - we write

$$\nabla^2\Psi = v^{-2}\, \partial^2\Psi/\partial t^2 \qquad (4)$$

If **E** is the total energy and **V** the potential energy we have momentum $\mathbf{p} = (E - V)/2m$ and the **de Broglie** condition gives

$$\text{phase velocity} = \lambda v = E/p = E/\sqrt{2m(E-V)}$$

whence (4) becomes $\qquad \nabla^2\Psi = \{2m(E-V)/E^2\}\, \partial^2\Psi/\partial t^2$

and since the disturbance is a wave form of frequency v we can write

$$\Psi(q,t) = \psi(q)\, e^{-2\pi i v t} = \psi(q)\, e^{-(2\pi i/h)E(t)}$$

The (4) finally becomes

$$\nabla^2 \psi(q) + (8\pi^2 m/h^2)(E - V)\,\psi(q) = 0 \qquad (5)$$

[16.6] Schrödinger's wave equation

This wave equation also follows from the "eigenket" equation, viz.,

$$H\{q, -i\hbar \partial/\partial q\}\, \Psi(q,t) = -i\hbar \partial\Psi/\partial t \qquad (6)$$

where the function Ψ is the "time-dependent wave function" for the system and where the classical **Hamiltonian** for the system has each canonical momentum variable, p_r, replaced by the operator $-i\hbar \partial/\partial q_r$ - thus making the **Hamiltonian** into an operator in the **Dirac/Schrödinger** system.

By writing the wave function as the product $\Psi = \psi(q)T(t)$ we can obtain the **stationary states** of the system via the equation

$$H\{q, -i\hbar\partial/\partial q\}\psi(q) = E\psi(q) \qquad (7)$$

E being the **separation constant** in the analysis - and (7) shows that it is properly regarded as a measure of the **energy** of the system.

Since in the classical case

$$H = T + V = (1/2m)(p_1^2 + p_2^2 + p_3^2) + V(x,y,z)$$

and replacing each p_r by the operator $(h/2\pi i)\,\partial/\partial q_r$ this reduces to

$$\nabla^2 \psi(q) + (8\pi^2 m/h^2)(E - V)\,\psi(q) = 0$$

as before, and we notice that the potential function for the electron is simply $V = e/r$.

We can now find the solution of (5) in terms of spherical polar co-ordinates (r,θ,φ), when the equation becomes

$$r^{-2}\partial/\partial r(r^2 \partial\psi/\partial r) + (r^2 \sin\theta)^{-1}\partial/\partial\theta(\sin\theta\,\partial\psi/\partial\theta) +$$

$$(r^2 \sin^2\theta)^{-1}\partial^2\psi/\partial\varphi^2 + (8\pi^2 m/h^2)(E + e^2/r)\,\psi = 0 \qquad (7a)$$

Since this is separable we can write $\psi = R(r)S(\theta,\varphi)$ and then (7a) gives us

$$R^{-1}d(r^2 dR/dr)/dr + (8\pi^2 m/h^2)(Er^2 + e^2 r) +$$

$$(S\sin\theta)^{-1}\{\partial(\sin\theta\,\partial S/\partial\theta)/\partial\theta + (\sin\theta)^{-1}\partial^2 S/\partial\varphi^2\} = 0 \qquad (7b)$$

We are searching for solutions in which ψ is continuous, single-valued, finite as $r \to 0$, and such that $|\psi| \to 0$ as $r \to \infty$ (for all θ and φ) - for these ensure that the electron is bound to the nucleus.

Writing $\ell(\ell+1)$ for the separation constant in the above equation (7b) we get

$$d(r^2 dR/dr)/dr + (8\pi^2 m/h^2)(E^2 + e^2 r)R = \ell(\ell+1)R \quad (8)$$

$$(\sin\theta)^{-1}\partial(\sin\theta\, \partial S/\partial\theta)\partial\theta + (\sin^2\theta)^{-1}\partial^2 S/\partial\varphi^2 = -\ell(\ell+1)S \quad (9)$$

In (8) we can introduce another separation constant, s, so that writing

$$S(\theta,\varphi) = \Theta(\theta)\Phi(\varphi) \quad \text{we then get the two equations}$$

$$d^2\Phi/d\varphi^2 = -s^2\Phi \quad (10)$$

and $\quad (\sin\theta)^{-1}d(\sin\theta\, d\Theta/d\theta)/d\theta + \{\ell(\ell+1) - s^2(\sin^2\theta)^{-1}\}\Theta = 0 \quad (11)$

To satisfy the boundary conditions (10) must give us

$$\Phi = A e^{is\varphi} \quad (s = 0, \pm 1, \pm 2, \dots)$$

and since (11) is the differential equation giving **Legendre polynomials** [v. Appendix-E [34]]

it has solutions $\quad \Theta = P^s_\ell(\cos\theta)$

where $\ell = 0, 1, 2, \dots$; and $|s| \leq \ell$.

To solve (8) we can write it in the form

$$\rho^2 d^2 R/d\rho^2 + 2\rho dR/d\rho + \{\beta\rho - \rho^2/4 - \ell(\ell+1)\}R = 0 \quad (12)$$

where $\rho = r\alpha$; $\alpha^2/4 = -(8\pi^2 m/h^2)E$; $\beta = (8\pi^2 m e^2)/h^2 \alpha$

This equation gives us a solution in terms of **Laguerre polynomials**, [v. Appendix-E [34]]

viz.,
$$R(\rho) = e^{-\frac{1}{2}\rho} \rho^{\frac{1}{2}(p-1)} L_k^p(\rho) \quad (0 \leq p \leq k\,; \text{ integers})$$

provided $\quad p^2 - 1 = 4\ell(\ell+1) \quad (13)$

and $\quad k - (p-1)/2 = \beta = 8\pi^2 m e^2/h^2 \alpha. \quad (14)$

Now $4\ell(\ell+1) = (2\ell+1)^2 - 1$ so that (13) gives $p = 2\ell + 1$

so then (14) gives $\quad k - \ell = 8\pi^2 m e^2/h^2 \alpha$.

Then $\quad \alpha^2/4 = -(8\pi^2 m)/h^2 \cdot E = 16\pi^4 m^2 e^4/h^4 (k-\mu)^2$

whence the (negative) energy levels form a point spectrum and are given by

$$E_n = -2\pi^2 m e^4/n^2 h^2 \quad (n = k - \ell)$$

The complete eigenfunctions for the electron in orbit are

$$\psi^s_{n,\ell} = A^s_{n,\ell}\, R_n^\ell(\rho)\, P_\ell^s(\cos\theta)\, e^{is\varphi}$$

where the $A_{n,\ell}{}^s$ are constants and $r = \rho/\alpha$. Normalisation of this function must be taken over the range

$$0 \leq r < \infty,\ 0 \leq \theta \leq \pi,\ 0 \leq \varphi \leq 2\pi$$

This eventually gives a value for the constants $A_{n,\ell}{}^s$ via the relation

$$A^s_{n,\ell} = (na_0)^{-3}\{(2\ell+1)(\ell-s)!(n-\ell-1)!\} \div \pi n(\ell+s)![(n+\ell)!]^3$$

the a_0 being the Bohr radius in the ground state - i.e. the one given by $n=1$, $\ell=s=0$.

The consequence of all this is that the eigenkets turn out to be

$$\psi^s_{n,\ell} = A^s_{n,\ell}\, e^{-\frac{1}{2}\rho} \rho^\ell L^{2\ell+1}_{n+\ell}(\rho) P_\ell^s(\cos\theta) e^{is\varphi}$$

with $n = 1, 2, \ldots;\ 0 \le \ell < n;\ -\ell \le s \le +\ell$; s, ℓ both integers.

[16.7] The Harmonic Oscillator

Schrödinger's equation for the Harmonic Oscillator gives the same energy levels as that already discussed, using matrix mechanics, in Chapter-15 - but also see Chapter-17.

Suppose a particle moves on a straight line under the central force $-\omega^2 x$ per unit mass. The **Schrödinger's** equation for the motion is

$$d^2\psi/dx^2 + (8\pi^2/h^2)(E - \tfrac{1}{2}\omega^2 x^2)\psi = 0 \qquad (15)$$

If we write $\xi = \sqrt{(2\pi\omega/h)}\, x$ this becomes

$$d^2\psi/d\xi^2 - (\xi^2 - \theta)\psi = 0 \qquad (16)$$

where

$$\theta = (4\pi/\omega h) E \qquad (17)$$

We need solutions of (16) in which $\psi(\xi)$ is continuous and single-valued, so we can write

$$\psi(\xi) = \exp[-(\tfrac{1}{2}\xi^2)]\, H_n(\xi)$$

provided $\theta = 2n + 1$ ($n = 0, 1, 2, \ldots$) and where the functions H_n are the **Hermite polynomials** [v. Appendix-E [34] [64]].

This gives the eigenvalues (energy quanta) as

$$E_n = (n + \tfrac{1}{2}) h\nu$$

where as before, the normalised wave functions being

$$\psi_n(\xi) = (2^n n!\sqrt{\pi})^{-\frac{1}{2}} \exp[-(\xi^2/2)]\, H_n(\xi)$$

in the range $-\infty < \xi < +\infty$.

Chapter-17 /PART-1   Cross-Ratios in a Field Theory

[17.1] Extending the Projective definition of metric.

In a Field theory we need to discuss, in terms of the Calculus, what can be associated with a point - that is to say, with a **vanishing neighbourhood**.

Since we have seen the important role played by the concept of Cross-Ratio in Physics - since it ultimately only depends on mappings into the fundamental projective geometry - it seems desirable to extend its application to measures other than "point" or "line".

If we have two values of what is in effect the "same sort of thing" then we have a situation similar to the range of "points" on a line Λ. This can occur whenever the two values are attributed to measures with **identical dimensions**.

Thus although, originally, the cross-ratio idea arose with respect to a range of four "points" (which have zero dimensions), it was simultaneously applied to a "range" (which is called a "pencil") of four lines (with dimensiions [L]) on a point λ.

We now intend to apply it to any quantities which have identical dimensions, and this will include variables such as **energy** (of various forms), **action, momentum** etc., and to remind ourselves of the general nature of the method we shall refer to the entities as **poins**.

Thus we contemplate a **poin** as, eg, point, line, momentum, energy, action, or whatever ; a "range" will now refer to a suitable quadruple set of such poins.

Since we are concerned with Field Theory we are also concerned with "values at a point", and this means that we must deal with **differentials** - defined in a vanishing neighbourhood.

At such a poin "P", and in the limit, any **finite departure** from "P" (however small) must be regarded as a move towards "infinity", ∞. Thus we expect, for example, the measure of a variable to correspond to the traditional projective metric, viz., $(1/2i) \ln (P\ O\ Q\ U)$, to become something like

$$\text{value of measure} = k \ln (?\ 0\ ?\ \infty)$$

where the ?'s replace measures of generating elements which are dimensionally indistinguishable and where 0 and ∞ are taken as the standard "poins" of reference. In some cases the references 0 and will be the "double poins" for that variable.

[17.2] Consequences for a Quantum Mechanics

Since we are assuming values in $\mathbb{C}$ we have

$$k \ln (?\ 0\ ?\ \infty) = k \{\ln r\ +\ i(\theta\ +\ 2n\pi)\}$$

and, if we are seeking quantised measures (as we are in any Quantum Mechanics), à la **Planck**, we see that this suggests that k must be $h/2\pi i$ (h being **Planck's** constant) - since this isolates the values as multiples of h, viz. nh. We also find that the significance of the need for "normalisation" in Quantum Mechanics follows from the projective measure, as above, by taking the modulus, r, as unity - for this gives us $\ln r = 0$ in the evaluation.

So now the "projective measure" we have introduced becomes

$$\text{value of measure} = (h/2\pi i) \ln (?\ 0\ ?\ \infty) = h\theta/2\pi + nh$$

When the "range" is harmonic (eg in the case of an involution) then we get the quantised measures as

$$(h/2\pi i) \ln (-1) = (h/2\pi i) \{i (\pi + 2n\pi)\}$$

giving the values $(\frac{1}{2} + n) h$ for the quantised measures of energy.

The same result follows by considering those cross-ratios which are harmonic - such as the property possessed by reflexive measures of base elements.

For then, since we shall have $(?\ 0\ ?\ \infty) = -1$, the value of $(h/2\pi i) \ln (?\ 0\ ?\ \infty) = (\frac{1}{2} + n)$ as before - we need only take the cross-ratio as $(T\ 0\ V\ \infty)$, where T and V are the relevant forms of the energy functions. Since their sum is constant they are mutually interchangeable in the mathematics. That is to say we get the common case where

$$(T\ 0\ V\ \infty) = (V\ 0\ T\ \infty) = -1$$

giving ther harmonic condition - each ratio being -1.

[c.f. the analysis of the **Harmonic Oscillator** in Chapters-15, 16]

So the **quantised** basic **Harmonic Oscillator** is a demonstration of the involutionary range of its measures (of eg kinetic and potential energy functions) when these **act as Base Elements** on the scales $\mathfrak{C}_\alpha$ which are used for their observations.

We may return to the **Metaphysical Question, MQ**, viz.,

[MQ] → "Is quantisation a property of "reality" or is it a property inherent in the Scales (Experimental setups) we use in our observations?"

Reflecting on this brings us to the **Metaphysical Answer, MA**, viz.,

[MA] → "As long as our experimental techniques (the $\mathfrak{C}_\alpha$) are such as to use the same Base Elements in observations of the Generating Elements, then we shall never notice the difference - so **MQ** cannot be asked, since it is in conflict with the underlying syntax of the observation process"

[17.3] Localised Cross-Ratios

Moving to the localised cross-ratio in a continuum we accordingly consider the value of $(d\xi\ 0\ d\eta\ \infty)$ where ξ and η are defined and differentiable at an arbitrary point of some manifold **M**.

This evaluates to $(d\xi\ 0\ \eta\ \infty) = d\xi/d\eta$

showing that in a vanishing neighbourhood of a poin P in **M** the **localised cross-ratio** exhibits the the gradient properties of the dual (tangent) space. It also justifies our choice of 0 as a double poin - since $d\xi$ and $d\eta$ must both vanish together, where the derivative is concerned. Also, as mentioned above, any finite move from the poin "P" (however small) must be regarded as "infinite" as the other double poin, and since the unerlying geometry is a projective space we see that, if k is any constant in $\mathbb{C}$, kP ≡ P so that

$$k(d\xi\ 0\ d\eta\ \infty) = (kd\xi\ k0\ kd\eta\ k\infty) = d\xi/d\eta$$

And since we have seen that cross-ratios can be multiplied by the relation

$$(A\ P\ Q\ B) \cdot (A\ Q\ B\ R) = (A\ P\ B\ R)$$

it follows that, eg,

$$(d\xi\ 0\ d\eta\ \infty) \cdot (d\eta\ 0\ d\tau\ \infty) = (dNILd\eta) \cdot (d\eta/d\tau) = d\xi/d\tau$$

as required by the differential calculus.

The cross-ratio also possesses an **additive** property in any one of its members, in the sense that, eg,

$$(d\xi\ 0\ d\eta_1+d\eta_2\ \infty) = (d\xi\ 0\ d\eta_1\ \infty) + (d\xi\ 0\ d\eta_2\ \infty)$$

It also follows that whenever $(d\mathbf{f}\ 0\ d\mathbf{g}\ \infty) = -1$ we can write

$$d\mathbf{f} + d\mathbf{g} = 0$$

and this will integrate into $\quad \mathbf{f} + \mathbf{g} = $ a constant.

Example-1 As it has already been seen in Chapter-9 **Bernoulli's equation** becomes an illustration of this in Fluid Mechanics.

For using a localised harmonic range, with poins ρdE and dp, we have

$$(\rho dE\ 0\ dp\ \infty) = -1$$

and this reduces to $\quad (1/\rho)dp + dE = 0$

Whence we get the usual form of **Bernoulli's** equation, viz.,

$$\int (1/\rho)dp + E = \text{constant}$$

where E will be $\frac{1}{2}(\text{vel})^2 + V$; V being the potential of any applied field (eg gravity) and the integration being taken along a **stream line**.

Example-2 The kinetic energy **T** and the potential energy **V**, in any conservative holonomic dynamical system is an example of the involution pattern in the dynamics, whence the harmonic condition $\quad (d\mathbf{T}\ 0\ d\mathbf{V}\ \infty) = -1$

gives us $\qquad dT + dV = 0$

and the **energy equation** $\qquad T + V =$ a constant.

Example-3 In a **Hamiltonian** system any pair of canonically conjugate co-ordinates p,q are given, in terms of the **Hamiltonian Function, H** by the equations

$$\dot{q} = \partial H/\partial p \quad \text{and} \quad \dot{p} = -\partial H/\partial q$$

and this means that, in the wedge space Λ^2 (c.f. **Pfaff's** linear 1-form) the local cross-ratio

$$(dHdt\ 0\ dqdp\ \infty) = \partial H/\partial q \cdot \partial t/dp = -\dot{p}/\dot{p} = -1$$

and since, in Λ^2, dpdq = - dqdp the corresponding cross-ratio, viz.,

$$(dHdt\ 0\ dpdq\ \infty) = \pm 7\ \dot{q}/\dot{q} = -1$$

so that the **action** variables form an involution with the localised double points 0 anmd ∞.

Conversely we can say that **Hamilton's Equations** for the dynamical system are a consequence of these involution "ranges", noticing that (eg) dHdt and dqdp both have dimensions of **action**.

Example-4 Referring again to fluid mechanics we can associate the equation of continuity (v. Chapter-9) with an harmonic range where the pairs of (localised) **poins** are

$$d(\rho u)dt \quad \text{and} \quad d\rho dx$$

ρ being the density function and u being the velocity in the x-direction.

These are legitimate poins since they are dimensionally identical via

$$[d(\rho u)dt] = ML^{-3}LT^{-1}T = ML^{-2}$$

and $\qquad [d\rho)dx] = ML^{-3}L = ML^{-2}$

Then $\qquad (d(\rho u)dt\ 0\ d\rho dx\ \infty) = -1$

gives $\qquad d(\rho u)dt + d\rho dx = 0$

or $\qquad \partial(\rho u)/\partial x + \partial \rho/\partial t = 0$

In 3-dimensional space the additive property of the cross-ratio means we can immediately extend this to $\qquad \partial \rho/\partial t + \text{div}\ (\rho \underline{V}) = 0$

[17.4] Planck's Postulate and Action variables

When dealing with a Quantum Mechanics we depend upon the implementation of the underlying **Planck's Postulate**, which amounts to

> When dealing with a dynamical system at a level in which quantisation is expected it is necessary to obtain results in which the dynamical "Action" variable takes values nh; $n \in J^+ \cup \{0\}$ - h being **Planck's constant**

Denoting an action variable by **A** then we expect two measures A_1 and A_2, occurring in such a situation, to be instances of Base Elements forming an involution with double points 0 and ∞ - the values being reflexive measures in the vanishing neighbourhood at a point P.

This will give us, eg, $(dA_1 \; 0 \; dA_2 \; \infty) = -1$

so that, as we have seen above, $\quad dA_1 + dA_2 = 0$

Now since **Hdt** is one such version of Action we get

$$Hdt + dA = 0$$

or, $\quad H = - \partial A/dt$

with a consequential quantised measure for energy **H** of

$$H = - (h/2\pi i)(\partial/\partial t)(A)$$

We also notice that considering the dimensions of **H** and of **A**

$$[\text{Energy}] = [\text{Action}] \times [T]^{-1}$$

and since $[T]^{-1}$ is the dimension of **frequency**, whenever the measures of **Action** are to be "nh" then the measures of **Energy** will be "nhv".

This fits in very well with the language of Wave Theory.

Looking back to **Dirac's** symbolism in Chapter-16 we can see that the **linear operator** representing **energy** in **Dirac space** will be

$$E_{op} \equiv ih/2\pi \; \partial/\partial t.$$

It also follows that if we consider the other two versions of **Action**, viz., Hdt and **pdq** then we get

$$(Hdt \; 0 \; pdq \; \infty) = -1$$

resulting in $\quad Hdt + pdq = 0$

which in turn gives $\quad p = - Hdt/dq.$

Then the **momentum operator** in **Dirac's Space** follows from

$$p_{op} \equiv - E_{op} \; dt/dq \; \partial/\partial t$$

or $\quad p_{op} \equiv (h/2\pi i) \; \partial/\partial q .$

[17.5] The operator **crop**

In the spirit of **Dirac's** enquiry we may introduce a C)ross-R)atio OP)erator, **crop**, and define it as

$$\text{crop} \equiv (\eth \; 0 \; \eth \; \infty)$$

with a symbol for it, viz., ℭ.

We also use the notation $\mathbb{C}^n$ for the operator $(\eth^n\ 0 \neq 1^n\ \infty)$ with n = 0, 1, ...

When   n = 0 we take $\eth^0$ as the identity operator, say 1, and get

$$\mathbb{C}^0 = (P\ 0\ Q\ \infty)\quad \text{the usual cross-ratio}$$

When   n = 1 we get $\mathbb{C}^1\ (\equiv \mathbb{C}) =$ **crop**, as above

When   n = 2 we get $\mathbb{C}^2 = (\eth^2\ 0\ \eth^2\ \infty)$

and so on.

We take $\eth \equiv \partial,\ d$, according to the differentiable circumstances.

When there is only <u>one operand</u> we interpret the differentiation w.r.t an understood parameter θ, otherwise we will expect <u>two operands</u>, say (ξ, η), and write

$$\mathbb{C}(\xi,\eta) = (\eth\ 0\ \eth\ \infty)(\xi,\eta) \equiv (d\xi\ 0\ d\eta\ \infty) = d\xi/d\eta$$

whilst if $\xi = f(\eta, \zeta, ...)$ we write

$$\mathbb{C}(\xi,\eta) = (\eth\ 0\ \eth\ \infty)(\xi,\eta) \equiv (\partial\xi\ 0\ \partial\eta\ \infty) = \partial\xi/\partial\eta.$$

and if, say $\xi = f(\tau)$ and $\eta = \eta(\tau)$, τ being an independent variable we write

$$\mathbb{C}(\xi,\eta) = (\dot{\xi}\ 0\ \dot{\eta}\ \infty) = \dot{\xi}/\dot{\eta}$$

the dot denoting differentiation with respect to τ.

<u>Example-5</u> This also gives us a means of finding the momentum operator, for if we consider two closely related variables with dimensions those of **Action**, viz., **dA** and **pdq** where **p,q** are a pair of canonically conjugate variables in our local dynamical system, then we expect there to be a constant K (in $\mathbb{C}$) such that **pdq** = KdA, and this requires that in our vanishing neighbourhood $p = K\ \partial A/\partial q$

Applying the operator **crop** ($\mathbb{C}$) to the pair **A, q** we get

value of **A** = $K \ln (\partial A\ 0\ \partial q\ \infty)$

which must be of the form $K \ln (re^{i\theta})$

and which gives, at the poin "P", a value of **A** as $K\{\ln r + i(\theta + 2n\pi)\}$.

Appealing to **Bohr's** correspondence principle we put   K = h x (factor in $\mathbb{C}$) and this immediately suggests that we take $K = h/2\pi i$, as before,

and this requires that $p = \partial A/\partial q \equiv (h/2\pi i)\ \partial/\partial q$ acting on **A**.

[17.6] Schrödinger's wave equation

We can see how this follows, from a consideration in **Dirac Space**, of the eigenket equation, viz.,

$$H_{op}|\Psi\rangle = (T_{op} + V_{op})|\Psi\rangle = E_{op}\Psi\rangle$$

and now the **Hamiltonian** operator will be obtained by taking the usual classical function **H** for the kinetic energy **T** and replacing each momentum variable $\mathbf{p}_r$ by its operator version $(h/2\pi i)\, \partial/\partial \mathbf{q}$, and the operator $\mathbf{E}_{op}$ by $-(h/2\pi i)\,(\partial/\partial \mathbf{t})$.

This is demonstrated in Chapter-16 via the solution of **Schrödinger's** equation for the case of the H-atom.

[17.7] Newton's Law of Motion

This enduring law of motion for a particle of unit mass, moving on the x-axis under a force **F** in that line, becomes

$$\mathbb{C}^2\,(\mathbf{x},\, \mathbf{t}) = \mathbf{F}.$$

Similarly the equation of **wave-motion**, in one dimension, is the expression of the invariance of a cross-ratio, viz., the equality of two cross-ratios

$$\mathbb{C}^2\,(\xi,\, \mathbf{x}) = \mathbb{C}^2\,(\xi,\, \mathbf{ct})$$

c being a constant with dimensions those of velocity.

PART 2

Objects are just Holes in Space

PROLOGUE to Part-2

In this part of the thesis we pursue the foundations of Physics by calling in some mathematical concepts which were introduced and developed during the 20th century, viz. the **topological notions** associated with various abstract spaces.

These notions place an emphasis on the generalised idea of a **hole** in a **space**, and this idea is defined algebraically by the introduction of **cycle** and **boundary**. Thus in an intuitive sense the usual concept of Space (the 3-D **Euclidean Space**, E^3) requires it to have <u>no holes</u> at all. This can be expressed by saying that every (intuitive) **cycle** is also the **boundary** of some bit of the space.

From this sort of beginning there sprang the concepts of **cycle** and **boundary** in other more abstract and general spaces [often predicated by mathematicians studying "pure mathematics"].

But a Physics begins with notions of **objects** in **spaces** - which latter are actually defined in some **Field Theory** (or **function space**).

So, in this Part-2 we pursue the link between these requirements and those of **cycle** and **boundary**, and in doing this we shall call on many of the results already established in Part-1.

Furthermore we shall need the **dual concepts** of **cocycle** and **coboundary** in order to make use of **mappings** (or **functions**) naturally required in the Physics. For if some object (usually a **particle**) is to be associated with a **cycle** (or **hole**) in some space then properties of that object need to be described by functionals - which will be mappings into that field of coefficients, say **F**, which carries the measures of the **generating elements**.

These functions will be mappings on the chain complexes which cover the point-set topological complexes. They are therefore in the **dual space** to C_* - and are called **cochains**, giving rise to **coboundaries** and **cocycles**. The latter are the natural carriers of the Laws of a Physics and this is where we introduce the ubiqitous **Cocycle Law** in a **Physics**.

In all of this we need to examine any special role which is played by the **base elements** in the **scales** used in the **Physics**.

By appealing to the mathematics of the Complex Plane, the Calculus of Variations, and the Exterior (Wedge) Algebra, we can illustrate the occurence of this **Cocycle Law**.

Since these discussions come under the headings of **Homology** and **Cohomology** in the subject of **Algebraic Topology**, we shall need to introduce some of the terminology found in those subjects.

Chapter-1 /PART-2 Holes (Objects) in Fields

[1.1] Base Elements - seeing objects

Suppose that A,B,C ... are Physicists, studying phenomena and taking observations on various Scales.

But for the sake of comparison suppose that

    A uses the Base Element $\mathcal{Z}_L$ (optical instruments)

    B uses the Base Element $\mathcal{Z}_Q$ (electric charges)

    C uses the Base Element $\mathcal{Z}_F$ (ultrasound signals)

    D uses the Base Element $\mathcal{Z}_X$ (X-rays)

    E uses the Base Element $\mathcal{Z}_H$ (magnetic field ; eg in MRI or Particle accelerator)

    F uses the Base Element $\mathcal{Z}_G$ (gravitational field)

In the first instance we notice that these Fields are associated with **sources** and **sinks**.

Thus the signal $\mathcal{Z}_G$ provides a field associated with **mass**, $\mathcal{Z}_Q$ with an electrically charged object, and so on. And in the functional space defined by such fields these **sources** or **sinks** are recognised as **singularities**.

In terms of an algebra such a thing as a **particle** has always been represented by a **point** [v. Physics-3 in Part-1] - insofar as its position in 3-space (usually **Euclidean E**^{3}) is concerned - and this has been fruitful in all the Physics which is based on the particle concept. But, for example, the various **fields** associated with the particle (gravity, electric potential, magnetic potential, sources and sinks in fluid mechanics, wave functions in atomic theory) have naturally required that the **point** occupied in the algebra by the abstract particle must be identified as a **singularity** in the relevant <u>Function Space</u>.

Thus, in the case of the inverse-square force with potential function V(r) - taking the particle to be at the origin (r = 0), then V being proportional to 1/r it cannot be defined at the origin (the **source** or **singularity** in the "function-space" defined by V).

This means that the field V cannot "see" the origin although it can (other things being equal) "see" all the other points in the space. But that also means that, as far as V is concerned, the origin is a "no-go" piece of the space - so it must "see" the origin as an **object** which it cannot penetrate.

This "object" therefore manifests itself (in the space defined by the function V) as a **hole** in the perceived space. Perhaps we can think of it as a **pinhole** in any planar description of the field ? [simply illustrated in Fig. 1.1]. The **pinhole** may also be regarded as surrounded by a (vanishing) circle (or by any curve homemorphic to a circle). This "circle" will be the boundary to an **empty disc** (or its homeomorphism).

The space free of any such singularities is the domain of (eg) any field vectors.

[1.2] An example

Another example is obtained by considering the magnetic field vector **H** which is generated by an electric current **I** flowing in a long straight wire (a filament). Using the usual cylindrical co-ordinates, the field at a radial distance **r** from the wire (the z-axis) is given by $\mathbf{H} = 2\mathbf{I} \div \mathbf{r}$ (in e.m.u) and so, again, we see that there is a "cylindrical hole" in the function space defined by **H** ; a singularity where the field vector **H** is not defined (viz., where $\mathbf{r} = \mathbf{0}$). The mapping associated with this **hole** is the electric current 2**I**.

So the difference (if any) between the two cases (in 1.2 and 1.3) is the **nature of the hole**, viz. whether it is a **zero-dimensional hole** (the particle picture) or a **one-dimensional hole** (the line picture). In the either case the hole may well be defined mathematically by being bounded by any 1-dimensional line or 2-dimensional surface.

When the above function **V(r)** is defined in a 3-space the **pinhole** becomes a **spherical hole** - and this can appear in a Physics without the requirement of a vanishing radius.

[1.3] Furthermore, when we consider the way in which the Base Elements explore the world of observations (of what we have called the **generating elements** {**E**}) they will see very different pictures.

The chief point is that some "things" are **opaque** to some signals and not to others - it being unnecessary to elaborate on it. [v. Fig. 1.2].

So what one Scale sees as an "object" with only surface (external) features another sees as an "object" with internal features. The Physics (plural) which they describe require a different mathematical language for their descriptions - and this leads to a search for some sort of theorising for their reconciliation.

That is to say, for example, that when one sees a **particle** (containing no structure) another sees a **particle** containing (probably) other **particles** - for whenever the Base Element β finds its "limit" it will always interpret the observation as a **particle** (with a surface which is opaque, that is to say which cannot be "penetrated"). [v. Part-1 §[14.4]]

In any of the above cases, whether the Base element seems to identify a **source/sink** or whether its role is to monitor the objects of generating elements, the analysis leads us to search for objects associated with "holes" in the relevant field of observations.

[1.4] Values attached to Pinholes

In a typical Physics we then find it essential to attach a **value** to this pinhole - such as "graviational mass" m, "electric charge" q, "magnetic pole" μ, "fluid source" s ... and so on.

So various kinds of particle physics are described via a Field Theory which uses values/mappings to appropriate pin**holes** (which are identified with **mathematical singularities** in the fields).

[1.5] Analysis procedure

Our analysis is therefore based on

(a) setting up a structure which can exhibit **holes** (often via **singularities**)

(b) associating mappings on these **holes** compatible with our observations of

either **sources/sinks** or measures of **generating elements**

In any event such a procedure does not depend on confining ourselves to the traditional **Euclidean** spaces, or to the use of the real number system $\Re^{\#}$ (as opposed to the use of the rationals, $\Re$).

[1.6] Some concepts of topology

(a) We find that if we wish to remain independent of the traditional **Euclidean** metric space then we need only turn to the modern mathematical notion of a **topological space**, which has developed with just that problem in mind.

This is defined as any set of "points", **X**, together with a collection of some (or all) of its subsets, $\Im$, called its **open sets**. We write the topological space as $\{X, \Im\}$.

$\Im$ always contains the whole set **X** as well as the empty set $\otimes 8$. When these are the only sets in $\Im$ the space is said to have the **discrete topology**.

In **Euclidean** space these open sets are things like (a< x <b) , eg (0,1) etc. or, in the plane, things like (a< x <b, c< y <d) ; the inside of the unit sphere is the open set

$$\{(x,y,z) \text{ where } (0 < r < 1) \text{ and } r^2 = x^2+y^2+z^2\} \quad \text{...etc}$$

all of which definitions are expressed in terms of a **distance function**.

For example, the traditional way of defining the continuity of a function $f(x)$ at a point x_0 says "given an $\varepsilon > 0 \; \exists$ a δ such that whenever $x \in (x_0-\varepsilon, x_0+\varepsilon)$ then $f(x)$ lies in the set $(f(x_0)-\delta, f(x_0)+\delta))$"

Clearly this may easily be translated into appropriate open sets on the real line.

So, given topological spaces $\{\mathbf{X}_1, \Im_1\}$ and $\{\mathbf{X}_2, \Im_2\}$ and a mapping

$f : \mathbf{X}_1 \to \mathbf{X}_2$ we can say that $f(x)$ is continuous at x_0 if,

given an open set $V \in \mathbf{X}_2$ the set $f^{-1}[V]$ is open in $\mathbf{X}_1$.

[Note : For in-depth analyses v. eg Ref. Appendix-E [87], [90], [91]]

An important related concept is that of a **homeomorphism** between two topological spaces. This is an onto mapping h: $\mathbf{X} \to \mathbf{Y}$ which preserves the open sets in the two spaces.

This means that topologically the two spaces are identical.
It may be intuitively understood as a deformation of the space $\mathbf{X}$ which is continuous and which is accomplished <u>without tearing</u>. Perhaps it justifies the old adage that

"A topologist cannot distinguish between a ring-doughnut and a coffee cup"
the handle of the cup being the hole in the middle of the doughnut.

Thus a disc in $\mathbf{E}^2$ is homeomorphic to a filled-in triangle ; a sphere is homeomorphic to a filled-in tetrahedron ; and so on.

Measures expressed by real numbers, $\Re^{\#}$, are always approximations and, being found only in the rationals, $\Re$NIL should rightly be shown as **open intervals**, such as

$(a < x < b)$ with $a, x, b \in \Re$.

This means that measures in Physics really define a **topological space**, being more general than the **Euclidean** space, $\mathbf{E}^3$, which is one with the open sets being defined via the usual **metric**.

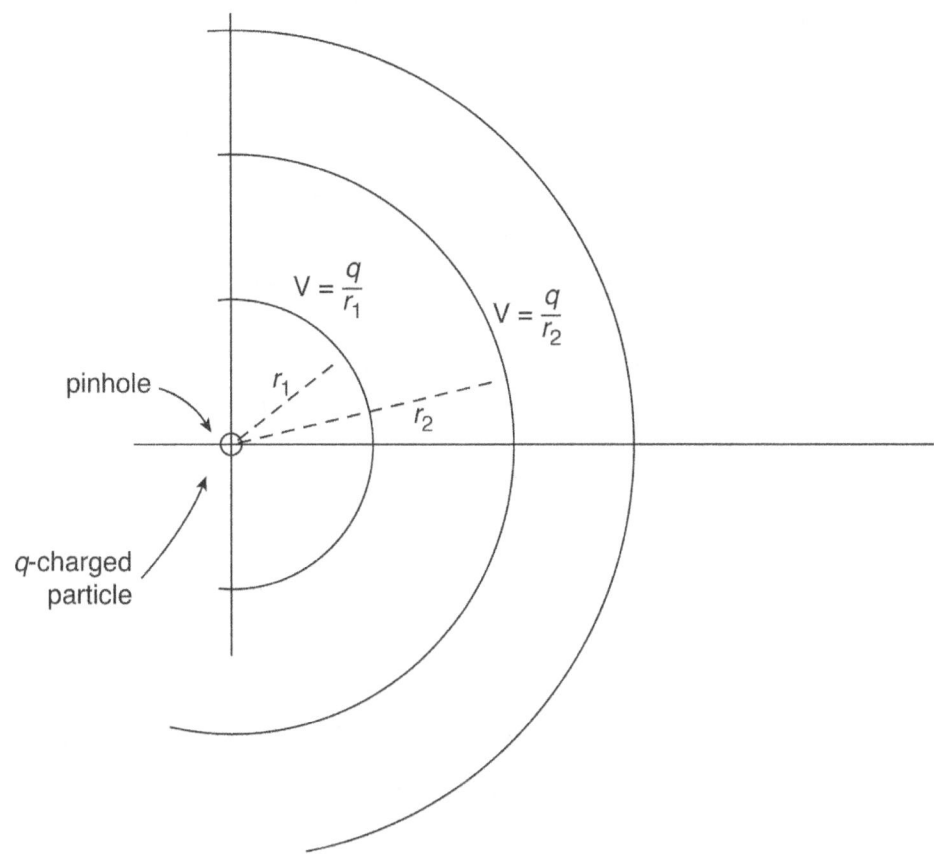

pinhole = singularity in V(r) field
charge (q) mapped into pinhole
particle = point-object at pinhole in V-field

Fig 1.1

162 MATHEMATICAL PHYSICS

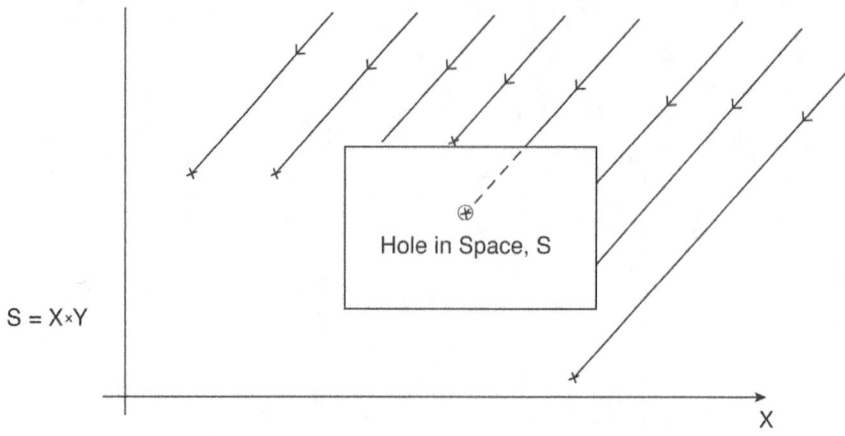

Box = Object = Hole in Space as seen by Base Element (Light)

∗ = accessible point;   ⊛ = inaccessible point

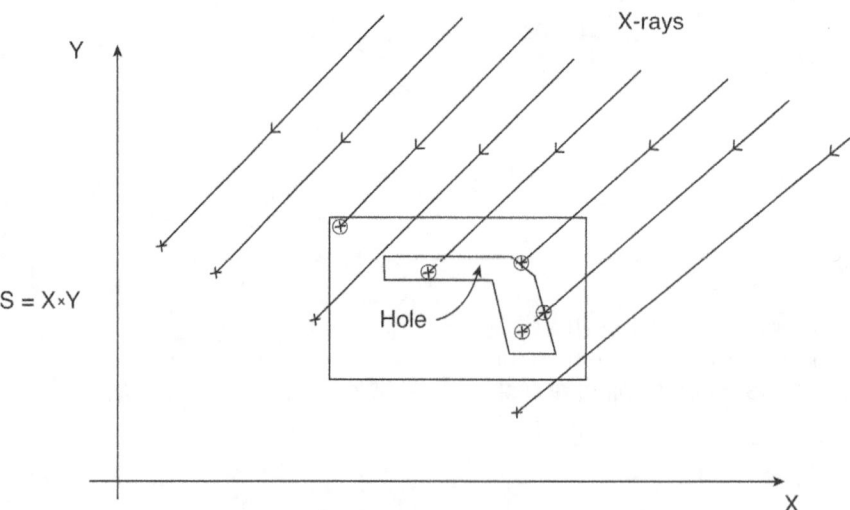

Gun = Object = Hole in Space as seen by Base Element (X-rays)

∗ = accessible point;   ⊛ = inaccessible point

Fig 1.2

Chapter-2 /PART-2   Chain Complexes & Homology Groups

[2.1] Simplices and Chains

We begin by laying out the historical essentials of what are called **Chain Complexes** - which we shall denote by the letter **K**. To be more precise we will denote this by the symbol **K.** since it consists of a **graded sequence** of structures, viz., $\{K_p\}$ where the subscript "p" takes integral values $0,1,2,\ldots,n$ (say) - "n" being the finite maximum and is called the dimension of **K**.

We first consider a finite set

$$V = \{v^i;\ i = 1 \ldots k\}$$

and a collection **K** of its subsets. We denote any one of these subsets by the symbol σ_p when it contains $(p+1)$ distinct elements of **V**, and we call such a subset a **p-simplex** (or a **simplex** of order p). If σ_q is a **q-simplex** defined by a $(q+1)$ subset of the $(p+1)$ elements of the points which define the first we say that σ_q is a **face** of σ_p, and we write $\qquad \sigma_q < \sigma_p$

Clearly this relation $<$ is a partial ordering on the collection **K**.

This collection is called a **simplicial complex** if and only if

(1) each singleton set $\{v^i\}$ is a member of K, each being a σ_0

(2) whenever $\sigma_p \in K$ and $\sigma_q < \sigma_p$ then $\sigma_q \in K$.

The set **V** is called the **vertex set** and, whenever appropriate, when σ_p is defined by vertices (such as) $v^1, v^2, \ldots v^{(p+1)}$ we can write

$$\sigma_p = \langle v^1 v^2 \ldots v^{(p+1)} \rangle$$

When the order in which the vertices are so written is irrelevant we say that the simplex is **unoriented** - otherwise (being the cases in which we are primarily interested) an **orientation** is attributed to the simplex and we denote the σ_p by $+\sigma_p$ when the sequence of its vertices is an **even permutation** of the defining set, and by $-\sigma_p$ when it is an **odd permuation**. In such a case we see that the vertices of a σ_p must be distinct for otherwise we would have $\langle v_i v_i \rangle = - \langle v_i v_i \rangle$, whence $\langle v_i v_i \rangle = 0$.

A natural and systematic precedure for introducing an orientation is to settle on a positive sequence for the whole set **V** and then to let this induce an orientation for each σ_p.

In any event most of the global properties of a complex **K**, such as the **homological** structure (v. later) are invariant under a change of orientation.

Also we must point out that none of these definitions require the vertices $\{v^i\}$ to be points in the traditional **Euclidean** metric space - although it is natural (as historically it has been) to use this assumption by way of illustration.

[2.2] Convex polyhedra in E^n.

We can obtain a representation of a complex **K** in terms of a connected **convex polyhedron** a **Euclidean** space - in, say, E^2 - as follows.

Let the vertices $\{v^1, v^2, v^3\}$ become the points (P_1, P_2, P_3) - being the vertices of a plane triangle $P_1P_2P_3$ together with the "inside".

Then the **convex set** is $\{P\}$ where

$$P = t_1P_1 + t_2P_2 + t_3P_3$$

where each $t_i \geq 0$, and where $t_1 + t_2 + t_3 = 1$

We can take this as a 2-simplex σ_2 and then the complex K consists of

the 3 0-simplices $\langle P_1 \rangle$ $\langle P_2 \rangle$ $\langle P_3 \rangle$

the 3 1-simplices $\langle P_1P_2 \rangle$ $\langle P_2P_3 \rangle$ $\langle P_3P_1 \rangle$

the 1 2-simplex $\langle P_1P_2P_3 \rangle$

This simple case, for the triangle $X_0X_1X_2$, is drawn in Fig. 2.1 and shows an inbuilt orientation.

Generally a p-simplex σ_p is represented by a convex polyhedron (with all its faces) with (p+1) vertices in the **Euclidean** space E^p, and a complex K as a collection of such polyhedra in a suitable space E^h.

We also notice that this complex will be equally applicable to any space which is homeomorphic to such **Euclidean** structures.

[2.3] Dowker complexes arise in a combinatorial context, being defined by binary matrices (each element of such a matrix being either 0 or 1).

Given a typical binary matrix like :

$$\begin{bmatrix} 0 & 1 & 1 & 1 & 0 & 1 & 1 & 0 \\ 1 & 1 & 0 & 0 & 1 & 1 & 0 & 1 \\ 0 & 0 & 0 & 1 & 1 & 1 & 0 & 1 \\ 1 & 1 & 1 & 0 & 0 & 1 & 1 & 0 \end{bmatrix}$$

where the columns are denoted by X_c and the rows by Y_r, then each row, say, can be identified as a simplex. For example the simplex Y_1 is defined by the vertices

$$\langle X_2, X_3, X_4, X_6, X_7 \rangle$$

and so on for the simplices Y_2, Y_3, Y_4, Y_5.

This illustrates the point that the vertices fully define a simplex - there is no requiremnt that the (**Euclidean**) points in between need be considered - although in §[2.1] above this has in fact been so. Such an interpretation defines a **simplicial complex**, say **K(Y,X)**, of simplices. The "connections" are represented by the sharing of 1's between the rows of the matrix. Thus Y_1 shares the vertices X_2, X_6 with the simplex Y_2.

Similarly, by looking at the transpose of such a matrix we can obtain a complex whose vertices are the **Y**'s and whose simplices are the **X**'s. This we would denote by **K(X,Y)**, and may call it the **conjugate complex**.

Dowker showed, in the reference above, that the **homology groups** of **K(Y,X)** and **K(X,Y)** are isomorphic.

[Some applications of such complexes will be discussed in Part-4 of this work].

[2.4] Simplices in Exterior Algebra

In a similar way we can see that the ideas of simplex and complex can be associated with the members of an **exterior algebra**, Λ^n, à la **de Rham**.

In this case, and using differentials as the defining elements in Λ^n,

a "0-simplex" (or vertex) will be denoted by a single dx_r

a "1-simplex" will be denoted by a pair of vertices, eg. $dx_r \wedge dx_s$

a "2-simplex" will be denoted by a triple, eg. $dx_r \wedge dx_s \wedge dx_t$

and so on. The subscripts must be unequal and the rules of Λ^n require

$$dx_r \wedge dx_s = - dx_s \wedge dx_r \quad ; \quad dx_r \wedge dx_r = 0$$

In all cases of a complex the sharing of faces expresses our intuitive idea of a **chain** of contiguous simplices, and if C_p denotes the totality of simplices $\sigma_p^{(i)}$ we can consider it as an additive (**Abelian**) group with, say, coefficients in J. This gives rise to the idea of a **chain complex** (on **K**) which is dimensionally graded, viz.,

$$C^* \equiv C_0 \oplus C_1 \oplus C_2 \oplus C_3 \oplus \ldots \oplus C_n$$

the sign $\oplus$ indicating the direct sum of the separate additive **Abelian** groups.

We can think of each group C_p as being a **module/vector space** with a finite basis determined by the distinct simplices, $\sigma^{(i)}_p$.

Referring to Fig. 2.1 we find a 0-chain

$$c_0 = <X_0> + <X_1> + <X_2> \quad \text{(a member of } C_0)$$

and a 1-chain $\quad c_1 = <X_0,X_1> + <X_1,X_2> + <X_2,X_0>$

In Fig. 2.2 we find, eg. a 2-chain in

$$c_2 = <X_0,X_1,X_2> + <X_2,X_3,X_0>$$

which can be written as $\quad \sigma^{(3)} + \sigma^{(1)}$.

With an orientation identified on the simplices we naturally have, eg.

$$- <X_0,X_1> = + <X_1,X_0>$$

[Note : This antisymmetry matches that to be found in the example of Λ^n above.]

The orientation (the ordering of the vertices) will be such that, eg. the directions on a common edge between two neighbouring triangles will be opposites.

[2.5] CW-complexes

These were introduced by **Whitehead** (Ref: Appendix-E [89]) and illustrate the generality of these concepts by their application to many types of **topological spaces**. In this it is clear that the notions need not be restricted to **Euclidean** metric spaces, though that is where they had their beginnings.

From our point of view, with an eye on the importance for the foundations of Physics, we expect to be able to generalise the notion of "point" (in its **Euclidean** sense) and, as we have introduced in Part-1 (v. Chapter-17, §[17.1]), we refer to the relevant elements as **poins**.

In the creation of a CW-complex the idea is that we can begin with a set of vertices X^0 ; then choose a set of one (or two) vertices and adjoin a collection of 1-cells to them to give a space X^1 ; then adjoin a collection of 2-cells to obtain a space X^2; with this structure adjoin a collection of 3-cells to obtain a space X^3, and so on.

[Note : A 1-cell is any space which is homeomorphic to the open region U^1 viz. the
open interval $(0 < x < 1)$ - together with its bounding end-points.

A 2-cell is any space which is homeomorphic to the open region U^2 where

$$U^2 = \{(x,y) \in \Re \times \Re \; ; \; x^2+y^2 < 1\} \; = \text{ inside of a disc}$$

Similarly a 3-cell is any space which is homeomorphic to the open region U^3 where

$$U^3 = \{(x,y,z) \in \Re \times \Re \times \Re \; ; \; x^2+y^2+z^2 < 1\} = \text{ inside of a sphere}$$

and so on]

Adjoining (eg) a U^2 to a piece of the graph X^1 is to use one (or more) edges as the boundary of a homeomorph to the disc U^2.

[2.6] Čech complexes

These are usually, but not necessarily, associated with the open sets of a compact **Hausdorff** topological space $\{X, \mathfrak{I}\}$ in the following way.

Suppose that **U** is a collection of open sets (which need not be in $\mathfrak{I}$) which form a finite **covering** of **X**.

Then we regard members $\{u_r\}$ of **U** as vertices in the ensuing complex **K**; a simplex σ_p of **K** is then identified with any $(p+1)$-subset of **U** whenever the intersection of such u_r, for example $u_0 \cap u_1 \cap u_2 \cap \ldots \cap u_p$, is not the empty set ■.

It is usual to call such a complex the **nerve** of the covering **U** and since this need not be unique we can obtain other and "finer" coverings, **V**. Such a **V** will often be a refinement of the first cover [v. Fig. 2.2a], and it can be shown that given any finite simplicial complex κ then there exists a subcomplex of the nerve of **U** which is isomorphic to κ.

[v. Refs. Appendix-E [87] [93] [94] for detailed analyses of Čech complexes.]

[2.7] Triangulation

This is a carve up of a space by imaging it to be covered by a suitable complex; the word having been introduced in the initial studies of 2-dimensional manifolds (surfaces in geometry).

By this means the point-set topological space (the surface) can be transformed into an algebraic structure via its covering simplicial complex.

It is by this means that the complexes underlying sets of measures in a Physics can be given an algebraic representation which results in a manageable analysis - and can bridge the apparent divide between the discrete and the continuous. As we have seen in PART-1 of this thesis this has been necessary to enable Physicists to excape from the constriction imposed on them by constantly demanding that all measures must be found in the real number system $\mathfrak{R}^{\#}$.

We illustrate some simple triangulations in Fig. 2.3 where we use both **Euclidean** type "triangles" to get the complex **cover** of the space (surface) and, as an alternative the use of CW-complexes. The examples illustrate the spherical surface S^2, the torus, and the projective plane.

[2.8] Homology groups in a complex

The **homology groups** defined on a chain complex, C_*, give us the **holes** in the underlying complex **K**.

In order to do this we need the use of a **boundary operator** (which we shall denote by ∂) which, giving us the natural boundary of a simplex/cell, will identify those cells which are empty in **K** (i.e. holes).

This operator will be a homomorphism, preserving the additive properties of the **Abelian** groups $C_0, C_1, C_2, \ldots C_p, \ldots C_n$ defined above.

Considering the sequence of C_p's

$$C_* = C_0 \xleftarrow{\partial_1} C_1 \xleftarrow{\partial_2} C_2 \xleftarrow{\ldots} \xleftarrow{\partial_p} C_p \xleftarrow{\partial_{p+1}} C_{p+1} \xleftarrow{\ldots} \ldots$$

each boundary operator ∂_p operates on its C_p to give us the boundary in the group C_{p-1}. This boundary will be the algebraic sum of some appropriate σ_{p-1}'s.

Given a simplex $\quad \sigma_p = \langle P_0 P_1 P_2 \ldots P_i \ldots P_p \rangle$

the boundary of this is defined as

$$\partial_p \sigma_p = \Sigma_i (-1)^i \langle P_0 P_1 \ldots P_{i-1} P_{i+1} \ldots \rangle \qquad i = 0 \ldots p$$

which is a signed sum of the (p-1)-faces of σ_p, obtained by **omitting successive** σ_i's.

The most important property of this operator is its **nilpotency**, that is to say, that $\quad \partial_{p-1}(\partial_p) = 0_{p-2}$

usually just written as $\quad \partial^2 = 0$.

This follows by the straightforward expansion of $\partial_{p-1}(\partial_p)$.

Example-1 See Fig. 2.1 for the simple case of a **triangle**, σ_2

Example-2 See Fig. 2.2 for the hollow **tetrahedron** (as a triangulation of S^2, the surface of a solid sphere, S^2).

Since each ∂_p is a homomorphism (<u>down</u> the graded chain C_*) its kernel, $\ker \partial_p$, and the image, $\text{im} \partial_{p+1}$, are both normal subgroups of C_p. Denoting these respectively by Z_p and B_p (for "zero" and "boundary") we naturally form the **factor group**

$$H_p = Z_p / B_p$$

called the **pth homology group**. In such a factor group the additive identity element - the "zero" - is the group of boundaries, B_p.

[Note: A full discussion of the concept of **factor group** will be found in most works on group theory, but see eg Ref. Appendix-E [95]].

The nilpotency of the boundary operator ensures that each

$b_p \in \mathbf{B}_p$ is also $\in \mathbf{Z}_p$.

[See Fig. 2.3a which illustrates the nilpotency of ∂]

The members of $\mathbf{B}_p$ are called the **p-boundaries** of **K** whilst the members of $\mathbf{Z}_p$ are called the **p-cycles** of **K**, so each p-boundary is a "trivial" p-cycle.

The subgroup $\mathbf{B}_p$ acts as the additive zero in the factor group whilst those members $z_p \in \mathbf{Z}_p$ which are not p-boundaries, in $\mathbf{B}_p$, represent the genuine **holes** in **K**.

Generally an algebraic **Abelian** group, G, consists of the direct sum of two subgroups, viz. a **free group** (no relations between the basis elements) and a **torsion group**, Tor(G), where some relational constraints are found among the basis elements.

So we can come across some homology groups which look like

$$\mathbf{H}_p \cong \text{(Free group)} \oplus \text{Tor}(\mathbf{H}_p)$$

The rank of the Free group is called the **Betti number** β_p of **K** and, when p=0,

β_0 = the number of disjoint components of K.

In the following examples we make use of the fact that the spaces of Torus and Projective plane can be neatly obtained by the identification (glueing together) of points and edges of a plane rectangle ; this is indicated by the colouring of the edges.

<u>Example-3</u>   See Fig. 2.4 for examples of CW-complexes and the associated $\mathbf{H}_*$ for a Torus, the Projective Plane, the Klein Bottle, and the Sphere S^2.

<u>Example-4</u>   See Fig. 2.5 for a Čech covering and its $\mathbf{H}_*$ for a Torus

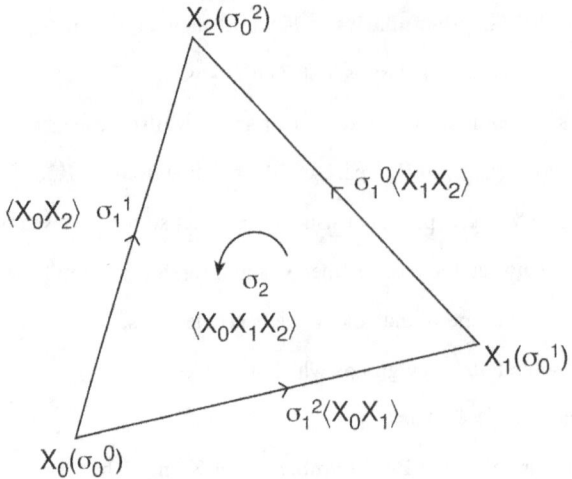

$\partial\{X_i\} = 0$ but eg $X_1 = X_0 + \partial\sigma_1^2$ so X_0 generates Z_0

$\partial\sigma_1^0 = \langle X_2 \rangle - \langle X_1 \rangle = \sigma_0^2 - \sigma_0^1;\ \partial\sigma_1^1 = \sigma_0^2 - \sigma_0^0;\ \partial\sigma_1^2 = \sigma_0^1 - \sigma_0^0$

So $\partial(\sigma_1^0 - \sigma_1^1 + \sigma_1^2) = 0;\ \sigma_1^0 - \sigma_1^1 + \sigma_1^2 \in Z_1$

But $\partial\sigma_2 = \sigma_1^0 - \sigma_1^1 + \sigma_1^2$, so $\sigma_1^0 - \sigma_1^1 + \sigma_1^2 \in B_1$

It follows that $H_0 \cong J$ and $H_1 = \dfrac{Z_1}{B_1} \cong 0$

Fig 2.1

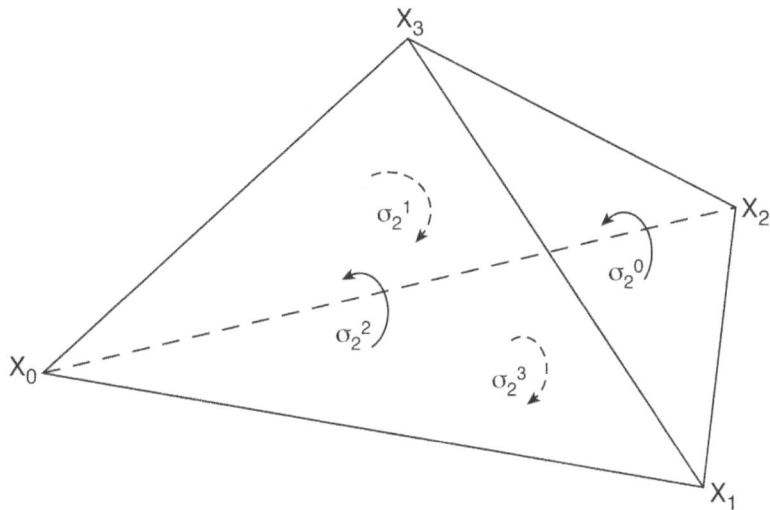

Hollow Tetrahedron = Homeomorphic to Sphere S^2

Notation: σ_2^i is the face opposite vertex X_i

Since $\partial \langle X_i \rangle = 0$ and eg $\langle X_1 \rangle = \langle X_0 \rangle + \partial \langle X_0 X_1 \rangle$, $\langle X_0 \rangle$ generates Z_0

C_3 is empty so B_2 is empty, but $\partial(\sigma_2^0 + \sigma_2^1 + \sigma_2^2 + \sigma_2^3) = 0$

Hence Z_2 is generated by the 2-cycle $Z_2 = (\sigma_2^0 + \sigma_2^1 + \sigma_2^2 + \sigma_2^3)$.

There are no 1-cycles, $Z_1 \cong 0$

$H_x = H_0 \oplus H_1 \oplus H_2 \cong J \oplus 0 \oplus J$ which is therefore $H_x(S^2)$.

Fig 2.2

Čech nerves V refining the U-cover of x-axis

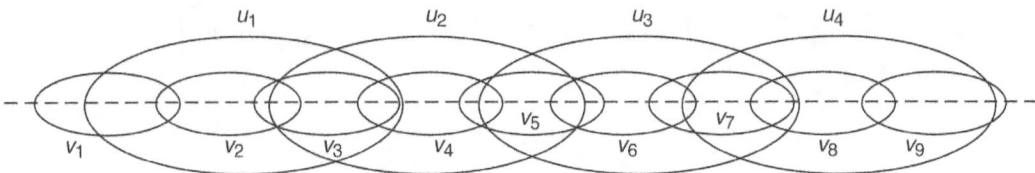

$U(u_1\ u_2\ \dots\ u_4)$ refined by $V(v_1\ v_2\ \dots\ v_9)$

Physical measures only give open sets in ℝ (rationals) which correspond to a Čech cover (a Nerve ≅ complex) "Refining" means measuring "more accurately"

Fig 2.2a

Coverings by Triangulation; Projective Plane P² ; Torus

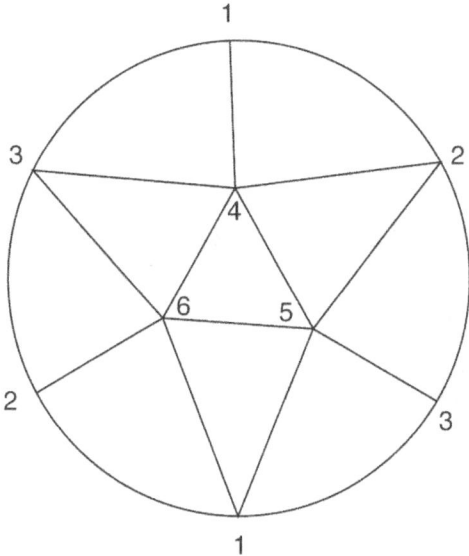

P² is obtained by identifying opposite ends of diameters

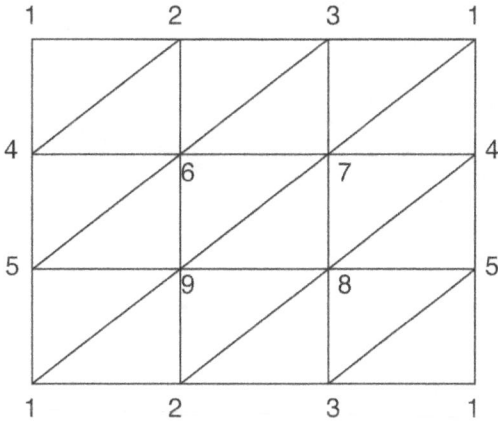

A CW-complex can be formed to simplify this Torus and is obtained by identifying opposite sides

Fig 2.3

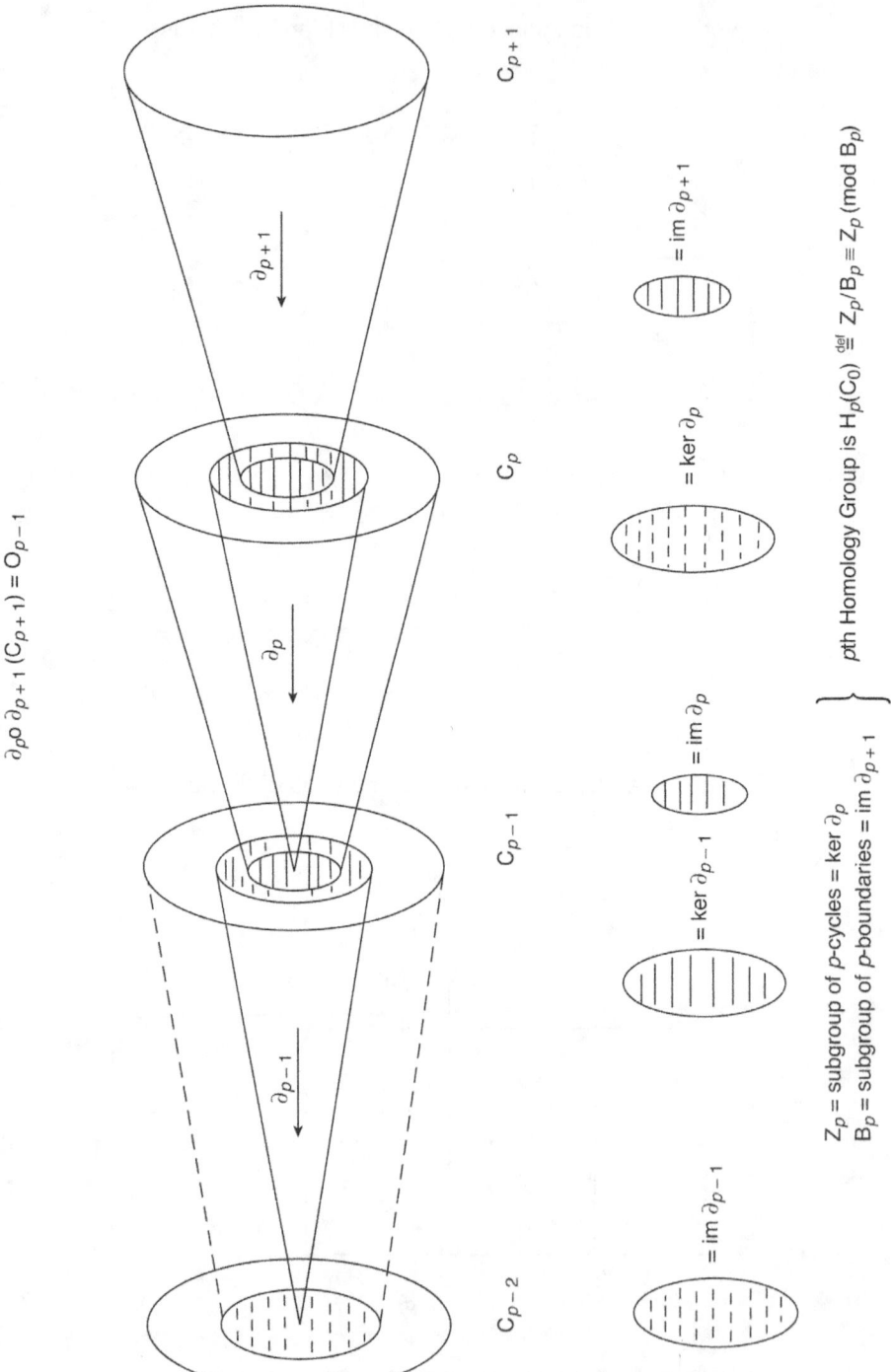

Fig 2.3a

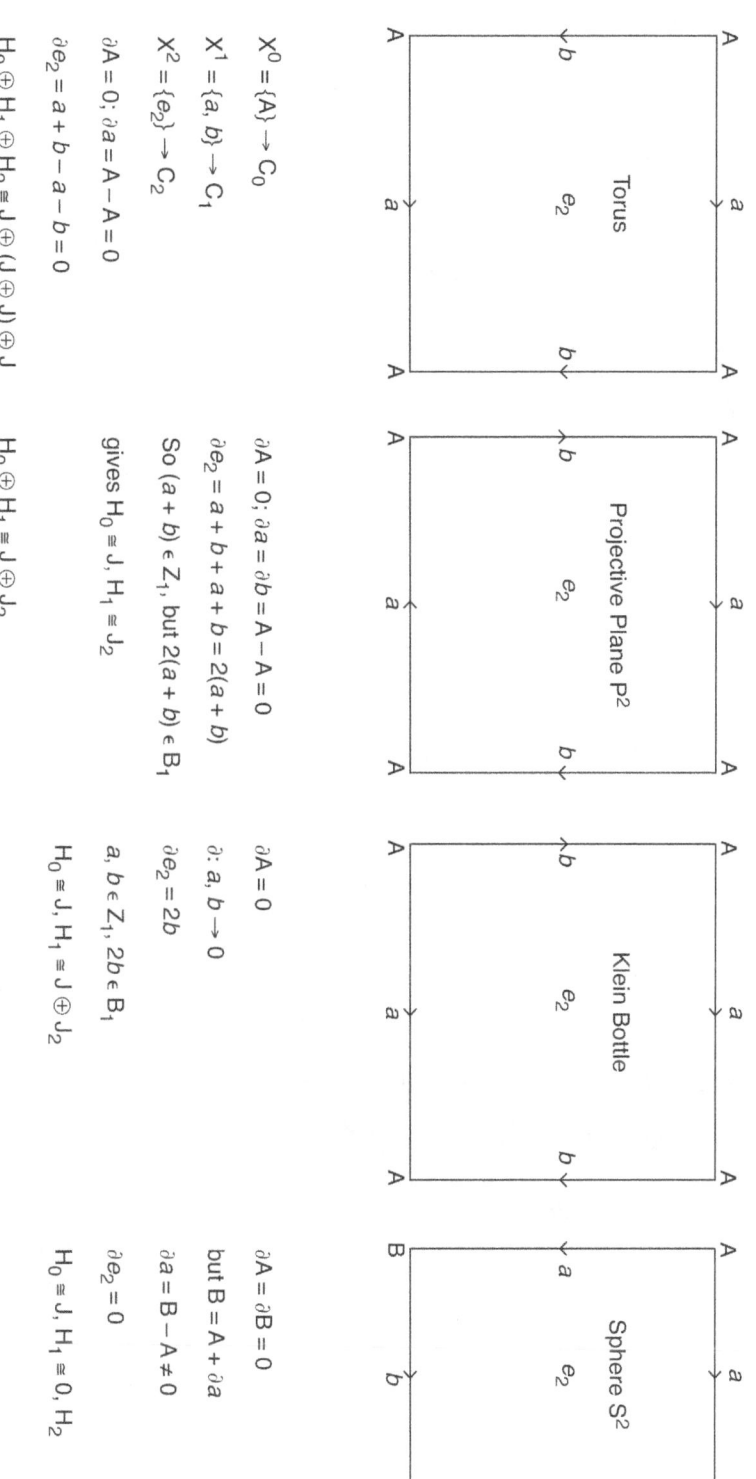

Fig 2.4

CW-complexes via identification of oriented edges of a rectangular sheet

Torus

$X^0 = \{A\} \to C_0$
$X^1 = \{a, b\} \to C_1$
$X^2 = \{e_2\} \to C_2$
$\partial A = 0; \partial a = A - A = 0$
$\partial e_2 = a + b - a - b = 0$
$H_0 \oplus H_1 \oplus H_2 \cong J \oplus (J \oplus J) \oplus J$

Projective Plane P^2

$\partial A = 0; \partial a = \partial b = A - A = 0$
$\partial e_2 = a + b + a + b = 2(a+b)$
So $(a+b) \in Z_1$, but $2(a+b) \in B_1$
gives $H_0 \cong J$, $H_1 \cong J_2$

Klein Bottle

$\partial A = 0$
$\partial: a, b \to 0$
$\partial e_2 = 2b$
$a, b \in Z_1, 2b \in B_1$
$H_0 \cong J, H_1 \cong J \oplus J_2$

Sphere S^2

$\partial A = \partial B = 0$
but $B = A + \partial a$
$\partial a = B - A \neq 0$
$\partial e_2 = 0$
$H_0 \cong J, H_1 \cong 0, H_2$

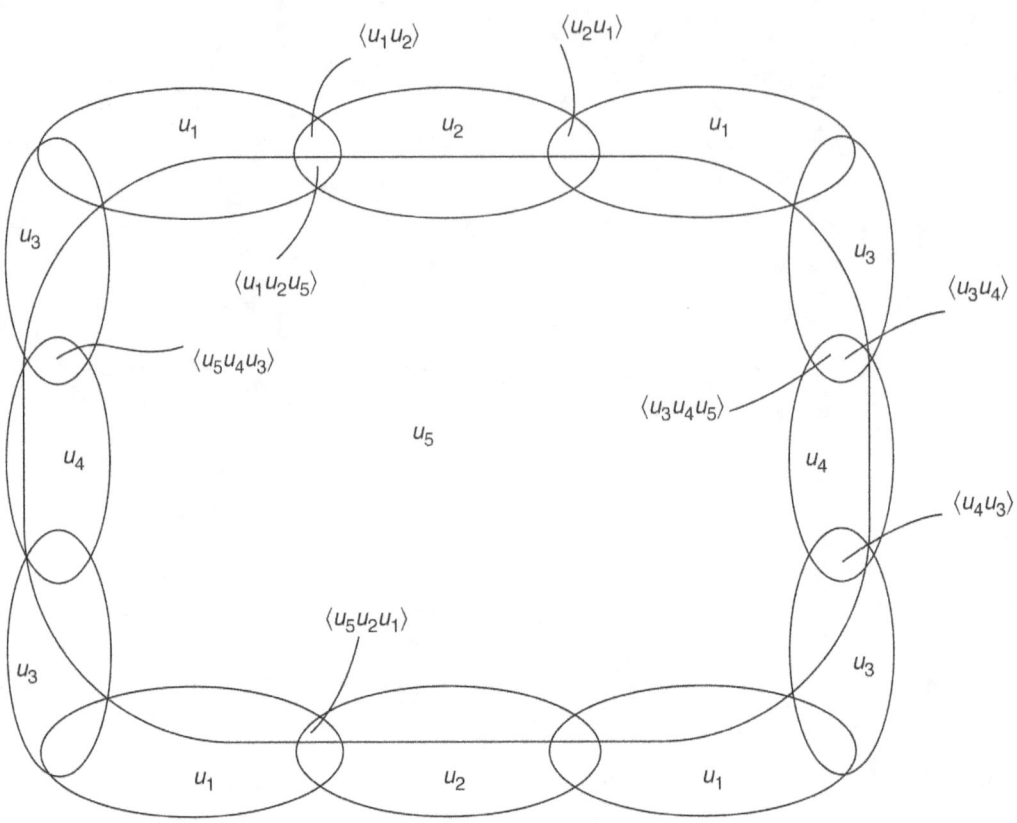

$Z_1^{(1)} = \langle u_1 u_2 \rangle + \langle u_2 u_1 \rangle \qquad \partial Z_1^{(1)} = \langle u_2 \rangle - \langle u_1 \rangle + \langle u_1 \rangle - \langle u_2 \rangle = 0$

$Z_1^{(2)} = \langle u_3 u_4 \rangle + \langle u_4 u_3 \rangle \qquad \partial Z_1^{(2)} = \langle u_4 \rangle - \langle u_3 \rangle + \langle u_3 \rangle - \langle u_4 \rangle = 0$

$Z_2 = \langle u_1 u_2 u_5 \rangle + \langle u_5 u_2 u_1 \rangle + \langle u_3 u_4 u_5 \rangle + \langle u_5 u_4 u_3 \rangle$

$\partial Z_2 = \langle u_2 u_5 \rangle - \langle u_1 u_5 \rangle + \langle u_1 u_2 \rangle + \langle u_2 u_3 \rangle - \langle u_5 u_4 \rangle + \langle u_5 u_2 \rangle + \text{Similar}$
$\qquad = 0$

$H_x = H_0 \oplus H_1 \oplus H_2 \cong J \oplus (J \oplus J) \oplus J$

Fig 2.5

Chapter-3 /PART-2   Cohomology and Measures/Observations

[3.1] The Classical acyclic backcloth

If we seek a view of measures of generating elements which does not depend entirely on notions of the continuum then we can find it via the **algebraic homology/cohomology** defined on graded groups of **p-chains**, as discussed in Chapter-2.

When these p-chains are defined on manifolds (which are locally homeomorphic to **Euclidean** space) they depend upon the ability to cover the space with non-overlapping **simplices** via the process of triangulation, or by some equally sufficient cover of point-sets. This approach gives rise to p-chains in a **complex** - simplex-based or cell-based.

The development of traditional Physics has been based on the idea that the underlying space available for carrying the measures of generating elements is **homologically trivial**, this being the **Euclidean** space E^3. That is to say the **homology groups** are

$$H_p(K) \cong 0$$

meaning that there are **no holes** in the space - which is naturally said to be **acyclic**.

[3.2] Mesh of Observations

If we consider a set of observations (which we shall agree to call **space-time** and using the word "point" without prejudice) then we may obtain a representation of the **Cartesian** product $S \times T$ by a diagram shown in Fig. 3.1 (Part-2). Here, S carries the co-ordinate (**x** say) whilst T carries the time **t** measures.

Generally we would more likely use the co-ordinates (x,y,z,t) and have a mesh based on observations of points in $S \times S \times S \times T$ (or S^3T). In relativistic notation the time **t** would be replaced with ct, "c" being the velocity of the signal.

Yet again we might consider the space associated with a general **Hamiltonian** dynamical system where there are n independent co-ordinates q_r, together with the co-ordinate t.

Now the absolutism of the **Newtonian Physics** regarded a mesh (such as Fig. 3.1) as embedded in a **Euclidean** plane $\Re^\# \times \Re^\#$, that the plane was objective, and that phenomena (measures) were "carried" by it. It would certainly be effectively regarded as a triangulation of an underlying continuum and would constitute a simplicial decomposition of that space - whereby the points in the diagram would be the 0-simplices and the pairs of points would be the 1-simplices. Naturally the p-simplices would be defined by a collection of (p+1) points of the mesh.

But we must now notice that, eg, a 1-simplex is adequately defined by a pair $P_1 = (s_1,t_1)$ and $P_2 = (s_2,t_2)$ and does not require there to be any other points (of the assumed continuum) between them.

This is equivalent to appealing to the **Dowker** complex [v. Chapter-2 above] if we wish only to rescue the notions of homology groups of the backcloth.

The acyclic nature of the backcloth, which was there before, is still there. Nor does it depend on whether the sets of points S or T are finite or infinite.

But since the co-ordinate values (x,y,z,ct) are observed by the Physicist in the rational numbers $\Re$ it can still be regarded as the "Physicist's continuum" - but not the "Mathematician's continuum". So this does not exclude localised concepts (calculus concepts).

[3.3] Mappings on a chain complex

Now let us suppose that the backcloth mesh represents measures in a Physics. Then the extension of this backcloth to embrace a physics of **dynamics** will be achieved by (say) identifying a "particle" with its "point" in the mesh. The additional elements of the dynamical physics, such as **mass, velocity, acceleration, electric charge, magnetic moment** etc. may also be associated with suitable "p-simplices" of the mesh.

Now, for example, attributing mass to a particle (at a point) amounts to describing it as a mapping $\quad$ mass : $C_0 \to \Re^{\#}$

where C_0 is the zero-order chain group in the complex K. Normally this mass (being a point particle) would be confined to one 0-simplex (at any one time) in the chain.

Mappings on a chain complex, that is to say on the graded chain group C_* of a complex **K**, which are **linear functionals** with values in an **Abelian** group **G**, are naturally called **cochains** associated with **K** - and become the dual graded group

$$C^* = C^0 \oplus C^1 \oplus C^2 \oplus \ldots \oplus C^n$$

"n" being the dimension of $C_*(K,J_+)$, J_+ being the positive integers.

The zero-order **co-chain**, c^0, consists of values g_i in the coefficient group **G** tied to vertices (points/poins) in any finite covering of **K**.

Mass of a particle is to be an example of such a c^0 and, in a similar vein a point-charge in electrostatics will be associated with a point, giving

$$\text{point-charge} : C_0 \to \Re^{\#}$$

In a similar way we may find, eg, velocity measures as 1st-order co-chains via

$$\text{velocity} : C_1 \to \Re^\# \quad \text{(v. Fig. 3.2)}$$

A force, acting an a line (a 1-simplex) will require a 1-cochain

$$\text{force} : C_1 \to \Re^\#$$

A 1-dimensional current flow in Electromagnetism will also be

$$\text{current} : C_1 \to \Re^\#$$

Values of, eg, angular velocity would require a 2-chain, c^2, since the area of an implied triangle would be needed to represent an **orientation** to the rotation. Similarly a torque would be a c^2.

Oriented volume would require a c^3 (a homeomophic tetrahedron for the three dimensions).

In like manner we might well consider that all our measures in a Physics could be regarded as p-cochains on a suitable backcloth complex **K** (not always or necessarily that of $S \times T$).

[3.4] Duality between C_* and C^*

Very commonly in the history of Physics the mappings are taken in the field $\Re^\#$ when they are regarded as **cochains** defined on the covering of the backcloth **K**.

If we write a p-cochain as c^p then its value on a p-chain c_p can be written as

$$(c^p, c_p) \quad \text{a value in } \mathbf{G} \text{ (usually in } \Re^\#\text{)}$$

The definition requires a cochain to respect the additive structure of the C_* associated with it - so it must be a homomorphism, obeying the rule

$$(c^p, c_p + c'_p) = (c^p, c_p) + (c^p, c'_p)$$

meaning that we need only know the action of a c^p on those $\{\sigma_p\}$ which form a basis for the chain group C_p.

This is exactly parallel to the case of the vector space, V^*, which is the **dual** of a given vector space **V** (over the rationals $\Re^\#$, say).

Whilst we take the **Abelian** (additive) group C_p to have integer coefficients in **J** we can be more general in the case of the dual cochain group C^p and can consider its values on the basis $\{\sigma_i\}$ of C_p to be found in the chosen group **G** (usually the additive group in $\Re^\#$.

It follows that although an integral p-chain c_p can be regarded as a p-cochain the reverse is not necessarily so.

The duality inherent in these structures means that we must consider the group which is dual to the **homology groups** H_p, $p = 0, 1, \ldots n$, and to this end we require the dual to the boundary operators ∂_p. We naturally call these the **coboundary operators** and denote them by $\{\delta^p\}$, where $\delta^p : C^p \to C^{p+1}$ in the scheme

$$C^0 \xrightarrow{\delta^0} C^1 \xrightarrow{\delta^1} C^2 \xrightarrow{\delta^2} \ldots \to C^p \xrightarrow{\delta^p} C^{p+1} \xrightarrow{\delta^{p+1}} \ldots$$

It is easily verified that the coboundary operator δ is nilpotent ; $\delta^{p+1}(\delta^p) = 0$.

The duality between the operators ∂ and δ follow the notion of operator and adjoint operator found in the algebra of vector spaces so that, if we denote the value of c^p on the chain c_p by (c^p, c_p) we can identify that $(p+1)$-cochain δ^p as that cochain c^p satisfying
$$(\delta^p c^p, c_{p+1}) = (c^p, \partial_{p+1} c_{p+1})$$

The value of the coboundary δ^p(cochain c^p) on a c_{p+1} equals the value of that c^p on the boundary of c_{p+1}.

Thus, whilst the boundary operator only requires a knowledge of any particular c_p in the complex **K** (that is to say a **local** knowledge), yet the coboundary operator requires knowledge of the **neighbouring** structure associated with that c_p.

We can now define the **pth cohomology group** as the factor group

$$H^p = \ker \delta^p / \operatorname{im} \delta^{p-1} \quad \text{or} \quad Z^p / B^p$$

the elements in the group Z^p become the **p-cocycles** whilst those in B^p are the **p-coboundaries**. Any two p-cocyles $z^p{}_1$, $z^p{}_2$ which differ by a p-coboundary are said to be **co-homologous** - and the group of p-coboundaries B^p act as the additive "zero" in the factor group H^p.

[3.5] Examples from Vector Field theory

We consider the **Euclidean** space E^3 with co-ordinates $(x,y,z) \in \Re^{\#} \times \Re^{\#} \times \Re^{\#}$ and take a triangulation (of points, arcs, areas, volumes - defining a complex **K**) of a suitable piece of it.

A c^0 will be a function defined at points of **K**, with values in $\Re^{\#}$.

A c^1 will be defined on the arcs of **K**, with values in $\Re^{\#}$.

A c^2 will be defined on the areas/triangles of **K**, with values in $\Re^{\#}$.

A c^3 will be defined on the volumes/polyhedra of **K**, with values in $\Re^{\#}$.

If ξ is a suitable differentiable function defined throughout **K** then it naturally defines such a c^0, say $c^0(\xi)$.

If **v** is a vector field defined on the arcs of **K** then the associated c^1 can be regarded as the integral $\int_\alpha \mathbf{v}$ where α denotes an arc in **K**.

Furthermore, if **u** is a vector field defined on the areas/triangles of **K** then the associated

c^2 can be regarded as the integral $\int_\beta \mathbf{u} \cdot \mathbf{n}$ where β denotes an area (piece of surface) and $\mathbf{n}$ is the normal at a point on it.

Similarly a c^3 can be obtained as the volume integral

$$\int_\gamma \xi$$

using the above function ξ and where γ is a finite volume.

The **coboundaries** of these c^p, viz., δc^0, δc^1, δc^2, and δc^3 will be such that in each corresponding case :

(a) δc^0 will be a c^1 via $\int_a \delta c^0 = \int_{\partial a} c^0$, a being an arc ($P_1 \to P_2$)

But this equals $\xi(P_2) - \xi(P_1)$ and this identifies the δc^0 with the well-known vector gradient, **grad** ξ, since $\int_{arc} \mathbf{grad}\xi = \xi(P_2) - \xi(P_1)$ on this arc.

(b) δc^1 will be a c^2, a surface integral, and this is defined by

$$\int_\beta dc^1 = \int_{\partial\beta} c^1$$

Appealing to **Stokes's** theorem which tells us that $\int_{\partial\beta} \mathbf{v} = \int_\beta \mathbf{curl}\ \mathbf{v}$ we can see that δc^1 is determined by the operator **curl**.

(c) δc^2 will be a c^3, a volume integral, and this is defined by

$$\int_\gamma \delta c^2 = \int_{\partial\gamma} c^2 \qquad \gamma \text{ denoting a finite enclosed volume}$$

Appealing to **Gauss's** theorem, viz.,

$$\int_{\partial\gamma} \mathbf{v} = \int_\gamma \mathbf{div}\ \mathbf{v}$$

we can see that δc^2 is determined by the operator **div**.

Furthermore the nilpotency of the coboundary operator δ reflects the properties of a vector field, viz.,

curl grad $\xi = 0$ and   **div curl v** $= 0$

[3.6] The Ring of Cochains

We have seen that the cochains on a complex are homomorphisms from one graded **Abelian** group C_* to another C^*. That is to say they behave like linear functionals, say f, g, with the property $(f + g)(x) = f(x) + g(x)$ when each function maps into a group **G** and each behaves like $f(x + y) = f(x) + f(y)$.

The first shows that the set of functions $\{f\}$ together with the "plus sign" themselves form an additive group - our group of cochains.

If we now contemplate extending the image group **G** by allowing it to be an algebraic **ring** **R** (such as all the integers **J**) then we can introduce a "product" into the set of cochains and thus obtain the **ring of cochains**, say $\{C^*, \cup\}$, where $\cup$ is usually read as **cup**, the resulting "product" being the **cup-product**. Furthermore the identity in this ring will be the 0-chain e^0 which takes the value 1 on every vertex in the complex **K**.

The cup-product $\cup$ is a bilinear functional, with values in **R**, and is **associative**.

The result of the cup-product of a c^p with a c^q is to obey the rule

$$c^p \cup c^q \in C^{p+q}$$

and if $\sigma_{p+q} = \langle v_1 v_2 ... v_p v_{p+1} ... v_{p+q}\rangle$ is a typical simplex in **K** the cup-product is defined by

$$c^p \cup c^q (\sigma_{p+q}) = c^p\langle v_1 v_2 ... v_p\rangle * c^q\langle v_p v_{p+1} ... v_{p+q}\rangle$$

the * denoting the product in the ring **R**.

Behaviour under the coboundary operator is to be ruled by

$$\delta(c^p \cup c^q) = \delta c^p \cup c^q + (-1)^p c^p \cup \delta c^q \qquad (A)$$

and this results in us being able to show that the product induced in the Homology group H^* gives

(cocycle) $\cup$ (cocycle) = cocycle

(cocycle) $\cup$ (coboundary) = coboundary

In this sense then we have constructed the **cohomology ring**, say $R(H^*)$.

PART 2 - OBJECTS ARE JUST HOLES IN SPACE

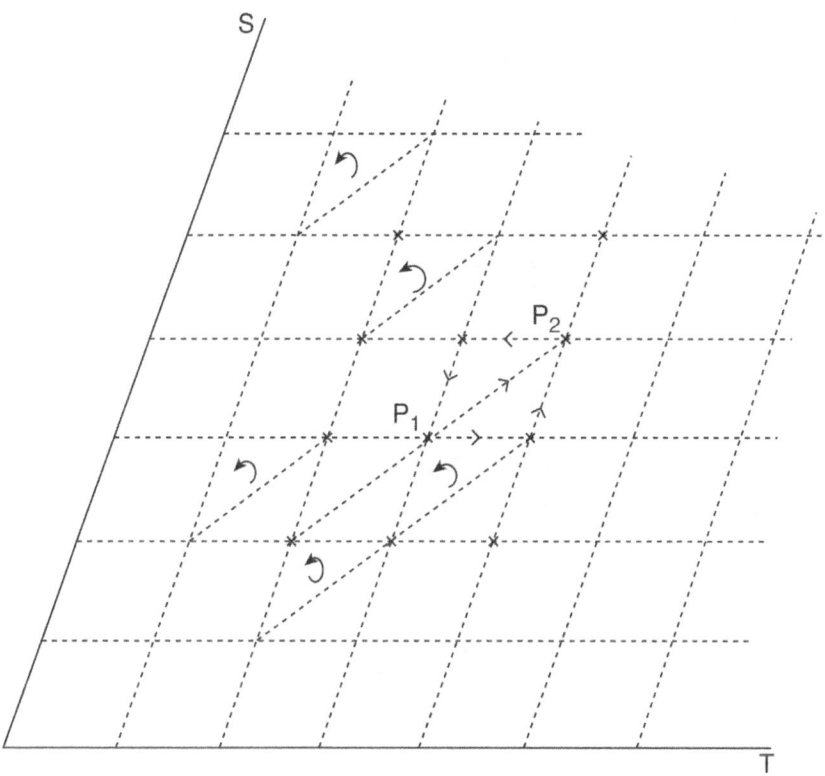

Mesh of "Points": S × T

Triangulation of the "Plane" with Orientation

Fig 3.1

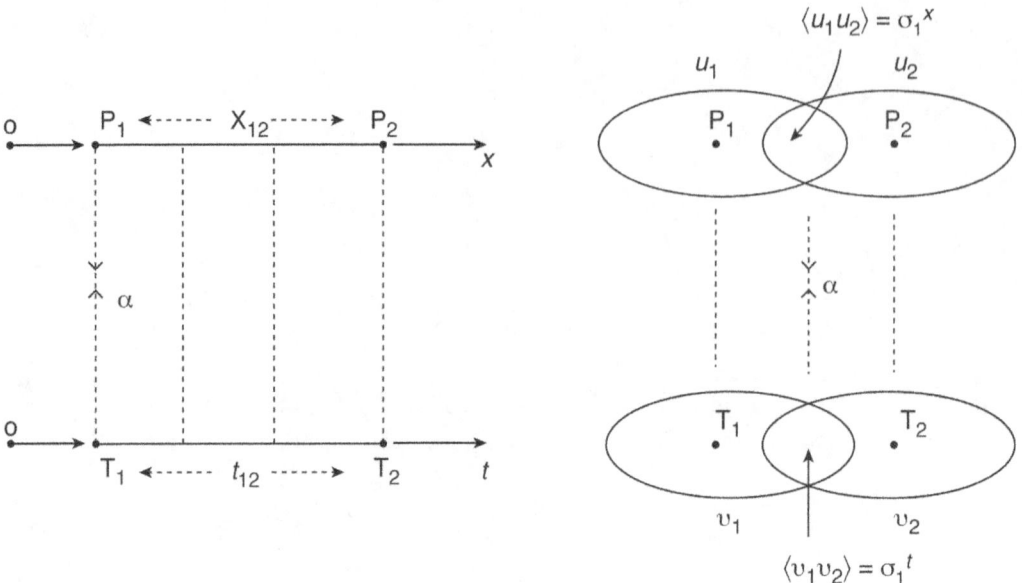

Bijection α "orders" position x-values in Čech cover of measures gives simplices $\langle u_1 u_2 \rangle$ & $\langle v_1 v_2 \rangle$ which correspond to x-interval σ_1^x & t-interval σ_1^t

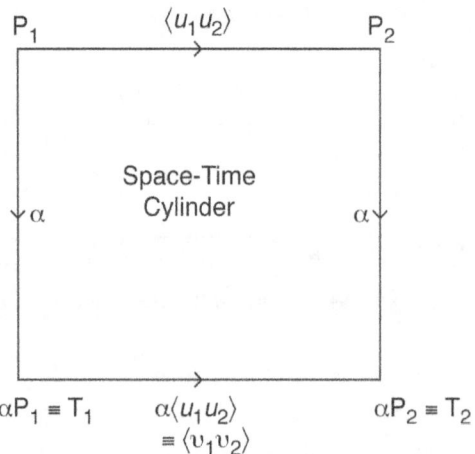

Represents a CW-complex covering velocity-space

Fig 3.2

Chapter-4 /PART-2   The CoCycle Law in Physics

[4.1] Heisenberg Uncertainty Principle & Measures in a Čech Complex

The usual argument in favour of the "Uncertainty Principle" is as follows.

Using the **Heisenberg** matrix mechanics in a general **Hamiltonian** dynamical system we take, as usual (see Chapter-4 in Part-1), the canonically conjugate variables **p** and **q** with matrix representations (in **Hilbert** space) $\check{P}$ and $\check{Q}$. Since they do not commute these matrices cannot be simultaneously reduced to diagonal form - so the eigenvalues for these variables cannot be simultaneously observed (that is to say, measured in $\Re^{\#}$).

If one measure (of, say **q**) is sharp then the other (of **p**) can only appear as an average with a probability distribution. We therefore let the "uncertainty" of a variable be the **root mean square** (deviation from the mean) of the possible values.

If ψ is the appropriate wave function (eigenfunction) for the variables then we write the average values (expectation values) as

$$\overline{p} = \int_\Omega \psi^* \check{P} \psi \, dq \quad \text{and} \quad \overline{q} = \int_\Omega \psi^* \check{Q} \psi \, dq$$

and the uncertainties will be Δp, Δq where

$$(\Delta p)^2 = \int_\Omega \psi^* (\check{P} - \overline{p})^2 \psi \, dq$$
$$= \int_\Omega \psi^* (\check{P} - \overline{p}I)^2 \psi \, dq \quad (A)$$

and similarly $(\Delta q)^2 = \int_\Omega \psi^* (\check{Q} - \overline{q}I)^2 \, dq \quad (B)$

where **I** denotes the unit matrix in the **Heisenberg** matrix meachanics.

Now if we write $\int_\Omega (\check{P}\psi)^* (\check{Q}\psi) \, dq$ as $a + ib \quad a,b \in \Re^{\#}$
together with $\int_\Omega \psi^* \check{Q}\check{P} \, dq$ as $a - ib$

then, by using the **Jordan** commutation relation, it follows that

$$\int_\Omega \psi^* (\hbar I) \psi \, dq \quad \text{becomes} \quad -2b$$

Thence   (1/4) (lhs of the above) = b^2 which is $\leq a^2 + b^2$

and then the lhs is $\leq \int_\Omega \psi^* \check{P}^2 \psi \, dq$ **times** $\int_\Omega \psi^* \check{Q}^2 \psi \, dq$

But using results (A) and (B) above

we deduce $(\Delta p)^2 (\Delta q)^2 \geq (h/4\pi)^2 \left[\int_\Omega \psi^* \psi \, dq \right]^2$

giving $\Delta P . \Delta q \geq h/4\pi \quad (C)$

The discussion equally applies to "simultaneous" measures of energy **E** and time **t**, so that $\Delta E . \Delta t \geq h/4\pi \quad (C')$

[4.2] Non-Hausdorf space in Physics Measures

The **Heisenberg Uncertainty Principle** says that, for example, simultaneous measures of **momentum p** and **position q** cannot both be sharp. For example, if we want $\Delta p = 0$ it means that $\Delta q \to \infty$. But this means, in the above topological space, that the values of **p** and **q** cannot be separated in the topology. But this is the condition that the **topological space of measurements** <u>cannot be</u> a **Hausdorff space**.

The **Heisenberg** formula, viz.

$$\Delta p \Delta q = \hbar/2 \; (= h/4\pi)$$

really expresses a relation between open sets $\{p\}$, $\{q\}$, $\{\hbar\}$ which states

$$\{p\} \cap \{q\} = \{\hbar\}$$

where $\{\hbar\}$ is not the empty set.

In the language of topology a **Hausdorff space** is called a T_2-space, that is to say that given two distinct points P, Q we can always find open sets A,B such that

$$P \in A, \; Q \in B, \text{ and where } A \cap B \text{ is the empty set (written as } \square \text{)}$$

To cope with our measures in Physics we can only always rely on a weaker separation axiom, viz. that of a T_1-space.

In such a space we only require, of the above sets A,B, that

$$P \in A, \; Q \text{ not-}\in A \text{ whilst } P \text{ not-}\in B, \; Q \in B$$

and A and B need not be disjoint (c.f a T_2-space).

Fortunately, for Physics measures in **Euclidean** space, every metric space is a T_1-space (although it is also a **Hausdorff** space).

[4.3] A Theorem due to Laurent ; Laurent expansion

We shall see that a **Laurent** series in the complex plane, $\mathbb{C}$, gives us an insight into the question of the **Cocycle Law** in Physics, when we come to look at its role in a Field Theory. To this end we remind ourselves of the **Laurent** expansion of a function $f(\zeta)$ in the complex plane.

We use the notation $\zeta = \xi + i\eta$ with $\xi, \eta \in \Re^{\#}$ for the complex variable.

Laurent's theorem states that if the function f is analytic throughout the annulus between two concentric circles, with centre $\zeta = a$, then in this region **R** f can be expanded as

$$f(\zeta) = A_0 + A_1(\zeta - a) + A_2(\zeta - a)^2 + \ldots + A_n(\zeta - a)^n + \ldots$$
$$A_{-1}(\zeta - a)^{-1} + A_{-2}(\zeta - a)^{-2} + \ldots + A_{-m}(\zeta - a)^{-m} + \ldots$$

where the coefficients $A_r \in \mathbb{C}$. (v. Appendix-E [36] [37])

When $m \to \infty$ the point $\zeta = a$ is called an **essential singularity**, otherwise it is called a **pole of order m** ; when $m = 1$ it is called a **simple pole**.

The coefiicients A_r are given by the formula

$$A_r = (1/2\pi i) \oint_C f(\zeta) (\zeta - a)^{-(r+1)} d\zeta$$

where C is any closed contour in **R** and where $r = 0, \pm 1, \pm 2, \ldots$

Each A_{-r} can be found by evaluating the limit, as $\zeta \to a$, of the expression

$$A_{-r} = \lim [1/(r-1)!] \, D^{r-1} \{(\zeta - a)^r f(\zeta)\} \; ; D \equiv d/d\zeta, \; r \neq 0$$

The particular coefficient A_{-1} is called the **residue** of $f(\zeta)$ at $\zeta = a$ and is given by the limit, as $\zeta \to a$, viz., $\quad A_{-1} = \lim \{(\zeta - a) f(\zeta)\}$.

The region **R** in which a function $f(\zeta)$ is analytic can circumnavigate a singularity by constructing a closed curve as C^+, shown in Fig. 4.1 . As $\varepsilon \to 0$ the curve C^+ becomes the simple circle shown as C_0, centred at the singularity $\zeta = a$.

[4.4] Laplace's equation, $\nabla^2 V = 0$,

We can derive this equation as a cocycle via a $\Gamma(1\text{-}1)$ relation.

Reminding ourselves of a $\Gamma(1\text{-}1)$ in our earlier discussion of a Physics (v. PART-1) we place the $\Gamma(1\text{-}1)$ in the complex field $\mathcal{C}$ and write it in the commonly occuring form associated with the ubiquitous potential field V and the inverse square law viz.,

$$\Gamma(1\text{-}1) \equiv Vr - aV - br - c = 0$$

But this gives the V-function as

$$V = (br + c)/(r - a)$$

which in turn becomes

$$V = b + (c + ab)/(r - a) \qquad (1)$$

a simple example of a **Laurent series**, with $A_0 = b$ and $A_{-1} = (c + ab)$.

The fact that when the homography is non-degenerate the matrix

$$M = \begin{bmatrix} 1 & -a \\ -b & -c \end{bmatrix} \neq 0$$

means that the **residue** (which is $A_{-1} = -\det(M)$, at the **pole** $r = a$ is $\neq 0$.

If we consider the circle C_0 as covered by observations - either through a standard triangulation (as in E^2) or via a Čech complex (v. Fig. 4.1) - then we can regard cochains on such a complex as **integrals** over appropriate chains, and since, eg , the boundary **C** will be a 1-chain so a 1-cochain on it will be an integral of the form

188   MATHEMATICAL PHYSICS

$$c^1 = \oint_C \mathbf{grad}(V(\zeta)) \cdot \underline{ds} \quad ; \underline{ds} \text{ being a vector piece of arc,}$$

and this immediately reminds us of the **residue** A_{-1} in the **Laurent** expansion of the $\Gamma(1\text{-}1)$ in (1) above.

This c^1 is determined by the vector $\mathbf{grad}\,(V)$ and so its **coboundary** δc^1 will be defined on a c_2 (the area of the region denoted by **R**) and where

$$(\delta c^1, c_2) = (c^1, \partial c_2)$$

Taking this boundary operator, which sends c^1 to a c^2, as determined by **div** applied to a vector field (v. Chapter-3 §[3.5]) we can write

$$\delta c^1 = \mathbf{div} \oint_C \mathbf{grad}\,(V) \cdot \underline{ds}$$
$$= \oint_C \mathbf{div}\,\mathbf{grad}\,(V) \cdot \underline{ds}$$

If this is to be a 1-cocycle $z^1 \in Z^1$ in the cohomology structure then we require that, at all points in **R**, **Laplace's** equation holds - which means

$$\nabla^2 V \equiv \mathbf{div}\,\mathbf{grad}\,V = 0$$

This will be a **Law in the Physics** - the **Cocycle Law**.

[4.5] Poisson's equation, $\nabla^2 V = 4\pi\rho$, as a coboundary statement

We introduce a scalar function $\rho(x,y,z)$ defined throughout a closed region Ω of $\mathbf{E}^3$ which is to represent a **charge/mass density function**. Assuming that we have some suitable covering of Ω (eg a **Euclidean** triangulation, or a Nerve provided by a Čech cover) then we can consider a 3-cochain, say c^3, defined throughout Ω.

This c^3 can be taken as the integral determined by ρ, viz.,

$$c^3 = \int_\Omega \rho\, d\Omega$$

and since we can take a σ_3 as homeomorphic to a sphere S^3 we can, for purposes of calculation, take the volume element $d\Omega$ as $d\{(4/3)\pi r^3\}$ - which is $4\pi r^2\, dr$.

It follows that $\quad c^3 = \int_\Omega \rho\, d\Omega = \int_\Omega 4\pi r^2 \rho\, dr$

Since the inverse-square law of force (for either the electrostatic/magnetic or gravitational fields) requires us to regard the "charge" as a "point-charge" we must take the charge as being centred in the small sphere. We can then incorporate this law of force by noting that on the surface of $d\Omega$ we get the **force-field** as determining the following c^3, viz.,

$$c^3 = \int_\Omega 4\pi \rho\, dr$$

But if the region contains the potential function $V(r)$ the force-field of attraction at the point $P(r)$ will be $\partial V/\partial r$ - generally **grad V**.

This will determine a 2-cochain, a c^2 say, which can be written as

$$c^2 = \int_S \mathbf{grad\ V} \cdot \mathbf{dS} \qquad S \text{ being the surface of } \mathbf{S}^3$$

Now we notice that if c^3 is in fact a **coboundary**, that is to say

$$c^3 = \delta c^2 = \int_\Omega \mathbf{div\ grad\ V}\ d\Omega$$

then, as in [4.5] above, we have throughout Ω

$$\int_\Omega \{c^3 - \delta c^2\}\ d\Omega = 0$$

giving the local relation

$$\nabla^2 V = 4\pi\rho$$

being the well-known **Poisson's equation**.

In these cases of the **Cocycle Law** we see the distinction, viz.,

  (a) **Laplace's equation** demonstrates the presence of a "true" p-cocycle, whilst

  (b) **Poisson's equation** demonstrates (only) the presence of a p-coboundary

The result (b) contains within it the assumption of the inverse-square law of the forces of attraction/repulsion in both the electrostatic and the gravitational fields of force - but also in many situations to be found in fluid mechanics (v. PART-1 for illustrations).

[4.6] Cocycles via the use of the Calculus of Variations

Referring to Fig. 4.2, and eg Appendix-E [8], we consider **Euler's** conditions for the integral $\quad I(\text{arc}) = \int_{\text{arc}} F(t, y, \dot{y})\ dt \quad$ to be <u>stationary</u> ; that is to say to find the arc between A and B for which this is true. Using δ_ε as the incremental operator on $\mathbf{I}$ (not to be confused with our coboundary operator δ !) we use the **Euler** equations for which $\quad \delta_\varepsilon\ I(\text{arc}) = 0 \quad$ neglecting powers of δ_ε beyond the first.

If the resulting arc, in the t-y plane, has the equation $\ y = \gamma(t)$ then **Euler's** analysis tells us that it satisfies the condition

$$\partial F / \partial y - D_t(\partial F / \partial \dot{y}) = 0 \qquad (E)$$

where D_t means the derivative d/dt.

In Fig. 4.2 this condition (E) is obtained by considering the incremental change brought about in the integral $\mathbf{I}$ as the arc of integration is changed from γ to a neighbouring arc, such as α. [The straightforward details are to be found in, eg, Appendix-E [8]

Now we naturally regard this problem to be embedded in a space homeomorphic to a triangulated piece of the standard **Euclidean** $\mathbf{E}^n$, for some finite n.

Then the arcs α and γ form the boundary of a 2-cell, say μ, and allowing for the obvious orientation shown in the diagram we have

$$\partial\{\mu\} = \alpha - \gamma \qquad (F)$$

Since the integral **I** is a 1-cochain determined by the function F we write it as c^1.

Then the coboundary of this c^1 is a c^2 ($= \delta c^1$) where, using (F),

$$(\delta c^1, \mu) = (c^1, \partial \mu) = (c^1, \alpha - \gamma)$$

But this is the incremental difference $I(\alpha) - I(\gamma)$ which is to vanish.

Since this is to apply to all such incremental differences we must have

$$\delta c^1 = 0 \qquad \text{the Cocycle Law.}$$

So the cocycle law, applied in this space (characterised by the function F), is equivalent to a **Law in Physics** given by the **Euler** equations of the form (E).

[4.7] Lagrange's equations for a general dynamical system

This classic result follows from **Hamilton's Principle** (Ref. Appendix-E [9]).

Given the **Lagrange** function $\quad L = T - V$

where **T** is the kinetic energy and **V** is the potential function and where the co-ordinates of the system are denoted by q_r, $r = 1 \ldots n$.

Then the behaviour of the system is given by the equations

$$\partial L / \delta q_r - D_t(\partial L / \partial \dot{q}_r) = 0 \qquad \text{for } r = 1 \ldots n$$

These are therefore the result of asserting that **the function L determines n 1-cocycles** in the system, viz., $\quad \delta\{\int_\gamma L \, dt\} = 0$

[4.8] Newton's Law of Motion

This follows from **Hamilton's Principle**, as in [4.7], by considering a particle of mass m moving along the x-axis under a force $X = -\partial V/\partial x$, the equation of motion via the functions $\quad T = (m/2)\dot{x}^2$, $L = T - V \quad$ and so

$$\partial(-V)/\partial x - D_t(m\dot{x}^2) = 0$$

which is simply $\quad \underline{m\ddot{x} = X}$

[4.9] Hamilton's Equations for a dynamical system

These equations are a natural consequence of the **Lagrange** equations already found.

Suppose there are n independent co-ordinates for the sytem, viz., $q_1, q_2, \ldots q_n$ so that the **Lagrange** function $L = L(q_1 \ldots q_n, \dot{q}_1 \ldots \dot{q}_n)$. Then define a set of n momentum co-ordinates $p_1, p_2 \ldots p_n$ by

$$p_r = \partial L/\partial \dot{q}_r \qquad (1)$$

and the **Hamiltonian** function **H** by $\quad H = \Sigma_r p_r \dot{q}_r - L$.

Since H is equally a function of the $\dot{q}_r$'s we get the following

(a) $\partial H/\partial p_r = \dot{q}_r + \Sigma_i p_i(\partial \dot{q}_i/\partial p_r) - \Sigma_i(\partial L/\partial \dot{q}_i)\partial \dot{q}_i/\partial p_r$

which by (1) reduces to $\quad \partial H/\partial p_r = \dot{q}_r \qquad (2)$

(b) $\partial H/\partial q_r$ becomes $- \partial L/\partial q_r$ which by (1) is $- d/dt(\partial L/\partial \dot{q}_r)$

so that $\qquad \partial H/\partial q_r = - \dot{p}_r \qquad (3)$

(2) and (3) being the canonical **Hamilton's** equations.

For a conservative system with potential function $V(q_1, q_2 \ldots q_n)$ we have

$$\Sigma_r p_r \dot{q}_r = \Sigma_r \partial T/\dot{q}_r \, \dot{q}_r = 2T$$

where **T** is the kinetic energy and where we appeal to **Euler's** theorem on homogeneous functions (in this case of degree 2).

This results in $\quad H = 2T - (T - V) = T + V$

so that **H** is a measure of the total energy of the system (at any given time).

[Note : A full discussion of the **Hamiltonian** function will be found in Appendix-E [9]]

Other illustrations of the significance of our treatment of the Calculus of Variations can be found in, eg, Ref. Appendix-E [8] and include such topics as

  (a) problems in Statics

  (b) laws in geometrical optics

  (c) laws of planetary motions

  (d) geodesics in general - eg in the **Lorentz** 4-space (x,y,z,ct).

and in most of the kernel of classical Physics.

A region R in the complex plane – avoiding singularity at $z = a$

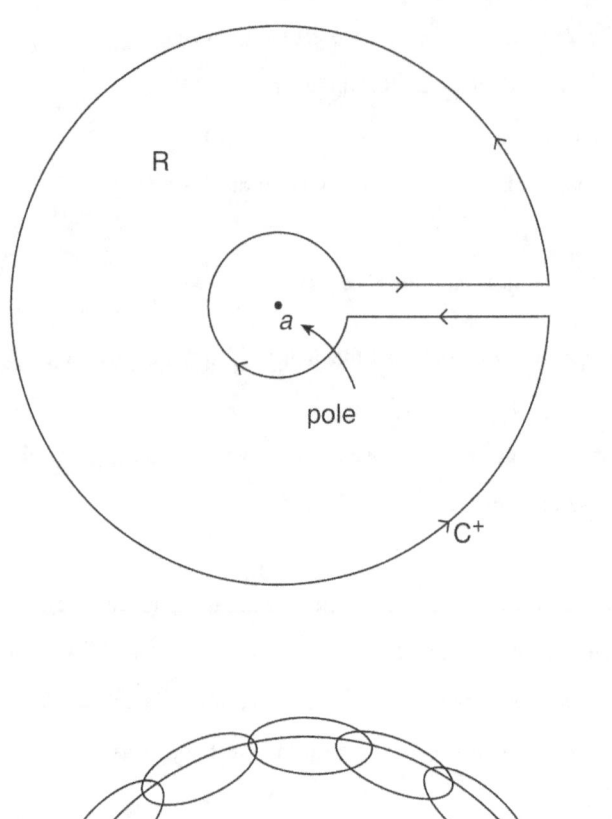

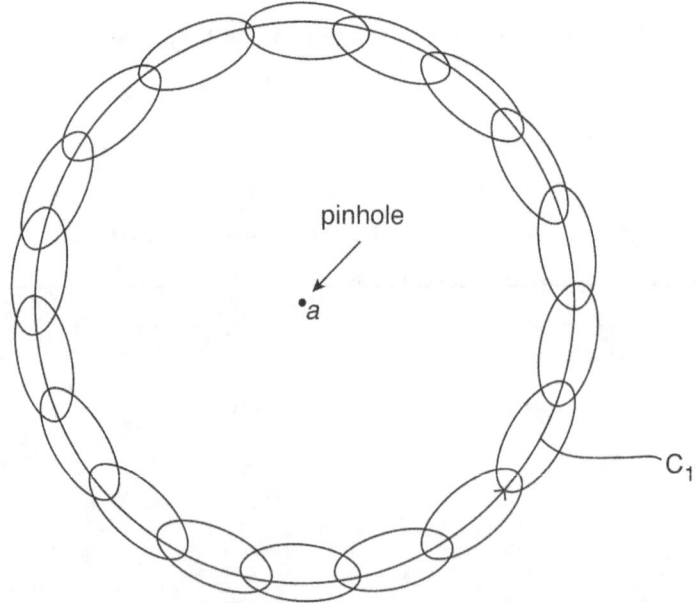

A possible Čech cover via physical measures of contour/circle C_1

Fig 4.1

Euler's condition for the integral $I_\gamma = \int_\gamma F(t,y,\dot y)dt$ **to be stationary**

[denoting this by $\delta_\varepsilon I_\gamma = 0$ and where $\dot y$ denotes dy/dt]

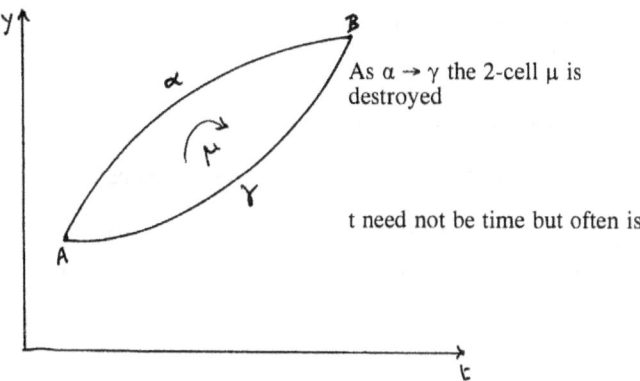

As $\alpha \to \gamma$ the 2-cell μ is destroyed

t need not be time but often is

Let the curve neighbouring to γ be the curve α - so that we can take its definition as
$$y = \gamma(t) + \varepsilon\beta(t) \text{ such that } \beta(A) = \beta(B) = 0 \text{ and let } \varepsilon \to 0$$
Then $\delta_\varepsilon I_\gamma = \int_A^B [F(t,\gamma+\varepsilon\beta,\dot\gamma+\varepsilon\dot\beta) - F(t,\gamma,\dot\gamma)]dt$ which, by **Cauchy's** Mean Value Theorem reduces to $\varepsilon \int_A^B \{\beta\partial F/\partial\gamma + \dot\beta\partial F/\partial\dot\gamma\}dt + O(\varepsilon^2)$
Integraton by parts gives $\delta_\varepsilon I_\gamma = \varepsilon \int_A^B \beta E dt + O(\varepsilon^2)$ where E denotes
$$E = \partial F/\partial\gamma - d/dt\{\partial F/\partial\dot\gamma\}$$
and so to the first order of approximation we deduce that I_γ is stationary on the curve γ when it satisfies the condition $\qquad \partial F/\partial\gamma - d/dt(\partial F/\partial\dot\gamma) = 0$

Fig 4.2

Chapter-5 /PART-2 CoCycles in Exterior Algebra, Λ^n

[5.1] A $\Gamma(1\text{-}1)$ as a cocycle

We have already shown what a powerful role a $\Gamma(1\text{-}1)$ can plays in what we might call a "finite/macro Physics", but now we can see how it finds a place as a cocycle - thus emphasising the role of our Cocycle Law.

The general homography over a field **F**, viz.,

$$axy + bx + cy + d = 0$$

with a,b,c,d $\in$ **F**, can be written, by using homogeneous coordinates, as

$$\Gamma(1\text{-}1) \equiv axy + bxs + cyt + dst = 0 \qquad (A)$$

in which x,y have been replaced by x/t, y/s respectively. This can be characterised by a 2×2 matrix **M**

$$\mathbf{M} = \begin{bmatrix} a & b \\ c & d \end{bmatrix}$$

since (A) is a scalar condition on a 2-dimensional vector space, say, $\mathbf{V}_2$, viz.

$$\Gamma(1\text{-}1) \equiv \mathbf{XMY}^T = 0 \qquad (B)$$

with $\mathbf{X} = (x,t)$ and $\mathbf{Y} = (y,s)$, the superscript T denoting the transpose of the vector so that $\mathbf{Y}^T$ becomes a column vector. The condition for (A) to be a non-degenerate $\Gamma(1\text{-}1)$ is that $\det(\mathbf{M}) = ad - cb \neq 0$.

We can now associate the homography (A) with an **exterior product space** $\Lambda \mathbf{V}_2$

$$\Lambda \mathbf{V}_2 = \Lambda^0 \mathbf{V}_2 \oplus \Lambda^1 \mathbf{V}_2 \oplus \Lambda^2 \mathbf{V}_2$$

in which $\Lambda^0 \mathbf{V}_2 = \mathbf{F}$.

If $\mathbf{V}_2$ is spanned by the vectors $\hat{e}_1 = (1\ 0)$ and $\hat{e}_2 = (0\ 1)$ then

$\Lambda^1 \mathbf{V}_2 = \mathbf{V}_2$ and $\Lambda^2 \mathbf{V}_2$ is 1-dimensional with basis $\hat{e}_1 \wedge \hat{e}_2$

The elements of $\Lambda^2 \mathbf{V}_2$ are **pseudoscalars** so that the symbol "0" in (A) may be viewed as $0 \in \Lambda^0 \mathbf{V}_2$ or as $0 \in \Lambda^2 \mathbf{V}_2$; zero is either a **scalar** or a **pseudoscalar**.

If we take $0 \in \Lambda^0 \mathbf{V}_2$ we can express the homography (A) as a 1-cycle statement in an homology theory based on the boundary operator ∂ in

$$\Lambda^0 \mathbf{V}_2 \xleftarrow{\partial_1} \Lambda^1 \mathbf{V}_2 \xleftarrow{\partial_2} \Lambda^2 \mathbf{V}_2$$

with $\partial_1(\partial_2) = 0$ (nilpotency).

For this to be so we define ∂_1 as

$$\partial_1 \equiv \mathbf{XM} = (ax+ct,\ bx+dt)$$

so that   (A) can expressed by the statement

$$\text{"}Y^T \text{ is a 1-cycle, } \partial_1 Y^T = 0\text{"} \quad \text{(scalar)}$$

provided ∂_2 is defined to ensure $\partial_1(\partial_2) = 0$.

This follows by taking the usual

$$\partial_2(\theta^1 \wedge \varphi^1) = \partial_1(\theta^1) \wedge \varphi^1 - \theta^1 \wedge \partial_1(\varphi^1)$$

where $\theta^1, \varphi^1 \in \Lambda^1 V_2$.

The homography in effect says :

> **Given a vector X the corresponding Y is such that Y^T is a 1-cycle in the homology on ΛV_2 with boundary operators ∂_1, ∂_2 defined as above.**

For the **cohomology representation** we first define the matrix M^*

by $\quad MM^* = (\det M)\begin{bmatrix} 0 & -1 \\ 1 & 0 \end{bmatrix}$

so that M^* "sees" the space ΛV_2 from the "other end", the matrix

$$\begin{bmatrix} 0 & -1 \\ 1 & 0 \end{bmatrix}$$

being a representation of the basis $\hat{e}_1 \wedge \hat{e}_2$.

This gives $\quad M^* = \begin{bmatrix} -b & -d \\ a & c \end{bmatrix}$

and we then define a coboundary operator δ_1 by

$$\delta_1 = (M^* X^T) \wedge$$

whence (A) becomes $\quad \delta_1 Y \equiv (M^* X^T) \wedge Y = 0$

Taking δ_0 to be
$$\delta_0 : k \to k(M^* X^T) \quad k \in F$$

ensures $\delta_1(\delta_0) = 0$ the zero map.

Now (A) is equivalent to the statement

> **Given a co-vector X^T the corresponding co-vector Y is a 1-cocycle in the cohomology on ΛV_2 with coboundary operators as above.**

[5.2] Cocycle laws in De Rham Cohomology

de Rham cohomology has already shown itself to be another suitable candidate for this sort of analysis, since it is based on the **exterior algebra** Λ^n.

The graded bases, Λ^p, are vector spaces of dimensions ${}^nC_p$ and may be considered to be spanned by symbols such as (typically) $u_1 \wedge u_2 \wedge u_3 \wedge \ldots \wedge u_p$ or, in this case by expressions like $dq_1 \wedge dq_2 \wedge dq_3 \wedge \ldots \wedge dq_p$, and the theory is peculiarly designed for being applicable to differentiable manifolds (spaces which are locally identifiable with pieces of **Euclidean** space E^k - and which have a well-defined **tangent space** at each point of itself).

The **de Rham** theory deals with what are called **differential forms**, being expressions of the form (eg) $a_1(q)dq_1 + a_2(q) + \ldots a_p d(q)$ this being a p-form, and where the $a_r(q)$ are suitably differentiable functions of the components q_i of the vector **q**.

Whilst such p-forms consist of functions associated with the "p-simplices" defined in the wedge space Λ^n they do not behave like homomorphisms on an additive group of the simplices - unless, eg. the functions $a_i(q)$ are identical - although we can contemplate forming "proper" cochains by evaluating suitable **integrals** over pieces of the underlying space.

These p-forms yet play a role in the **de Rham** theory which is exactly parallel to that of cochains on a set of simplices K, for they are defined in a graded direct sum of spaces Λ^p which possesses a nilpotent "boundary operator" $\delta : \Lambda^p \to \Lambda^{(p+1)}$.

Because of this correspondence the analysis is referred to as **de Rham Cohomology**. One of these bases in Λ^p consists of a linear set of these "(p-1)-simplices" - and there are ^nC_p of them in the subspace Λ^p, p taking the values $0, 1, 2, \ldots n$, and where $^nC_0$ consists simply of suitable functions, eg., $f(q_1, \ldots q_n)$, these taking values in the field (or ring) which constitutes the range of the mappings.

The **co-boundary operator** in the **de Rham** cohomology is the one already introduced, viz., that defined by $\quad \delta^p : C^p \to C^{(p+1)}$

with $\quad\quad\quad \delta c^0 \equiv dc^0 \quad\quad$ "d" denoting (calculus) differentiation

and $\quad\quad\quad d^2 c_0 = 0. \quad\quad$ nilpotency.

In detail, and comparing it with the definition (A) in §[3.6] above, we define this operator in such a way that, eg.,

If λ is a p-form and μ is a q-form then

$$\delta(\lambda \wedge \mu) = \delta(\lambda) \wedge \mu + (-1)^p \lambda \wedge (\delta\mu) \quad\quad (B)$$

For example, taking the simple 1-form $\lambda = a_1(x,y,z)dx$ (B) will give us

$$\delta(\lambda) = [(\partial a/\partial x)dx + (\partial a/\partial)dy + \partial a/\partial z)dz]dx + 0 \quad (\text{since } d(dx) = 0)$$

which is the 2-form $(\partial a/\partial y)dy \wedge dx + (\partial a/\partial z)dz \wedge dx \quad$ (since eg $dx \wedge dx = 0$).

Then corresponding to the idea of a formal cocycle we consider p-forms, λ^p, which are such that $\quad \delta\lambda^p = 0 \quad$ (referred to in the literature as **closed** forms)

and those for which a (p-1)-form, μ, exists and where

$$\lambda = \delta\mu \quad \text{(referred to in the literature as \textbf{exact} forms).}$$

With this understanding we shall refer to these as **cocycles** and **coboundaries**. The cohomology groups H^p naturally being defined as the factor groups Z^p/B^p.

[5.3] Hamilton's equations

For a dynamical system which is defined by n independent co-ordinates q_r and their corresponding n independent momenta p_r follow by identifying a "**de Rham**" cocycle, viz., a 1-form λ where

$$\lambda = \Sigma_r(p_r dq_r + q_r dp_r) - H(q,p,t)dt \quad \text{(t being time)}$$

H being the **Hamiltonian** function for the holonomic system.

For, to illustrate the calculations, we take the case of n = 2, and so

$$\lambda = (p_1 dq_1 + q_1 dp_1) + (p_2 dq_2 + q_2 dp_2) - H(q_1,q_2,p_1,p_2,t)$$

and recalling the definition of the p_r, viz.,

$$p_r = \partial L/\partial \dot{q}_r = \partial T/\partial \dot{q}_r$$

Also the independence of the p's and q's ensures that $\partial p_r/\partial q_r = \partial q_r/\partial p_r = 0$ for all r.

Then we have

$$\delta\lambda = [(\partial p_1/\partial t\, dt) \wedge dq_1 + \partial q_1/\partial t) \wedge dp_1)] + \text{(similarly for } q_2 \,\&\, p_2)$$
$$- (\partial H/\partial q_1\, dq_1) \wedge dq_1 - (\partial H/\partial p_1\, dp_1) \wedge dp_1 +$$
$$\text{(similarly for } q_2 \,\&\, p_2) - \partial H/\partial t\, dt \wedge dt.$$

If this λ is a "**de Rham**" cocycle it must vanish, so

$$(\dot{p}_1 - \partial H/\partial t)\, dt \wedge dq_1 + (\dot{p}_2 - \partial H/\partial t)\, dt \wedge dq_2) +$$
$$(\dot{q}_1 + \partial H/\partial t)\, dt \wedge dp_1 + (\dot{q}_2 + \partial H/\partial t)\, dt \wedge dp_2 = 0 \text{ (in } \Lambda^2)$$

where we have used the properties: $du \wedge du = 0$ and $du \wedge dv = -dv \wedge du$.

These give the **Hamiltonian equations** viz.,

$$\dot{p}_r = \partial H/\partial t \quad \text{and} \quad \dot{q}_r = -\partial H/\partial t$$

via the **Cocycle Law** in the **de Rham** cohomology.

Furthermore, if we require that λ shall be a "**de Rham**" coboundary we can look for a function $-W(\mathbf{q},\mathbf{p},t)$ such that $\lambda = \delta W$, that is to say, eg,

$$p_1 dq_1 + q_1 dp_1 - Hdt$$
$$\equiv \partial W/\partial q_1\, dq_1 + \partial W/\partial p_1\, dp_1 + \partial W/\partial t\, dt$$

So then $\quad p_1 = \partial W/\partial q_1 \quad q_1 = \partial W/\partial p_1 \quad \text{and} \quad H = -\partial W/\partial t$.

the equation $\quad \partial W/\partial t + H = 0 \quad$ beings the well-known **Hamilton-Jacobi** equation.

[5.4] Maxwell equations for the electromagnetic field.

Maxwell field equations for the electromagnetic field, in cgs/emu, are :

$$\text{div } \mathbf{D} = 4\pi\rho$$

$$\text{div } \mathbf{B} = 0$$

$$\text{curl } \mathbf{H} = 4\pi \mathbf{J} + \partial \mathbf{D}/c\partial t \; ; \quad \mathbf{H} = \text{magnetic field vector}$$

$$\text{curl } \mathbf{E} = - \partial \mathbf{B}/c\partial t$$

together with the subsidiary relations

$$\mathbf{D} = \kappa \mathbf{E} \qquad \mathbf{D} = \text{electric displacement vector}$$

$$\mathbf{B} = \mu \mathbf{H} \qquad \mathbf{B} = \text{magnetic induction vector}$$

$$\mathbf{J} = \sigma \mathbf{E} \qquad \mathbf{J} = \text{electric current vector}$$

$$\mathbf{B} = \text{curl } \mathbf{A} \qquad \mathbf{A} = \text{magnetic potential vector}$$

$$\mathbf{E} = - \partial \mathbf{A}/c\partial t - \text{\textbf{grad} } V \text{ where}$$
$$\mathbf{E} = \text{electric field vector \&}$$
$$V = \text{electric potential function}$$

$$\text{\textbf{div} } \mathbf{A} = -(\kappa\mu/c) \partial V/\partial t - 4\pi\sigma\mu V \text{ where}$$
$$\kappa = \text{dielectric constant \&}$$
$$\mu = \text{magnetic permeability}$$
$$\text{for the medium concerned.}$$

Now if we introduce 2-forms α, β, γ as follows, viz.,

$$\alpha = (E_1 dx \ldots E_3 dz)cdt + (B_1 dydz \ldots B_3 dxdy)$$

$$\beta = - (H_1 dx \ldots H_3 dz)(cdt) + (D_1 dydz \ldots D_3 dxdy)$$

$$\gamma = (J_1 dydz \ldots J_3 dxdy) - \rho dxdydz$$

then the **cocycle statement** $\quad \delta\alpha = 0 \quad$ readily yields the equations

$$\text{\textbf{curl} } \mathbf{E} = -(1/c)\partial \mathbf{B}/\partial t \quad \text{and} \quad \text{\textbf{div} } \mathbf{B} = 0$$

whilst the **coboundary statment** $\quad 4\pi\gamma + \delta\beta = 0$

yields the equations

$$\text{\textbf{curl} } \mathbf{H} = (4\pi/c)\mathbf{J} + \partial \mathbf{D}/c\partial t) \quad \text{and} \quad \text{\textbf{div} } \mathbf{D} = 4\pi\rho$$

Thence $\quad \delta\gamma = 0$ yields $\quad \text{div } \mathbf{J} + \partial\rho/\partial t = 0$

Furthermore, if we seek a vector potential **A** (for the magnetic field) and its associated scale function A_0 then we obtain them by regarding α as a coboundary - so we look for a 1-form λ where $\quad \alpha = \delta\lambda$.

Putting $\quad \lambda = A_1 dx + A_2 dy + A_3 dz + A_0 cdt$

we get that the statement $\delta\lambda = \alpha$ implies

$$\text{\textbf{curl} } \mathbf{A} = \mathbf{B} \quad \text{and} \quad \text{\textbf{grad} } A_0 - \partial \mathbf{A}/c\partial t = \mathbf{E}$$

all of which amount to the general form of **Maxwell** equations in any medium.

In **free space** we must take

$$\mathbf{D} = \mathbf{E} \quad \mathbf{B} = \mathbf{H} \quad \mathbf{J} = 0 \text{ and } \rho = 0$$

giving the equations

$$\text{curl } \mathbf{E} = -\partial \mathbf{H}/c\partial t \quad \text{with } \text{div } \mathbf{E} = 0$$

and $\quad \text{curl } \mathbf{H} = \partial \mathbf{E}/c\partial t \quad \text{with } \text{div } \mathbf{H} = 0$

These in turn easily produce the equations for electromagnetic waves, viz., by appealing to the vector algebra expansion

$$\text{curl curl } \mathbf{F} = \text{grad div } \mathbf{F} - \nabla^2 \mathbf{F}$$

so that, eg, $\text{grad div } \mathbf{H} - \nabla^2 \mathbf{H} = -\nabla^2 \mathbf{H}$ (because $\text{div } \mathbf{H} = 0$) and this becomes $\partial(\text{curl } \mathbf{E})/c\partial t$ which in turns equals

$$-\partial^2 \mathbf{H}/c\partial^2 t$$

whence we get $\quad \nabla^2 \mathbf{H} = \partial^2 \mathbf{H}/c^2 \partial^2 t \quad$ the equation of **wave motion**.

The field equations likewise result in the same wave equation for **E**.

[5.5] The **Hodge** $*$-operator in Λ^n space.

When we are given a vector space, **L**, of dimension n and with an orthonormal basis of unit vectors $\{u_r ; r = 1 \ldots n\}$ then the graded "wedge space" $\Lambda^n \mathbf{L}$ is such that

$$\Lambda^n \mathbf{L} = \Lambda^0 \oplus \Lambda^1 \oplus \ldots \oplus \Lambda^p \oplus \Lambda^{p+1} \oplus \ldots \oplus \Lambda^n$$

and since ${}^n C_p = {}^n C_{n-p}$ it follows that $\dim \Lambda^p = \dim \Lambda^{n-p}$.

This suggests that there can be a mapping between members of Λ^p and those of Λ^{n-p}, and this is achieved by a mapping (epimorphism) introduced by **Hodge** and denoted by

$$*_p : \Lambda^p \to \Lambda^{n-p}$$

with homomorphic properties, viz.,

(i) $*(\alpha + \beta) = *\alpha + *\beta$ and when $f \in \Lambda^0 \quad *(f\alpha) = f(*\alpha)$

(ii) $**\alpha = *(*\alpha) = -\alpha$

(iii) $\alpha \wedge *\beta = \beta \wedge *\alpha$

(iv) $\alpha \wedge *\alpha = 0$ if and only if $\alpha = 0$

We normally drop the subscript "p" in $*_p$ when the context is obvious.

[Note : v. Appendix-E [40], [96], [97] for a fuller discussion of "$*$"]

We also require that there should be an **inner product** defined on the vector space **L** which will be a generalisation of the "scalar product" found in standard vector algebra over the **Euclidean** space E^3. Such an inner product is generated by a metric on the space **L**, and in a general **Riemannian** space with a metric tensor $\mathbf{g} = (g^{ij})$ the inner product (α, β) of two vectors α and β is taken as

$$(\alpha, \beta) = \Sigma_{i,j} g^{ij} \alpha_i \beta_j$$

the g^{ij} being found in the matrix $\mathbf{g}$.

This results, in the **Riemannian** space of n dimensions, in the metric form

$$ds^2 = \Sigma_{ij} g^{ij} dx_i dx_j$$

If σ = a basis for the space Λ^n then a basis σ_p for Λ^p will consist of all p-selections from the dx's in σ and so $*\sigma_p$ will be an (n-p)-form in Λ^{n-p}, which has a basis consisting of all (n-p)-selections, with attached factor, from the set difference $\sigma - \sigma_p$), the factor being taken from the inner product matrix $\mathbf{g}$.

Since we are only interested in the **Lorentz** 4-space, with co-ordinates x_i, $i = 1 \ldots 4$, we consider the case of $\mathbf{g}$ = unit matrix, viz.,

$$\mathbf{g} = (g^{ij}) \quad \text{where } g^{ij} = 1 \text{ with } i=j, \text{ otherwise } g^{ij} = 0.$$

But when we wish to express our results in the usual **Euclidean** co-ordindates x,y,z,t we must introduce the relations

$$x_1 = x, \quad x_2 = y, \quad x_3 = z, \quad x_4 = ict \quad \text{where } i^2 = -1.$$

Then the **Lorentz** metric is written as

$$ds^2 = dx^2 + dy^2 + dz^2 - c^2 dt^2$$

[Note : The 4-space introduced by **Minkowski** differs only in his ds^2 being minus(the above) to ensure that it is always +ve ; but the following analysis is equally applicable]

Restricting ourselves to the Physics associated with the **Lorentz** 4-space and taking the 1-form $\quad \lambda = a_1 dx_1 + a_2 dx_2 + a_3 dx_3 + a_4 dx_4$

then the **star-operator** will carry it over to a 3-form (n-1 = 3) which will be spanned by a basis of vectors like $dx_1 \wedge dx_2 \wedge dx_3$ - and cycling round for the others. [v.Fig. 5.1]

In this mapping under $*$, and using the properties listed above, we see eg that

$$*(a_i dx_i \text{ - in } \Lambda^1) \rightarrow a_i (*dx_i \text{ - in } \Lambda^3) \text{ which in turn is identified}$$

as $\quad (\pm 1) a_i dx_j \wedge dx_h \wedge dx_k \quad$ where (i,j,h,k) are in cyclic order

and similarly for the other terms in Λ^1 - the signs being taken from the matrix $\mathbf{g}$.

[5.6] Equations of **Wave Motion** in **Lorentz** 4-space.

We consider a scalar function $\psi(x_1,x_2,x_3,x_4) \in \Lambda^0$ in the **Lorentz** 4-space, together with the graded wedge space and its associated boundary operators (à la **de Rham**), viz.,

$$\Lambda^0 \oplus \xrightarrow{\delta_0} \Lambda^1 \oplus \xrightarrow{\delta_1} \Lambda^2 \oplus \xrightarrow{\delta_2} \Lambda^3 \oplus \xrightarrow{\delta_3} \Lambda^4$$

To show that we are operating in the 4-space we shall indicate the usual operators "grad" and "div" by **GRAD** and **DIV** respectively.

Then, eg, $\quad \delta\psi = \mathbf{GRAD}(\psi) \quad$ giving us a 1-form, λ, whose components will be

$$\partial\psi/\partial x_1, \ \partial\psi/\partial x_2, \ \partial\psi/\partial x_3, \ \partial\psi/\partial x_4$$

which we will deal with as

$$\psi_x, \ \psi_y, \ \psi_z, \ \psi_t \qquad (A)$$

but noting that since $x_4 = ict$ then for any function f we have

$$\partial f/\partial x_4 = (\partial f/\partial t) \cdot (\partial t/\partial x_4) \quad \text{which equals } (ic)^{-1} \cdot (\partial f/\partial t)$$

and this means that the 4th component of **GRAD** (ψ) in (A) is $(ic)^{-1}\psi_t$.

If we map this into Λ^3, with which it will be equivalent under $*$, then

$$\delta(*\lambda) = \mathbf{DIV}(*\lambda)$$

which $\quad = (\psi_{xx} + \psi_{yy} + \psi_{zz} - c^{-2}\psi_{tt})\,d\Omega$

in the usual notation, as in (A), and writing the element of volume as $d\Omega$.

And so whenever $*\lambda$ **is a 3-cocycle** in the wedge space defined on the **Lorentz** 4-space

$$\mathbf{DIV\ GRAD}\ \psi = 0 \quad \text{or its equivalent}$$

$$\nabla^2\psi - c^{-2}\psi_{tt} = 0$$

being the standard equation of **wave motion**, c being the velocity.

So we can now see that the equation of **wave motion** for the function ψ is equivalent to the $\quad$ 3-cocycle statement $\quad \delta(*\lambda) = 0$

where $\quad \lambda$ is the 1-form $\lambda = \psi_x dx + \psi_y dy + \psi_z + (ic)^{-1}\psi_t$.

which again illustrates the **Cocycle Law**.

[5.7] The Schrödinger wave equation

This is a natural consequence of the Cocycle Law - which generates any equation of **wave motion**, it being based on a postulate by **de Broglie**, viz., [v. PART-1 §[16.5] & Appendix-E [74]], that a particle of mass m moving with a velocity v should be assigned a wavelength λ where

$\lambda = h/mv = h/p$; $h =$ **Planck's** constant and $p =$ momentum.

This was verified experimentally by **Davisson** and **Germer**, who bombarded nickel crystals with a beam of electrons, as well as by **G.P.Thomson**, who studied diffraction patterns produced by a beam of electrons passing through a thin metal foil.

In **Schrödinger's** treatment the equation of wave motion, to be attributed to atomic particles, is to be
$$\nabla^2 \Psi = (1/v^2)\partial^2 \Psi/\partial t^2$$
where $\Psi = \Psi(q,t)$ and q is a position co-ordinate.

Then if **E** is the total energy of the particle and **V** the potential energy in the field we have, since the K.E. $T = p^2/2m$, $p = \sqrt{2m(E-V)}$, and this gives the phase velocity of the particle-wave as $\lambda v = E/p = E/\sqrt{2m(E-V)}$.

The wave equation then becomes
$$\nabla^2 \Psi = [2m(E-V)/E^2]\partial^2 \Psi/\partial t^2$$
and since the disturbance is to be governed by a wave function of frequency v we can obtain the stationary condition via the time-independent function $\psi(q)$ where
$$\Psi(q,t) = \psi(q)e^{-2\pi i v t} = \psi(q)\exp(-2\pi i E t/h)$$
whence the final wave equation for $\psi(q)$ becomes
$$\nabla^2 \psi(q) + (8\pi^2 m/h^2)(E-V)\psi(q) = 0.$$

[Note : see PART-1 Chapter-16 for our earlier analysis]

HODGE $*$ operator on $\Lambda^n L$, viz. $*: \Lambda^p L \to \Lambda^{n-p} L$

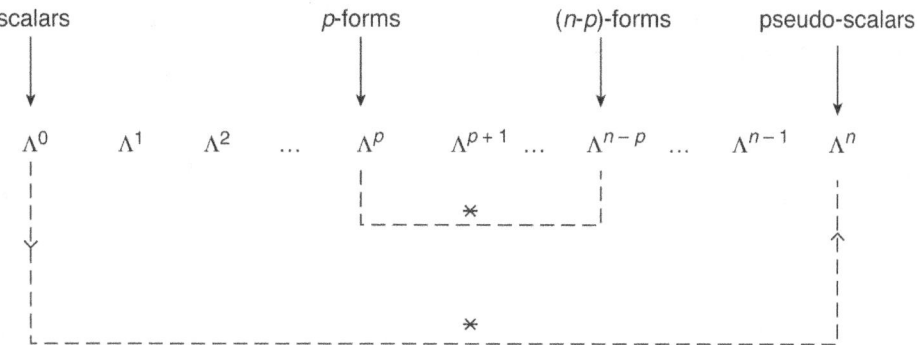

In Lorentz 4-space the matrix $g = \begin{bmatrix} 1 & 0 & 0 & 0 \\ 0 & 1 & 0 & 0 \\ 0 & 0 & 1 & 0 \\ 0 & 0 & 0 & ic \end{bmatrix}$

The metric is $ds^2 = dx^2 + dy^2 + dz^2 - c^2 dt^2$

When $\lambda = \Psi_x dx + \Psi_y dy + \Psi_z dz + \dfrac{1}{ic}\Psi_t$

we get $*\lambda = \Psi_x dx_2 dx_3 dx_4$ + similar (cyclically)

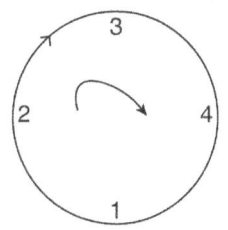

$*: 1 \to 234 \quad 2 \to 341 \quad 3 \to 412 \quad 4 \to 123$

Fig 5.1

PART 3

Dirac's Anti-Particles

PROLOGUE to Part-3

This is an approach to the experimental phenomena of **fundamental particles**.

The phenomena are entirely based on experimental data - which are derived from laboratories housing evermore elaborate (and expensive) **particle accelerators**. These are to be found all over the scientific world, and they require the creation of very strong **magnetic fields** which propel streams of particles (such as electrons and protons) in large circular orbits and with ever-increasing energies until they bombard other nuclei in, for example, bubble chambers. The resulting display of nuclear fireworks can then be interpreted in terms of other (and new) particle "bodies".

This process can probably be seen as the final triumph of the **particle concept** which has been a constant theme throughout the modern history of Physics, and which we have discussed in appropriate mathematical languages in the earlier Parts of this work.

An intriguing aspect of this process has finally appeared in the past twentieth century, viz., the abandonment of strictly algebraic descriptions of the experimental (Scale) measurements and their replacement by what we might call more "homely" language.

Thus we see that the many new particles which the experiments throw up are attributed properties called, eg. "strangeness", "charm", "colours" (various), "up" and "down".

These developments put themselves outside the scope of this work - unless and until some suitable **mathematical/algebraic** language becomes obviously relevant.

Nevertheless we see, in this PART-3, that the underlying properties of **mass** and of **momentum/spin** have been successfully deduced by the algebras which can be associated with the **Lorentz-Einstein** theory of Special Relativity.

As for the rest, we give a sketchy outline with references for further reading and information.

Chapter-1 /PART-3   Gravity and the Cohomology Ring

[1.1] Direct product of complexes

Given two complexes, $\widetilde{L}$ and $\widetilde{K}$, we can form their Cartesian product $\widetilde{L} \times \widetilde{K}$ and combine their components by the **direct (tensor)** product - denoted by $\otimes$. We can then consider the resulting **homology** structures by appealing to the **Kunneth** formula, viz.

$$H_p(\widetilde{K} \times \widetilde{L}) = \Sigma_{s,t} (H_s(\widetilde{K}) \times H_t(\widetilde{L}))$$

where, for each value of p, s and t are chosen so that $s+t = p$, and where we are assuming that the structures $\widetilde{K}/\widetilde{L}$ are torsion-free.   [v. Appendix-E [87],[88],[90],[94]]

The corresponding formula for the cohomology structures becomes

$$H^p(\widetilde{K} \times \widetilde{L}) = \Sigma_{s,t} (H^s(\widetilde{K}) \times H^t(\widetilde{L})) \quad (1)$$

with $s+t = p$.

[1.2] Gravity defined by two finite bodies

We begin with the idea that the massive bodies which are to manifest the usual features of gravitation shall be observable **2-holes** in actual-space - no harm done by contemplating the 2-hole as that of a hollow sphere, S^2.

The value of **mass** will then be regarded as a 2-cocycle defined on the "surface" of the 2-hole.

When the system consists of a <u>single such body</u> it will be associated with a space with the familiar homology structure :

$$H_0 \cong J, \ H_1 \cong J, \ H_2 \cong J$$

J being the additive group of the integers.

At this stage we assume that the question of gravitation does not arise. However, when we consider two such bodies, the phenomena of gravitation become evident.

To allow for this extension we need to contemplate the **Cartesian** product of two suitable complexes, say, $\widetilde{K}$ and $\widetilde{L}$, and we shall denote cocycles in $\widetilde{K}$ by the symbols z^p whilst the cocycles in $\widetilde{L}$ will be denoted by μ^p.

By using the **Kunneth** formula we see that the cohomology structures of this **Cartesian** product will be :

$$H^0 \cong J, \ H^1 \cong J \oplus J, \ H^2 \cong J \oplus J \oplus J, \ H^3 \cong J \oplus J, \ H^4 \cong J$$

If the generators of the respective cohomology groups for $\widetilde{K}, \widetilde{L}$ are denoted by $\hat{z}^i$ and $\hat{\mu}^i$, for relevant values of "i", the generators of the above cohomology groups will be $\hat{z}^0 \otimes \hat{\mu}^0$ for H^0

$\hat{z}^0 \otimes \hat{\mu}^1$ and $\hat{z}^1 \otimes \hat{\mu}^0$ for H^1

$\hat{z}^0 \otimes \hat{\mu}^2$, $\hat{z}^1 \otimes \hat{\mu}^1$, $\hat{z}^2 \otimes \hat{\mu}^0$ for H^2

$\hat{z}^1 \otimes \hat{\mu}^2$ and $\hat{z}^2 \otimes \hat{\mu}^1$ for H^3

and $\hat{z}^2 \otimes \hat{\mu}^2$ for H^4.

The presence of the non-zero group H^4 indicates the presence of a gravitational field, and so we can generalise the classical **Newtonian** theory in which the potential "V" and the radius "r" (in the spherically symmetrical case) are related by

(A) $\quad\quad Vr + k = 0 \quad\quad$ (k being a constant $\in \Re^{\#}$)

by taking k to be $k\hat{w}$ - where $\hat{w}$ is to be the single generator of H^4 and k is to contain a universal gravitational constant.

This then requires us to place the "V" and the "r" in H^2, and the usual product in Vr to be the product found in the Cohomology Ring.

[Note: We remind ourselves that, in that algebraic ring, when $z^p \in H^p(K \otimes L)$ and $\mu^q \in H^q(K \otimes L)$

we have $\quad\quad z^p . \mu^q = (-1)^{pq} \mu^q . z^p$

and that $\quad\quad z^p . \mu^q \in H^{p+q}(K \otimes L) \quad]$

Introducing $\hat{e}_0$, $\hat{e}_1$, and $\hat{e}_2$ for the generators of H^2 we therefore replace (A) by writing $\quad V \rightarrow (V_0 \hat{e}_0 + V_1 \hat{e}_1 + V_2 \hat{e}_2)$

and $\quad\quad r \rightarrow (r_0 \hat{e}_0 + r_1 \hat{e}_1 + r_2 \hat{e}_2)$

For convenience we write

$$\underline{U} \equiv (V_1 \hat{e}_1 + V_2 \hat{e}_2)$$

$$\underline{E} \equiv (r_1 \hat{e}_1 + r_2 \hat{e}_2)$$

giving $\quad\quad (V_0 \hat{e}_0 + \underline{U}) . (r_0 \hat{e}_0 + \underline{E}) = -k\hat{w} \quad$ (B)

Multiplying, in the Cohomology Ring, on the right with $(r_0 \hat{e}_0 - \underline{E})$

we get
$$(V_0 \hat{e}_0 + \underline{U})(r_0^2 \hat{e}_0 \hat{e}_0 - |\underline{E}|^2)$$
$$= k\hat{w} (r_0 \hat{e}_0 - \underline{E})$$

where we have used the commutative property of the Ring product.

Now since there is only a single generator for the H^4, viz., $\hat{w}$, and since the $\hat{e}_i$ are to be found in H^2 whilst the tensor products $\hat{e}_i \hat{e}_j$ are in H^4, we must have

$$\hat{e}_i \hat{e}_i \equiv \hat{w} \quad \text{for } i = 0,1,2$$

Equating the coefficients of $\hat{w}\hat{e}_i$ we obtain

$$V_0 = -kr_0 \div (r_0^2 - |\underline{E}|^2)$$

and $$\underline{U} = k\underline{E} \div (r_0^2 - |\underline{E}|^2)$$

The function V_0 is a function of r_0 only and gives a spherically symmetrical potential function corresponding to the Newtonian case. So there will be a **constant of angular momentum**, h_0, and since the dimension of $|\underline{E}|$ will be that of length, [L], we see that by writing

$$|\underline{E}| = h_0/c$$ (c being the velocity of light) we reproduce the **Einstein** correction term to the **Newtonian** case for - assuming $|\underline{E}| \ll r_0$ and expanding V_0 - the **radial force** is

$$\underline{F} = -\partial V_0/\partial r_0 = -k/r_0^2 - 3kh_0^2/c^2 r_0^4$$

the second term of which is the **Einstein** correction term.

Also, to a first approximation, $|\underline{U}| = k|\underline{E}| \div r_0^2$ and, with $|\underline{E}| = h_0/c$, we see that the dimension of $|\underline{U}|$ is [k]/[L] as expected (as it is associated with V_0), the value being $$|\underline{U}| = kh_0/r_0^2 c^2.$$

This means that the gravitational field contains, at a distance of r_0 from the origin, an intrinsic **angular momentum component** - which decreases by a factor of $[r_0 c]^{-2}$ - so that it is only likely to be observable at very small values of r_0 (of the order of c^{-1}).

Since we are considering the gravitational field arising from a "universe" of two masses the effective origin can be taken as the centre of either of our imaginary spheres - so the field effectively associates this apparent "spin" with the surface of either (extremely small) particle mass.

Chapter-2 /PART-3 Relativistic Hamiltonian - Dirac's treatment

[2.1] Relativistic Hamiltonian, Dirac's treatment

Considering a particle of rest-mass m moving with a velocity $\underline{v}$ in a 3-dimensional space we have seen in previous sections, that the **Einstein/Lorentz** expression for the total energy of the particle is $\quad E = m\beta c^2 \quad$ where $\beta^{-1} = \sqrt{(1 - \underline{v}^2/c^2)}$

[Note: c = velocity of light, and $\underline{v}^2$ means the scalar product $\underline{v}.\underline{v}$]

and that the classical momentum vector $\underline{p}$, viz., $m(\dot{x}, \dot{y}, \dot{z})$ needs to be replaced by the relativistic 4-vector $\quad \underline{p} = m\beta(\dot{x}, \dot{y}, \dot{z}, c)$,

for it can be shown that relative to a second Scale $\mathfrak{C}^*$ this transforms under the **Lorentz** operator $\mathscr{L}$ via $\quad \underline{p}^* = \mathscr{L}\underline{p} \quad$ where

$$\mathscr{L} = \begin{bmatrix} \beta & 0 & 0 & \gamma \\ 0 & 1 & 0 & 0 \\ 0 & 0 & 1 & 0 \\ \gamma & 0 & 0 & \beta \end{bmatrix} \quad \text{writing } \gamma \text{ for } -\beta\underline{V}/c$$

We also notice that

$\underline{p}^2$ = square of the modulus of $\underline{p}$ = $m^2\beta^2\underline{v}^2 \quad$ (A)

It is also necessary to take the **Lagrange** function as

$$L = mc^2 - mc^2[1 - (\dot{x}^2 + \dot{y}^2 + \dot{z}^2)/c^2] - \Phi(x,y,z)$$

where $\Phi(x,y,z)$ is the potential function for the force-field the particle moves in.

The **Hamiltonian** function then becomes

$$H = \text{total energy } E = \dot{x}\partial L/\partial \dot{x} + \dot{y}\partial L/\partial \dot{y} + \dot{z}\partial L/\partial \dot{z} - L$$

giving $\quad H = mc^2\beta + \Phi(x,y,z) - mc^2$

and because the last term is a constant it is customary to take the **Hamiltonian** as

$$H = mc^2\beta + \Phi(x,y,z)$$

and in the absence of a field $\Phi(x,y,z)$ we can derive **Dirac's** expression via

$$H = mc^2\beta = mc\beta[c^2(1 - \underline{v}^2/c^2) + \underline{v}^2]^{\frac{1}{2}}$$

or $\quad H = mc(c^2 + \beta^2\underline{v}^2)^{\frac{1}{2}}$

giving, by (A), $\quad H = c[m^2c^2 + \underline{p}^2]^{\frac{1}{2}} \quad$ (B)

[2.2] The relativistic Hamiltonian operator in Dirac's treatment

If we wish to introduce **Dirac's** equation (B) into Quantum Theory we shall need to translate H into the form of a **linear operator** - where e.g. we shall need to replace momentum variable, p, by the standard operator $(h/2\pi i)\partial/\partial q$ for conjugate co-ordinate, q.

This means that (B) must be transformed into a **linear** expression in the p_i's, and **Dirac** achieved this by introducing coefficients α_i into (B) and writing

$$[p_1^2 + p_2^2 + p_3^2 + m^2c^2]^{1/2} \text{ as } \alpha_1 p_1 + \alpha_2 p_2 + \alpha_3 p_3 + \alpha_4 mc$$

This requires that

$$p_1^2 + \ldots + m^2c^2 = \alpha_1^2 p_1^2 + \ldots + \alpha_4^2 m^2 c^2 +$$
$$\ldots + (\alpha_1\alpha_2 + \alpha_2\alpha_1)p_1 p_2 + \ldots + (\alpha_3\alpha_4 + \alpha_4\alpha_3)p_3 p_4$$

and for this to be satisfied we need

$$\alpha_r \alpha_s + \alpha_s \alpha_r = 0 \quad \text{when } s \neq r$$

and

$$\alpha_r^2 = I_4 \quad r = 1, 2, 3, 4$$

I_4 being the unit matrix, viz.,

$$I_4 = \begin{bmatrix} 1 & 0 & 0 & 0 \\ 0 & 1 & 0 & 0 \\ 0 & 0 & 1 & 0 \\ 0 & 0 & 0 & 1 \end{bmatrix}$$

By taking the α_i's as operators defined by the matrices

$$\sigma_1 = \begin{bmatrix} 0 & 1 \\ 1 & 0 \end{bmatrix} \quad \sigma_2 = \begin{bmatrix} 0 & i \\ -i & 0 \end{bmatrix} \quad \sigma_3 = \begin{bmatrix} 1 & 0 \\ 0 & -1 \end{bmatrix} \quad \sigma_4 = \begin{bmatrix} 1 & 0 \\ 0 & 1 \end{bmatrix}$$

and with

$$\alpha_i = \begin{bmatrix} 0 & \sigma_i \\ \sigma_i & 0 \end{bmatrix} \text{ whenever } i = 1, 2, 3$$

and

$$\alpha_4 = \begin{bmatrix} \sigma_4 & 0 \\ 0 & -\sigma_4 \end{bmatrix}$$

These give

$$\alpha_1 = \begin{bmatrix} 0 & 0 & 0 & 1 \\ 0 & 0 & 1 & 0 \\ 0 & 1 & 0 & 0 \\ 1 & 0 & 0 & 0 \end{bmatrix} \quad \alpha_2 = \begin{bmatrix} 0 & 0 & 0 & i \\ 0 & 0 & -i & 0 \\ 0 & i & 0 & 0 \\ -i & 0 & 0 & 0 \end{bmatrix}$$

$$\alpha_3 = \begin{bmatrix} 0 & 0 & 1 & 0 \\ 0 & 0 & 0 & -1 \\ 1 & 0 & 0 & 0 \\ 0 & -1 & 0 & 0 \end{bmatrix} \quad \alpha_4 = \begin{bmatrix} 1 & 0 & 0 & 0 \\ 0 & 1 & 0 & 0 \\ 0 & 1 & -1 & 0 \\ 0 & 0 & 0 & -1 \end{bmatrix}$$

The **Hamiltonian** linear operator for the system then becomes, e.g.,

$$\tilde{H} = (h/2\pi i) \{\alpha_1 \partial_x + \alpha_2 \partial_y + \alpha_3 \partial_z\} + \alpha_4 mc$$

where e.g. we write ∂_x for $\partial/\partial x$ etc..

Then the quantum theory equation for the system is

$$\tilde{H}(x, y, z, t) \Psi = (ih/2\pi c) \partial_t \Psi$$

where $\Psi = \begin{bmatrix} \psi_1 \\ \psi_2 \\ \psi_3 \\ \psi_4 \end{bmatrix}$

[2.3] Quantum numbers for spin in the **Dirac** picture

[Note: We remind ourelves that if we wish to observe the eigenvalues of two operators simultaneously then it must be possible to reduce both of them simultaneously to diagonal forms - for then the eigenvalues are real numbers (the operators being **Hermitian** and it normally being rquired that measurements must be in $\Re^{\#}$). In that form they will **commute** and this is a property which they must possess before we transform them to their respective diagonal forms. In this context we are always looking for operators which commute with the **Hamiltonian** $\tilde{H}$. If such a one is denoted by $\tilde{K}$ we therefore require it to satisfy the equation $\tilde{H}\tilde{K} - \tilde{K}Hz = \tilde{0}$, $\tilde{0}$ being the zero operator.]

To examine the **angular momentum** associated with the particle we begin with the non-relativistic angular momentum **operators** (derived from the **curl** of vector algebra) with components

$$\tilde{M}_1 = (h/2\pi i)(y\partial_z - z\partial_y)$$

$$\tilde{M}_2 = (h/2\pi i)(z\partial_x - x\partial_z)$$

$$\tilde{M}_3 = (h/2\pi i)(x\partial_y - y\partial_x)$$

these being the rectangular components of the **operator**

$$\tilde{M} = (h/2\pi i)\,\underline{r} \times \nabla \qquad (1)$$

If this were the whole story we would expect these operators to commute with those of the relativistic **Hamiltonian** found in [2.2] above - but it turns out that this is not so.

For example, we first consider whether $\tilde{M}_1$ commutes with α_4, and noting

$$\alpha_4 \Psi = \alpha_4 \begin{bmatrix} \psi_1 \\ \psi_2 \\ \psi_3 \\ \psi_4 \end{bmatrix} = \begin{bmatrix} \psi_1 \\ \psi_2 \\ -\psi_3 \\ -\psi_4 \end{bmatrix}$$

so that, by (1), we get

$$(2\pi i/h)\tilde{M}_1 \alpha_4 \Psi = \begin{bmatrix} \tilde{M}_1\psi_1 \\ \tilde{M}_1\psi_2 \\ -\tilde{M}_1\psi_3 \\ -\tilde{M}_1\psi_4 \end{bmatrix}$$

with the same expression for the product $(2\pi i/h)\alpha_4\tilde{M}_1\Psi$

and this shows that the operators $\tilde{M}_1$ and α_4 commute - the same applying to the products of $\tilde{M}_2$, $\tilde{M}_3$ with α_4.

Straightforward calculation then shows that, eg,

$$\tilde{H}\tilde{M}_1 - \tilde{M}_1\tilde{H} \text{ reduces to } (-h^2/4\pi^2)(\alpha_2\partial_3 - \alpha_3\partial_2)$$

which is not the zero operator.

Similar results for $\tilde{M}_2$ and $\tilde{M}_3$ means that none of these operators commute with $\tilde{H}$ and so none of them can be brought to a diagonal form simultaneously with $\tilde{H}$.

For this to be achieved we need a modified **angular momentum** operator which will commute with $\tilde{H}$.

Dirac achieved this by introducing what amounted to a **spin** operator, viz. $\tilde{N}$ with components defined by

$$\tilde{N}_1 = -k\alpha_2\alpha_3 \quad \tilde{N}_2 = -k\alpha_3\alpha_1 \quad \tilde{N}_3 = -k\alpha_1\alpha_2$$

where $k = (hi/4\pi)$.

It is easy to show, eg, that

$$\tilde{H}\tilde{N}_1 - \tilde{N}_1\tilde{H} \text{ becomes } (+h^2/4\pi^2)(\alpha_2\partial_3 - \alpha_3\partial_2)$$

and that therefore it follows that the operators

$$\tilde{M}_1 + \tilde{N}_1 \quad \tilde{M}_2 + \tilde{N}_2 \quad \tilde{M}_3 + \tilde{N}_3$$

separately commute with $\tilde{H}$.

But it also follows that the $\tilde{N}_i$ do not commute among themselves.

We therefore take just one of these $\tilde{N}_i$ to find the eigenvalues which can be seen with those of $\tilde{H}$, and for this we (arbitrarily) choose $\tilde{N}_3$

We notice too that

$$\tilde{N}_3^2 = -(h^2/16\pi^2)\alpha_1\alpha_2\alpha_2\alpha_1 = (h^2/16\pi^2)I_4$$

I_4 being the unit matrix in 4-dimensions

The eigenvalues for this **extra angular momentum**, which is called the **spin** about a pre-assigned axis, must therefore be

$$\sigma_3 = \pm\tfrac{1}{2}(h/2\pi) \qquad (2)$$

[Note: **spin** is the accepted term in scientific circles, although it is a little unfortunate since the word traditionally means only the **angular velocity** of a particle whereas we have seen that, via **Dirac's** derivation, it has the dimensions of **angular momentum** - which assumes the presence of a particle with **mass**. But experimentalists have occasionally felt the necessity to attribute **spin** to a massless particle (eg the **photon**.]

[2.4] Spin and the Involution range in a Projective Space

Planck's constant "h" is a measure of **angular momentum**, normally expressed as Joule-Second and therefore with dimensions [Work]x[Time] or $ML^2T^{-2}.T$ which $= ML^2T^{-1}$, and this is the same as [angular momentum] which equals $MLT^{-1}.L$ which is ML^2T^{-1} again.

The quantity $\hbar$, which denotes $h/2\pi$, is therefore the **quantised unit of angular momentum**, viz.,

angular momentum per unit angle (the radian)

We have noted that the spins associated with "particle & antiparticle" are natural mates in an involution range which, with reference points "0" and "∞", means that the cross-ratio $(0\ s^+\ \infty\ s^-) = -1$

If we measure these spins on scales in a Projective Space and using the natural **projective metric** (Ref. earlier sections of this work), viz.,

projective distance between "a" and "b" ò $(1/2i) \ln (0\ a\ \infty\ b)$

then the observations of the spins follow from the expression

$$\hbar\ (1/2\pi i) \ln (-1)$$

But $\ln (-1) = \ln (1 \cdot \exp i(\pm \pi \pm 2n\pi))$ with n = 0, 1, 2, ...

which equals $\ln(1) + \ln (\exp i(\pm \pi \pm 2n\pi))$, and since $\ln(1) = 0$

the **spins** will therefore become any of the values

$$\hbar \cdot (\pm(1/2)\ \pm(3/2)\ \pm(5/2)\ ...\ )$$

the positive values being associated with the "normal" (+ve) particles and the negative values being associated with the antiparticles.

Chapter-3 /PART-3   Lorentz-Einstein 4-Space : $\Gamma(1\text{-}1)$ in $\mathbb{C}$

[3.1] Using Magnetic Field iH as Base Element $\mathfrak{Z}_L$

Our pevious discussion of the **Lorentz-Einstein** (1-1)-correspondence, between velocity Scales $\mathfrak{C}$ and $\mathfrak{C}^*$, assumed only that the velocity measures v and v^* (and all other parameters) were mapped into the real numbers $\mathfrak{R}^\#$.

We now extend this discussion to allow for the expression of the velocities, and the other parameters, in the complex algebra $\mathbb{C}$.

We do this by restricting the Base Element $\mathfrak{Z}_L$ to its component iH, where $i^2 = -1$, and we shall examine the results obtained under two possible hypotheses.

[3.2] Hypothesis-1

We suppose that <u>all velocity measures "v" become "iv"</u> whilst other parameters are of the general complex form "x + iy" - "x" and "y" being reals.

It follows that the basic $\Gamma(1\text{-}1)$ becomes, say,

$$vv^* + av + bv^* + d = 0$$

where $v \to iv$, $v^* \to iv^*$, $a \to a_0+ia_1$, $b \to b_0+ib_1$, $d \to d_0+id_1$, and a double point of the homography (characteristic of the Base Element) $c \to ic$, with $v, v^*, a_0 \ldots d_1, c \in \mathfrak{R}^\#$.

There is no loss of generality by taking the fundamental $\Gamma(1\text{-}1)$ as

$$vv^* + (a_0+ia_1)(iv) + (b_0+ib_1)(iv^*) + d_0+id_1 = 0_C = 0+i0$$

or $\quad vv^* + (ia_0-a_1)v + (ib_0-b_1)v^* + d_0+id_1 = 0_C \quad (1)$

resulting in the two conditions

$$vv^* - a_1v - b_1v^* + d_1 = 0 \qquad (2a)$$

$$a_0v + b_0v^* + d_0 = 0 \qquad (2b)$$

As before, we have pairs in this homography, viz.,

$\quad v = V, v^* = 0 \quad$ and $\quad v = 0, v^* = -V \quad$ which result in

$\quad a_1V = d_1 \quad$ and $\quad -b_1V = d_1 \quad$ from (2a)

as well as $\quad a_0V + d_0 = 0 \quad$ and $\quad -b_0V + d_0 = 0 \quad$ from (2b)

Then (2a) gives $\quad b_1 = -a_1 = -d_1/V \quad$ with a resulting $\Gamma(1\text{-}1)$

$$Vvv^* - d_1(v - v^*)/V + d_1 = 0 \qquad (3a)$$

whilst (2b) gives $\quad d_0(v - v^*) + d_0V = 0 \qquad (3b)$

The condition (3b) immediately gives us the relative velocities associated with the classical theories of **Galileo/Newton**, viz.,

$$v = v^* + V \qquad (4)$$

whilst (3a) gives, when we take a double point as $v = v^* = c$ (being the same as $iv = iv^* = ic$) $\qquad Vc^2 + d_1 = 0$

whence $\qquad Vvv^* + c^2(v - v^*) - Vc^2 = 0 \qquad (5)$

This being the standard $\Gamma(1\text{-}1)$ defining the **Lorentz-Einstein 4-Space**.

To find the relation between vectors (x, ct) in $\mathfrak{C}$ and (x^*, ct^*) in $\mathfrak{C}^*$ we write (5) in the homogeneous form

$$Vxx^* + c^2(xt^* - x^*t) - Vc^2tt^* = 0$$

and this is $\qquad x^*(Vx - c^2t) - c^2t^*(Vt - x) = 0$

allowing us to introduce the separation constant β by writing

$$x^* = \beta(Vt - x) \qquad ct^* = \beta(Vx/c - ct)$$

and so we can introduce the **Lorentz** matrix $\mathscr{L}$ where

$$\mathscr{L} = \begin{bmatrix} -\beta & \beta V/c \\ \beta V/c & -\beta \end{bmatrix}$$

so that we have

$$\begin{bmatrix} x^* \\ ct^* \end{bmatrix} = \begin{bmatrix} -\beta & \beta V/c \\ \beta V/c & -\beta \end{bmatrix} \begin{bmatrix} x \\ ct \end{bmatrix}$$

The value of β follows from equating the inverse of $\mathscr{L}$, viz., $\mathscr{L}^{-1}$ via $\mathscr{L} \equiv \mathscr{L}(V)$ and $\mathscr{L}^{-1} \equiv \mathscr{L}(-V)^{-1}$. Then since $\mathscr{L}\mathscr{L}^{-1}$ = Identity matrix I_2 we get $\qquad \beta^{-2} = (1 - V^2/c^2) \qquad (6)$

[3.3] Hypothesis-2

We now suppose that <u>only the velocity associated with the Base Element i**H** is a measure</u> in the complex algebra, $\mathbb{C}$, although the coefficients a,b,d are to be complex numbers, viz., $a \to a_0 + ia_1$, $b \to b_0 + ib_1$, and $d \to d_0 + id_1$.

In this case we replace "c" by "ic" in the usual discussion of the **Lorentz-Einstein** $\Gamma(1\text{-}1)$ and substitute for a,b,d. Then (1), in [3.2], still applies and so do equations (2a),(2b),(3a),(3b),(4) in [3.2].

Therefore the **Galileo/Newton** assumption also applies, via (4), but (5) becomes a new $\Gamma(1\text{-}1)$ - for the **Lorentz-Einstein** special relativity theory. viz.,

$$-Vvv^* + c^2(v - v^*) + Vc^2 = 0 \qquad (1)$$

Usig the homogeneous co-ordinates, in which we write (x t) for x and (x* t*) for x*, this becomes $\quad -Vxx^* + c^2(xt^* - x^*t) + Vc^2tt^* = 0 \quad$ (1a)

and then $\quad -x^*(Vx + c^2t) + c^2t^*(x + Vt) = 0 \quad$ (1b)

The separation constant, β, is now replaced with, say, γ and from (1b) we can write

$$x^* = \gamma(x + Vt) \quad \text{and} \quad ct^* = \gamma(Vx/c + ct)$$

The corresponding **Lorentz** matrix is now, say, $\mathcal{L}_\gamma$ where

$$\mathcal{L}_\gamma = \begin{bmatrix} \gamma & \gamma V/c \\ \gamma V/c & \gamma \end{bmatrix}$$

so that we have

$$\begin{bmatrix} x^* \\ ct^* \end{bmatrix} = \begin{bmatrix} \gamma & \gamma V/c \\ \gamma V/c & \gamma \end{bmatrix} \begin{bmatrix} x \\ ct \end{bmatrix}$$

Replacing V by -V in $\mathcal{L}_\gamma$ gives us the inverse operator $\mathcal{L}_\gamma^{-1}$ and by identifying the product matrix $\mathcal{L}_\gamma \mathcal{L}_\gamma^{-1}$ with the identity matrix I_2 we deduce

$$\gamma^{-2} = [1 + V^2/c^2] \quad (2)$$

corresponding to equation (6) in **[3.2]** above.

Chapter-4 /PART-3 Dirac's Theory of Anti-Particles

[4.1] Matter and Anti-Matter

We have seen (eg PART-1 Ch-3 §[3.2] ; Ref Appendix-E [13]) that the energy associated with a particle rest-mass m_0 is, say, E^+ where

$$E^+ = m_0 \beta_0 c^2 \qquad (A)$$

where $\beta_0 = \{1 - |v|^2/c^2\}^{-\frac{1}{2}}$ and v is the particle velocity w.r.t the Scale $\mathfrak{C}$ and naturally replaces V (the Scale $\mathfrak{C}^*$ being coincident with the particle in its motion.

Equation (A) means also that the "ordinary" **mass** is $m_0 \beta_0$ (M^+)

Now these results are by reference to the **Lorentz-Einstein** $\Gamma(1\text{-}1)$ defined over the reals $\mathfrak{R}^\#$.

If we pursue **Hypothesis-1** and refer to the discussion in [1.1] equation (A) becomes

$$E^- = -m_0 \beta_0 c^2 \qquad (A^-)$$

and the **mass** becomes $-m_0 \beta_0$ (M_β)

On the other hand, if we pursue **Hypothesis-2** the expression for Energy is

$$E^- = -m_0 \gamma_0 c^2 \qquad (B^-)$$

with consequent **mass** being $-m_0 \gamma_0$ (M_γ)

In either case we obtain the **Dirac** result viz., there exist **anti-particles** with (apparently) negative masses, and that these correspond (1-1) with the accepted **particles**. For example, the **anti-electron** corresponds to the **positron** which has the same numerical value for its mass, with other properties being similarly reflected (eg a positive charge numerically equal to that of the electron).

Expansion of the expressions, under either **Hypothesis**, gives the result

$$E^- = -m_0 c^2 + \tfrac{1}{2} m_0 |v|^2 + \text{etc.}$$

Since the rest-mass energies of the particle and anti-particle are respectively

$$m_0 c^2 \quad \text{and} \quad -m_0 c^2$$

it will only be possible to observe the anti-particle if we can inject at least an amount of energy

$$\Delta E = 2 m_0 c^2 \qquad \text{into the experimental observations.}$$

This is the rationale behind the massive toroidal accelerators, with their powerful magnetic fields, which have been constructed in modern times.

The experimental idea is to inject sufficient magnetic energy into the Toroidal apparatus as to cause an anti-particle to be "knocked out" of its world into the observable

Lorentz-Einstein world. Its mass then becomes an m⁺ and the "anti" properties appear as opposing **charge, spin** etc..

[4.2] Calculating masses of particles/anti-particles

Under our traditional methods of observation we can see and measure a particle mass as, say **m⁺**, but we cannot "see" its anti-article, say, **m⁻**. But the measures of **m⁺** are always in the context of Scales with Base-Element $\mathcal{B}_L$ (= light), and in that context we cannot observe its velocity as any value greater than "c" - and the consequences of this lead us to the derivation of the **Lorentz-Einstein** transformations.

But this leads us to the simplest postulate for this "blindness", viz.,

$$\text{an anti-particle moves with a velocity greater than "c"} \quad (P)$$

This, in turn, suggests that the Scales should have a Base Element $\mathcal{B}_{iH}$ and that **Hypothesis2** should control the transformations - for if $V \geq c$ then neither the usual **Lorentz** metric nor that of **Hypothesis1** cannot give observable (real) values.

So the separation constant γ_0 (= $1/\sqrt{\{1 + v^2/c^2\}}$) applies - it being understood, via the condition $\Delta E \geq 2m_0\gamma_0 c^2$, that the classically observable kinetic energy is $m_0 v^2/2 \geq \Delta E$.

So we require that $v^2 \geq 4c^2$.

We can obtain discrete (particle) masses of the anti-particles (which will appear in the Scales as numerically equal to those of the corresponding particles) by taking the velocity v to satisfy $\quad v^2 = 4nc^2 \quad$ with $n = 0,1,...$

If we refer the masses to a selected reference particle of mass, M_0, we see that it is related to particles of mass m_0 where

$$M_0 = m_0\gamma_0 = m_0 1/\sqrt{\{1 + 4n\}} \quad (Q)$$

Taking the reference particle to be a **pion**, in units of MeV/c², we have for the **pion** π^+ its accepted mass is 140, as is that of the **pion** π^-, whilst the **pion** π_0 has a mass of 135.

The corresponding possible particle masses will be

$$m_0 = (\textbf{pion mass}) * \sqrt{(4n + 1)} \quad (R)$$

There will be an infinite number of such particles. A table of some of the initial values with comparisons with observed values is given in Table-1.

[4.3] Particle/Anti-particle Involution properties

The properties of particles and anti-particles must be in a simple (1-1)-correspondence, so for example m^+ gives rise to m^- and vice-versa, and this suggests that the values (however they are observed) satisfy the cross-ratio relation

$$(0 \; m^+ \; \infty \; m^-) = (0 \; m^- \; \infty \; m^+) = -1 \qquad (1)$$

it being natural to take the double points of this involution as 0 and ∞.

This expresses, what we have already required, viz.,

$$m^+ + m^- = 0 \qquad (2)$$

and we would also expect that other properties of the particles/anti-particles will also exhibit the same involutionary relations.

Until we discuss such properties later two examples will suffice viz., "charge" and "spin" giving us charges q^+ & q^- and s^+ & s^-.

The charge on the neutron being 0 means that the charge on the anti-neutron is 0, whilst the respective spins are ½ and -½.

Table of Particle Masses with comparisons

Using m = 135*sqrt(4n + 1) and m = 140*sqrt(4n + 1)

Value of n	mass(135)	particle/mass/symbol	mass(140)	% difference
0	135 <---	π Ref pions π --->	140	
3	487	(meson 498) κ --->	505	1.39
4	557 <---	(meson 548) η	577	1.62
7	727	(meson 750) π --->	754	0.53
8	776 <---	(meson 782) ω	804	0.77
10	864	(meson 888) κ --->	896	0.89
11	906	(nucleon 939) ν --->	939	0.00
14	1019 <---	(baryon 1020) η	1057	0.10
18	1153	(baryon 1196) Σ --->	1196	0.00
19	1185 <---	(baryon 1189) Σ	1228	0.34
20	1215	(baryon 1250) η --->	1260	0.79
21	1245 <---	(baryon 1238) Δ	1291	0.56
22	1274	(baryon 1318) Ξ --->	1321	0.23
23	1302 <---	(baryon 1311) Ξ	1350	0.69
24	1330 <---	(baryon 1318) Ξ	1379	0.90
26	1383 <---	(hyperon 1385) Σ	1435	0.14
27	1409 <---	(hyperon 1405) Λ	1462	0.28
29	1460	(nucleon 1512) N --->1514		0.13
31	1509 <---	(baryon 1512) N	1565	0.20
32	1533 <---	(baryon 1533) Ξ	1590	0.00
35	1603	(baryon 1660) Σ --->	1662	0.12
36	1626	(nucleon 1688) N --->1686		0.12
38	1670 <---	(baryon 1676) Ω	1732	0.36
39	1692 <---	(baryon 1688) N	1754	0.24
45	1816 <---	(baryon 1815) Ω	1884	0.06
47	1856	(baryon 1920) Δ --->	1925	0.26

Table-1

Chapter-5 /PART-3   The Spectrum of Particles

[5.1] Experimental Techniques for Observing Particles

We need to appreciate that our knowledge of elementary particles is all based on experimental observations. There are not, as yet, any deductions based on a mathematical structure - such as that of the **Lorentz-Einstein** theories, although the work of **Dirac** (v. Chapter-4 in this PART-3) exceptionally derived the notion of **anti-particle** as well as that of **spin** from such a context.

However the twentieth century saw the development of experimental set-ups (Scales) whose results allowed the traditional enthusiasm for **particle Physics** to blossom in the search for an impressive array of experimental data.

These set-ups have run through a series of engineering projects involving, eg. electrostatic high voltage generators (notably the **van der Graaf** generator), cyclotrons, linear accelerators and the proton synchrotons - all designed to enable the experimentalists to accelerate various particles into a state of ever-higher energy levels so that they can be made to impinge on various targets. The resulting scattering displays are then detected by (eg) photographic plates and/or bubble chambers. Analysis of these observed particle paths, such as length, angular deviations, mirror-like reflections, curvature, together with clockwise/anticlockwise rotations, allows interpretation of lifetime, spin, mass/energy, and "normal" or "anti" nature, of the particle.

Such projects are now to be found in many countries around the world - at least in those countries which can afford the huge financial outlay - and the process looks to be developing in the future.

There is ample literature to give pictorial/engineering descriptions of the history (and certain future) of this process, [v. eg Appendix-E [100] [101] [102] [103]], but such descriptions are outside the remit of this work - since they do not yet involve any novel mathematical foundation.

The following sections give an outline of the observed data.

[5.2] Categories of Observed Particles

Particles are classified into two main types, viz., **leptons** (lighter) and **hadrons** (heavier). Examples of **leptons** are :- electrons, muons (formerly known as μ-mesons), or tau(s), associated neutrinos (these are presumed to be of zero rest-mass).

Examples of **hadrons** are :- baryons (protons etc) and mesons (pions etc v. §[4.2] above).

The quantum values of **spin** assciated with the **hadrons** allows them to be further classified into **fermions** and **bosons** ; the former have values which are an odd number of $½\hbar$ and which are subject to the **Pauli Exclusion Principle**, whilst the latter have values which are a whole number of $\hbar$. All **leptons** are **fermions**.

The **baryons** are all unstable, except the **proton**, and lifetimes - whilst varying - have values which appear to be incredibly small (often of the order of 10^{-20} secs).

These particles (together with their anti-particles) interact via **strong, weak,** and **electromagnetic** interactions. For example, the **strong** interactions accur between particles which bind together to form a nucleus (of some other particle). The **gravitational** interaction is generally negligible in comparison.

Another novel feature of the 20th Century thinking was the introduction of the so-called **gauge bosons**. These are "particles" which, by way of **mutual exchange**, act between other particles to produce the effect which otherwise was assigned to the attractive **field** to which individual particles (used to) respond. Thus, for example, the electric attraction between charged particles is regarded as the result of the mutual exchange of **photons** between them. One big advantage of this approach is that **two** particles are required to produce an interaction. [v. §[2.2] above re the gravitational phenomenon].

The mediator in the **strong** interactions is called a **gluon** and, for example, is what binds the various **quarks** which are now regarded as combining to form (eg) the **hadrons**.

These **quarks**, which are not expected to be observed in isolation, come with various properties which allow further classifications to be made. These properties are referred to as

  up, down, strange together with the **anti-up** etc..

[5.3] Safeguarding Conservation Laws

The history of conservation laws in Physics is what corresponds to the appearance of **invariants** in the mathematics of algebraic/geometry and the **conservation of momentum** has always played a significant part in any Physics.

In the present context we look to the quantum **spin** numbers to exhibit the corresponding conservation law.

Indeed to preserve this notion in inter-particle collisions/decays/etc it has been essential to propose the various additional properties, together with their new quantum numbers, which particle/anti-particle observations have elicited.

The introduction of quantum numbers to be associated with baryons (the baryon number, B) and strangeness (number S) permits conservation in a production reaction, such as,

$$\pi^+ + p \rightarrow \Sigma^+ + K^+ \quad \text{where}$$
$$B \rightarrow \quad 0 + 1 \quad \quad 1 + 0 \quad \text{and}$$
$$S \rightarrow \quad 0 + 0 \quad -1 + 1$$

Many such examples will be found in the literature - the various quantum numbers requiring only simple arithmetic for their expression.

[5.4] Websites for further information

The whole subject of elementary particles has grown extensively during the past (20th) and present (21st) century. For readers who wish to pursue it in earnest there are now a number of informative websites available on the internet.

Perhaps the most significant is the following :

http://pdg.lbl.gov/pdg.html

which gives data (with regular updating) under the following headings, viz.,

> gauge & Higgs bosons
> leptons
> quarks
> mesons
> baryons
> searches (various)
> tests (of conservation laws)

Discussion of the **standard model** of particle physics can be found on

http://www.benbest.com/science/standard.html

A website giving comprehensive information about other specialist websites covering all areas of the subject is

http://www.psigate.ac.uk

Specialist discussion of quarks and anti-quarks can be found on

http://hyperphysics.phy-ast.gsu.edu/hbase/particles/quark.html

An educational program can be found on

http://www.sciencepark.info/particle/fundamental.html

[5.5] Some ideas to explore in the spirit of this thesis

[a] The idea of "particle" can be associated with the **vertices** of a Complex- since they are to be observed as **points** in some complex of observations.

This suggests an association with a **Dowker** complex ?

In this case we must begin to associate a "particle" (such as an electron) with and "edge", rather than with a vertex - for then the other end of the edge may well carry the **positron** ?

So, if a particle is identified with one vertex of a tetrahedron with the opposite face (consisting of 3 particles/vertices) it can be associated with 3 smaller (?) particles contained within itself ?

[b] In the Ring of cochains, **R**, we know that $c^p \cup c^q$ is a c^{p+q}

Hence, for example, since $c_1^0 \cup c_2^0$ equals some c_3^0 it follows that any c^0 (such as the c_3^0) can be the same as (manifest itself as) any number of "particles" (the various c_i^0, as in the above).

[c] Use $\mathbb{C}$ and/or $\mathbb{Q}$ as the group **G** in the mapping $(c^p, c_p) \in$ **G**.
This will allow us to have special values in the Projective Metric (x 0 y ∞) ?

[d] Use the idea of an **anti-complex** $\overline{K}$ so that $\overline{K}$ can carry the **anti-matter** ?

This might allow us to use the union $K \cup \overline{K}$ to give us a new $K^\#$. Then an edge in this $K^\#$ can be a join of a vertex with an anti-vertex ?

Since any one particle has only one(?) anti-particle associated with it the join of K with $\overline{K}$ must only be as a set of edges ?

[e] In the wedge space $\Lambda^n(V)$ if we look for **spin** in, say Λ^p, might not **isospin** be found as ∗**spin** in Λ^{n-p} ?

[f] The various values of **spin** etc seem to occur at multiples of "½"$\hbar$ - $\hbar$ being the accepted symbol for h/2π, h being **Planck's** constant - and this suggests that they could be derived as part of the logarithm function found in the **projective metric** ?

[g] The various properties of Masses and Anti-masses will presumably satisfy some $\Gamma(1-1)$?

[h] Masses → Masses (otherwise) will be found only in K (above) whilst Anti-masses → Anti-masses will be found only in $\overline{K}$? Are there any instances of this phenomenon ?

PART 4

Structures, Times, and Events

PROLOGUE to Part-4

In this fourth Part we pick up from Part-2 §[2.3] the idea of a **Dowker** complex generated and defined by a binary matrix - in which the elements take only values 0 or 1.

It seems to be rather distant from the problems of Physics and Physicists - the first of which we have seen to be about measurements in $\Re$, and the second of whom have been seduced by the mathematicians real numbers $\Re^{\#}$. But in the observational science of Physics itself the measures can all be reduced to the countable integers **J** - as readers who have travelled with us so far may readily agree.

So now we are looking at what kind of science can be based on nothing other the integers $\{0,1\}$.

The age of digital orientation has now arrived - its chief messenger being the computer and all its associated technology. To deal with numbers, and all kinds of calculations, we now only need the two symbols 0 and 1 - all credit due to the amateur mathematician **George Boole** (1815 - 1864) as well as to the pioneer in computing viz. **Alan Turing**.

These reflections are meant to illustrate the extension of this mathematical revolution and, in fields other the merely arithmetical, we pay tribute to the mathematician **C.H.Dowker** who unearthed a fundamental theorem in Algebraic Topology (v. Appendix-C §[5]) using a binary matrix array for the representation of a **relation**.

The writer, together with other co-workers (v. Appendix-C), introduced the ideas of what has been called **Q-Analysis**, viz., the study of connectivities inherent in a relation between sets of entities - and which can always be represented by a matrix array of 0's and 1's - what is hereinafter called a **Dowker array**, and the ensuing **simplicial complexes** denoted by KY(X) and its conjugate KX(Y) are defined. And here again such ideas would not be practicable without the development of the (digitalised) computer system - which is needed to analyse large files of data in a reasonably short time.

Subsequently the idea of **p-Events** is introduced and the concept of an inherent sense of time, called **Komplex-Time**, is compared with what we normally think of as time, viz. **Clock-Time**.

0-Events giving rise to other p-Events leads to remarks on the probabilities induced thereby.

Readers who have experimented with the techniques in Q-Analysis will be very familiar with the contents of the first few chapters.

The approach to **Komplex-Time** is introduced in chapter-2 and this feeds through to later chapters and later topics.

Finally it is very much hoped that no esoteric knowledge of mathematics is required for the reading of these reflections - **except** for those interested. Appendix-B contains suggestions for an algebraic approach (using our familiar Exterior Algebra) with which they might wish to experiment.

Appendix-C contains references to some of the papers already published on the subject.

Chapter-1 / PART-4 Basic Set Properties

[1.1] Mathematical and Other Sets

We shall assume that the idea of a "set" is largely an intuitive one, but one which is compatible with the ideas first advanced by the mathematician **CANTOR**.

We therefore regard a set as any well-defined collection of items, whether finite (in number) or infinite, and by "well-defined" we shall use the **CANTOR** test, viz., that of any given item it is given that we can decide whether or not it is a member of the given set.

Mostly we shall be concerned with **finite sets**, but we shall need to refer to a few infinite ones.

Example-1

Take the finite set, denoted by **Boys**, and defined by its members

$$\text{Tom, Dick, Harry.}$$

We shall write   **Boys** = {Tom, Dick, Harry}

Nothing very remarkable about that !

Example-2

The set of all positive and negative integers (together with the number 0) is usually called the set **J** and written as

$$\mathbf{J} = \{0, \pm 1, \pm 2, \pm 3, \ldots \text{etc}\}$$

This is clearly a set with a countably infinite number of members, so it is impossible to list them all When it is desirable to distinguish them we shall write the set of all non-negative integers as

$$\mathbf{J}^+ = \{0, 1, 2, 3, \ldots \text{etc}\}$$

Example-3

The infinite set of all the **rational numbers** we shall write as

$$\mathfrak{R} \quad \text{which will mean}$$

all numbers of the form p/q, where p,q are in **J** but where $q \neq 0$.

If the item x is a member of some set **P** we write $x \in \mathbf{P}$, otherwise we write $x \notin \mathbf{P}$

So, for example,   Tom $\in$ **Boys**   but   George $\notin$ **Boys**

and $2/3 \in \mathfrak{R}$ but $4/6 \notin \mathfrak{R}$ because $4/6 = 2/3$.

We also need to take cognisance of the **null set** which has no members and we shall denote this by the symbol $\square$ (for the empty box).

A **universal set U** - which contains "everything relevant in the context" - can often be used.

[1.2] Subsets of Sets

When all the items of a set **A** are to be found in another set **B** we call **A** a **subset** of **B** and write **A** ⊂ **B** (**A** is contained in **B**) or that **B** contains **A** (written as **B** ⊃ **A**).

For the sake of the **algebra of sets** we also require that ☐ be contained in every other set - so that ☐ ⊂ **A** (for all sets **A**).

Items which are common to two sets **A** and **B** constitute the **intersection** of **A** and **B**, and the set produced thereby is written **A**∩**B**, whereas the set which contains all the members of **A** together with the members of **B** (without repetition) is the **union** of **A** and **B** and written as **A**∪**B**. Not surprisingly these symbols ∩ and ∪ are usually read as **cap** and **cup** - which are are equally and respectively denoted by AB (the "product") and A+B (the "sum")

The intersections and unions of sets are also represented by **Venn diagrams** - as shown on page 5.

[1.3] Hierarchy of Sets

From the above we can see that a set can arise via combinations of other sets - e.g. the simple case of the set **A**∪**B**, in which sets **A** and **B** behave as "items" (members) of a set **C**, say, and which equals {**A**,**B**}. This leads us to the idea of a **hierarchy of sets** viz. those formed by way of

items → (sets of items) → (sets of (sets of items)) → (sets of (sets of (sets of items))) → etc.

The items themselves can be regarded as **singleton** sets - each of which contains just the one item. For example the items in our set **boys** can themselves constitute the singleton sets

{Tom} {Dick} {Harry}

which fit it into the above hierarchy sequence.

Another example of a set of sets is derived from any given (finite) set **A**. It is called the **power set** of **A** and consists of the total subsets of **A** : we write this as ℘**A**. It also contains ☐.

Clearly the union of all the sets in ℘**A** is **A** itself.

Simple examples of ∪ and ∩ are shown in Fig.1 following.

The shaded areas are the subsets **A**∪**B**, **A**∩**B**, and **A**∩**B**∩**C**.

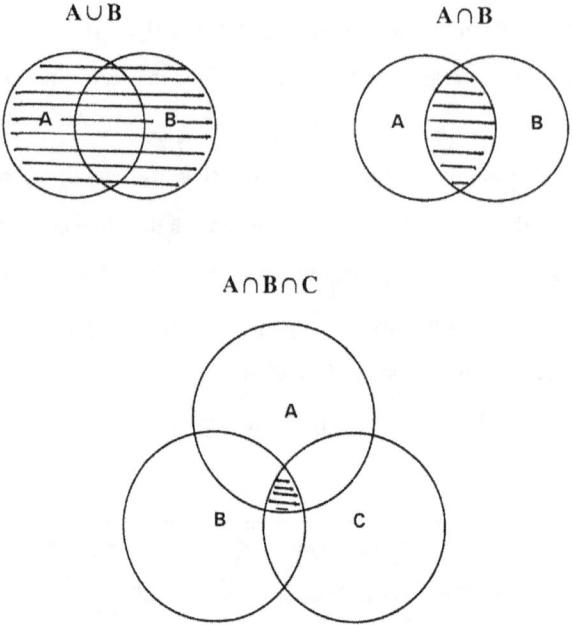

Fig.1   Venn Diagrams

[1.4] The Russell Paradox

This was famously posed by the mathematician **Bertrand Russell** in the following homely terms.

In a certain town (a) all the men are clean shaven
(b) the barber is a man and shaves all the men who do not shave themselves.

Question : Does the barber shave himself ?

The paradox -- >

(1) If the barber shaves himself then he must be a man who does not shave himself
(2) If the barber does not shave himself then he must be shaved by the barber - who is "himself".

A straightforward analysis of this unanswerable question was provided by **Russell** by appealing to

the above concept of a **hierarchy** of sets, as follows.

The "barber" is being appealed to (defined) in different places in the hierarchy - in the one case as

the **set** of men he shaves and in the other case as the **item** man/barber.

So the syntax inherent in the algebra of sets has been used improperly. The question is therefore or which cannot logically be asked. The hierachy involved is therefore

{items = all shaven men} → {the set of two sets, viz. self-shaved set and barber-shaved set}

and the "barber" is the name given to an **item** in the first set **as well as** to a **set** in the second set.

The lesson to be learned from this paradox is that we must respect the different **hierarchical levels**
One way of doing this is to indicate the levels by some nomenclature and we shall use the
following notation.

In any given hierarchy we shall denote

       the 1st level by **N**   (for "normal", if you like)

       the 2nd level by **N+1**

       the 3rd level by **N+2**   and so on.

The syntax then **forbids** us to make statements about entities which link the same name, but differe
definitions at, say, levels **N+r** and also at **N+s** ($r \neq s$).

The obvious thing is to remember to use diferent names for the entities - and to note the **N**-level at
which any such statements are being used.

[1.5] Dowker arrays

At the (N+1)-level, where the members are themselves N-level **sets** there can be a relation betwen
Such a relation can be described by a matrix-array whose elements are 0's and 1's. This is also call
a **binary array** but because of his early work with these arrays we shall call them **Dowker arrays**.
A simple example of this is the following relation, K, between sets X and Y which are defined by

       Y = {Y1, Y2, Y3, Y4, Y5, Y6, Y7}   and

       X = {X1, X2, X3, X4, X5, X6, X7, X8, X9}

The possible **Dowker-array** can have the following matrix form, viz.,

K	X1	X2	X3	X4	X5	X6	X7	X8	X9
Y1	1	0	1	1	0	1	0	1	1
Y2	0	1	1	0	1	1	1	0	0
Y3	1	0	0	1	0	1	0	0	1
Y4	0	1	1	0	0	0	1	1	0
Y5	1	0	1	1	1	1	0	1	1
Y6	0	0	0	1	0	0	1	1	0
Y7	1	1	1	0	1	1	0	0	1

This tells us, for example, that looking along the rows

       Y1 is related to {X1,X3,X4,X6,X8,X9}
       Y5 is related to {X1,X3,X4,X5,X6,X8,X9}

whilst if we look down the columns we see that

X1 is related to {Y1,Y3,Y5,Y7}   and
X6 is related to {Y1,Y2,Y3,Y5,Y7}

[1.6] **Equivalence relation** defined on a set

A relation, R, can be defined on a set **A** by writing xRy, where x $\in$ **A** and y $\in$ **A**.

Example-4

The idea of "similarity" is a relation on the set, **G**, of all plane figures in **Euclidean** geometry.

Denoting it by S we notice that for members x,y,z $\in$ **G** we have the following properties :

(a) xSx   (this is the property of being **reflexive**)

(b) if xSy then ySx (the property of being **symmetric**) - and

(c) if xSy and ySz then xSz (the property of being **transitive**).

These conditions on **G** define what we call an **equivalence relation**.

Of course we soon realise that not all relations on a set will be equivalence relations.

Example-5

The relation "is the mother of", say M, is certainly not an equivalence relation on the set of all human beings - because it does not satisfy any of the above tests (a),(b),(c).

Example-6

The relation "is greater-than or equal to", written as $\geq$, on the set of integers **J** satisifes the conditions (a) (b) and (c), but the more restrictive relation " > " does not satisfy (a) or (b), but does satisfy (c).

Then $\geq$ is an equivalence relation on **J** but " > " is not.

[1.7] An Equivalence relation defines a **partition** of a set

The significance of an **equivalence relation** lies in the following theorem, viz.,

An **equivalence relation**,R, defined on a set **A** separates the members of **A** into disjoint sets - constituting a partition of **A**. If we denote these subsets of **A** by

$B_1, B_2, B_3, \ldots . B_n$ we require the conditions :-

(1) **A** = Union of all the B_r,   r=1..n   and

(2) For any r$\neq$s $B_r \cap B_s = \square$ - the sets B_i are disjoint.

The proof of this follows from the three conditions (a), (b) and (c) above.

For if we assume that there is some x such that x$\in$**B1** and x$\in$**B2** then if **B1**$\cap$**B2** is not empty we have for any other y$\in B_1$ xRy and also yRz (for every z$\in$**B2**).

By the transitive property of R, xRz, and by the symmetric property zRx, whence it follows that $B_1 = B_2$.

So, choosing an $x \in B_1$ fully defines an exclusive subset of **A**. If this does not exhaust the set **A** then we choose a member $b \in A$ which is not in B_1 and discover another B_k which is disjoint from our first subset B_1.

This process means that the conditions (1) and (2) above are satisfied.

Having obtained such a partition we can see that the distinct subsets B_1, B_2 etc. may now be regarded as **members** of a set which is one step higher up the hierarchy, say,

$$C = \{B_1, B_2, B_3, \ldots B_n\}$$

and this set is called the **quotient set** of the pair [**A**, R] - sometimes called the **factor set** and often written as **A**/R since there is a sense in which **A** has been "divided up" by R.

[1.8] Cartesian product of sets

We can form the Cartesian product, written $A \times B$, by taking all pairs of elements such as (x,y) where $x \in A$ and $y \in B$, the order being significant.

This is exhibited in the geometrical plane first used by the French mathematician **Descartes** - where the points of the plane (the **Cartesian plane**) have such pairs as their co-ordinates. In that geometry the values of x and y are taken to be real numbers.

We can also remind ourselves that a relation R on a set **A** actually consists of pairs of its members which satisfy that relation.

So if members x and y are such that xRy then these members will be found in the Cartesian product **A** with itself, that is to say as members of $A \times A$.

This is why it is often useful to regard the relation R as a subset of $A \times A$ and so it is quite OK to write $R \subset A \times A$.

Example-7

If we let **A** be the set of all real numbers then we can contemplate the relation C as the subset of points in the Cartesian plane which form the unit circle.

This is the subset in $A \times A$ - being all pairs (x,y) which satisfy the relation C, viz., the circle represented by the relation

$$C: \quad x^2 + y^2 - 1 = 0$$

Chapter-2 /PART-4   Events and Times via a DOWKER Complex

[2.1] The Dowker complex

The idea of a simplicial complex has already been defined in chapter-2 of Part-2 and we begin with a few reminders of such a structure.

Given a finite set **A** we rename some of its properties as follows :

- (a) the items/elements of **A** are to be called **vertices** or **nodes**.
- (b) if there are $(q+1)$ vertices the power set $\mathcal{P}\mathbf{A}$ is to be called a **q-simplex**
- (c) the subsets of $\mathcal{P}\mathbf{A}$ are to be called the **faces** of the q-simplex
- (d) the q-simplex is to mean **all of the faces** of the q-simplex
- (e) the number "q" is to be called the **dimension** of the q-simplex.

It is usual to denote such a q-simplex by the Greek letter σ : if it is of dimension q it will then be denoted by σ_q.

This means that we regard

the vertices/nodes to be of zero dimension - the σ_0's

the sets of pairs to be of 1-dimension - the σ_1's

the sets of triples to be of 2-dimensions - the σ_2's

the sets of quadruples to be of 3-dimensions - the σ_3's, and so on.

Each of the subsets of this q-simplex is therefore an r-simplex, where $r <= q$.

Example-1

A 2-dimensional simplex may be visualised as a triangle - possessing
    3 vertices (0-simplices)
    3 edges (1-simplices)   and
    1 triangle (2-simplex)

It will be written as :-

a σ_2 whose faces are three σ_1's and three σ_0's.

If we denote the vertices by the letters A,B,C then we can sometimes use the notation

the σ_0's as $<A>$, $<B>$, $<C>$
the σ_1's as $<AB>$, $<BC>$, $<CA>$ and
the σ_2 as $<ABC>$

Now a **DOWKER** complex is a collection of such simplices, for various values of q, and they are more or less connected (joined together) by sharing their various faces.

It is defined by the following binary matrix array viz., the **Dowker**-array (v. above).

The resultant structure is a **DOWKER complex**, which we shall denote by the letter **K** and in simple cases is usually visualised by a geometrical resprepresentation.

In point-set topology the σ_2 would be identified as the **convex set** determined by the three vertices A,B, and C. That is to say, in homely terms, σ_2 can be thought of as the "solid"/"filled-in" triangle ABC.

[2.2] **If N is the maximum value of q occurring in K** it can be proved that the complex K can always be embedded in a Euclidean Space of dimension $(2N+1)$ - (v. Appendix-E [94]). This means that we must expect drawn representations of higher order complexes to be somewhat elusive - so we shall only draw the geometry of the following example, which is the complex defined by the **Dowker-array** :-

```
K  | X1 X2 X3 X4 X5 X6 X7 X8 X9 X10 X11
Y1 |  1  1  1  1  0  0  0  0  0  0   0
Y2 |  1  1  0  0  1  0  0  0  0  0   0
Y3 |  0  0  0  0  1  1  0  0  0  0   0
Y4 |  0  0  0  0  0  1  1  1  0  0   0
Y5 |  0  0  0  0  0  1  0  1  0  0   0
Y6 |  0  0  0  0  0  1  0  0  1  0   0
Y7 |  0  0  0  0  0  0  1  0  0  1   1
```

Its geometrical representation is shown in Fig.2 and gives us the opportunity to introduce the **two complexes** which are always available viz., those denoted by

$$KY(X) \quad \text{and} \quad KX(Y).$$

The first is obtained by regarding the X's as the vertices whilst the Y's are the names of the q-simplices. For example Y1 is the $\sigma_3 = <X1,X2,X3,X4>$ and can be drawn as the (solid) tetrahedron.

The second is obtained by regarding the Y's as the vertices and the X's as the simplices. In the first case we read the array row-by-row, and in the second case we read it col-by-col. In KX(Y) the maximum q-value is 2, given by $\sigma_2 = <Y4,Y5,Y6>$ (triangle).

In each case the complex falls into two distinct parts ; in general the number of distinct parts which a complex exhibits is called its zero-oder **Betti** number - following the theorem which **Dowker** proved in his original paper (v. Appendix-B [1]) in which he showed that the algebraic **homology groups** of KY(X) and KX(Y) are isomorphic - although we shall not need to use those particular results here, except to say that the isomorphism explains why the zero order **Betti** numbers are the same for both KY(X) and for KX(Y).

238 MATHEMATICAL PHYSICS

KY(X) complex

Two disjoint parts

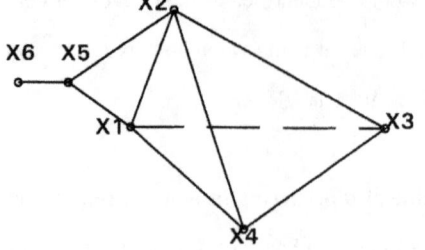
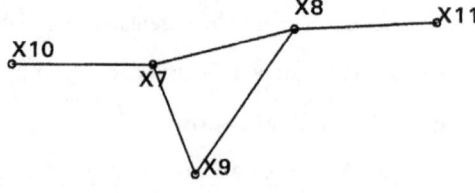

KX(Y) complex

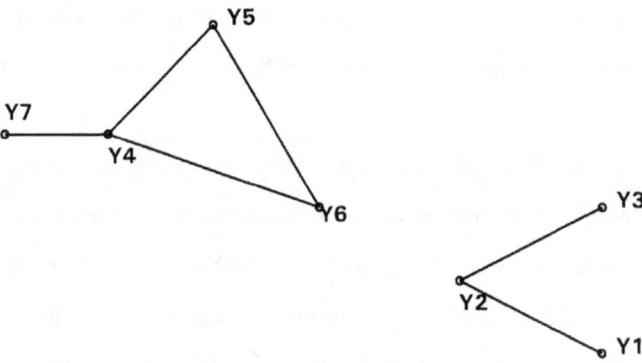

Fig.2 The Two Geometries of a Dowker Complex

[2.3] A q-connection relation

When two simplices in a complex K share a face of dimensiion q we shall say that they are q-connected, and a **q-chain**, c_q, is to be a sequence of such σ's ach of which shares a q-face with at least one member of the chain. By this means we can define a relation Γ on the simplices of K by saying that $\sigma^1 \, \Gamma \, \sigma^2$ if there exists a c_p, for some p, which contains σ^1, σ^2 somewhere within it ; they need not be immediately adjacent.

It is then clear that Γ is **reflexive** since σ shares all its faces with itself,

It is also **symmetric** since if $\sigma^1 \, \Gamma \, \sigma^2$ then $\sigma^2 \, \Gamma \, \sigma^1$, and it is **transitive** since if $\sigma^1 \, \Gamma \, \sigma^2$ and $\sigma^2 \, \Gamma \, \sigma^3$ then $\sigma^1 \, \Gamma \, \sigma^3$.

Thus the relation so defined is an **equivalence relation** on the set of all the simplices of the complex KY(X), as there is also an equivalence relation on the simplices of KX(Y).

[2.4]
The overall effect is that Γ produces a **partition** of all the simplices throughout K.

Of course, with larger arrays and complexes the disjoint components of such a partition will only be obtained via the use of computer software. The whole process of finding such a partition has been called **Q-Analysis** [v. Appendix-E, [111] [113] [114] [115]].

If we write K_0 for the disjoint components which are 0-connected (share vertices) and K_1 for the 1-connected components (share edges) ... up to K_N then we are showing all the disjoint pieces exhibited by K at each dimensional level.

That is to say we can write $K \equiv K_0 \oplus K_1 \oplus K_2 \oplus ... \oplus K_p \oplus ... K_N$ where the "tensor sum" $\oplus$ is here used merely to keep the components separate.

Thus the complex K can be seen as a **graded complex** - graded by the q-values.

In the diagram (Fig.2) we can see, by inspection, that in the complex KY(X) there are

    2 pieces in K_0 (**Betti** number is 2)

    5 pieces in K_1 (which do not share edges)
    and these are {Y1, Y2} {Y3} {Y4,Y5} {Y6} and {Y7}

    2 pieces in K_2 (which do not share triangles)
    and these are {Y1} and {Y4}

    1 piece in K_3 (a single tetrahedron)
    and this is {Y1}.

In the conjugate complex, KX(Y), we have

    2 pieces in K_0 (same **Betti** number as in KY(X))

    4 pieces in K_1 (which do not share edges)
    and these are {X1 ≡ X2} {X5} {X7, X9} {X8}

1 piece in K_2 (a single triangle)
and this is $\{X7\}$.

[2.5] Defining p-Events associated with a Complex

Insofar as the vertices and simplices of a complex K are relevant to some practical experience of a relevant observer (by a human or experimental instrument) we shall thereby refer to any one of its σ's as an **Event**. Since any such event will carry the structure of the σ_p which identifies it we shall say that it is a **p-Event**.

[Note: We use a parameter "p" since we do not wish to overburden the letter "q" : but of course the "p" is an example of a "q"]

So by these definitions we shall suppose that a **Dowker** array not only defines a complex, K, with its inbred **q-connectivities** but also, via any associated happenings it defines a collection of **p-Events**.

The Events in KY(X) will correspond to the Y's (the row names) whilst the Events in KX(Y) will correspond to the X's (the column names).

[2.6] Complexes representing Time-Moments and Time-Intervals

What we naturally call **Clock-Time** is an example of the definitions of **p-events**. If the X's are written as T's (say, ... T_0 T_1 T_2 ... T_n ...) and the Y's written as I's (say ... I_1 I_2 I_3 ... I_n ...) then the T's represent the **0-Events** ; these are the **clock-time-moments**, the vertices in the complex, and the I's become the **1-Events**, being the **clock-time-intervals**. A typical piece of this standard Clock-Time will have a **Dowker** array like the following.

A Piece of the infinite Linear Clock-Time

L	T0	T1	T2	T3	T4	T5	T6	T7	T8	T9	......	->
I1	1	1	0	0	0	0	0	0	0	0	......	->
I2	0	1	1	0	0	0	0	0	0	0	......	->
I3	0	0	1	1	0	0	0	0	0	0	......	->
I4	0	0	0	1	1	0	0	0	0	0	......	->
I5	0	0	0	0	1	1	0	0	0	0	......	->
I6	0	0	0	0	0	1	1	0	0	0	......	->
I7	0	0	0	0	0	0	1	1	0	0	......	->
I8	0	0	0	0	0	0	0	1	1	0	......	->
I9	0	0	0	0	0	0	0	0	1	1	......	->
↓				.... ad infinitum								->

In Fig.3 (v.seq.) we have here picked out a small finite piece of what is actually an infinite complex - its dimension being 1, so it can also be referred to as **Linear-Time**.

PART 4 - STRUCTURES, TIMES AND EVENTS **241**

A finite piece of Linear Clock-Time

o = clock moment

o———o = clock interval

The Complex for a cyclic Clock-Time

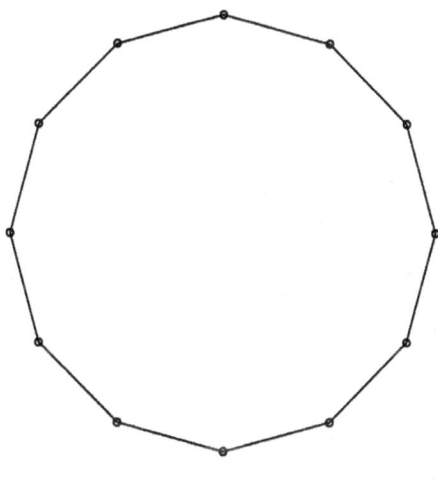

Fig.3

Clock moments are 0-simplices - Clock intervals are 1-simplices

The existence of this **linear** clock-time complex is our experience of "ordinary" time - which we take as a standard (scientific) clock-time. Our mechanical and electronic machines for recording the usual "time" are such that, to allow for the ongoing developments in electronic gadgetries (such as the computer on which this book is being typed), it has become necessary to be able to record ever smaller units for the time-intervals. For example the "second" is now defined in the scientific world as 9,192,631,770 periods of the radiation which arises from the transitions between two hyperfine levels of the ground state of the caesium-133 atom (I hear you say "have the scientists gone completely mad" !).

The time-interval, from the caesium-133 atom, looks very much like "simultaneous" and cannot possibly be discriminated by simple human observation.

At an everyday practical level however we record our paths along the linear clock-time by using what we might call collections of **cyclic clocks**. These contain a practical and finite set of successive linear time-intervals which repeat themselves. To this end we can associate our experiences of our **own p-events** in time concepts like

...... seconds, minutes, hours, days, weeks, months, years ......

and thereby fit in a **lifetime** for ourselves, and others, on this planet earth.

It is as if we take a finite piece out of the infinite linear time complex and wrap it around an imaginery circle - and this circle is repeated down the linear complex for whatever reason we think fit, such as a **lifetime**.

The complex for a **cyclic clock** is shown in Fig.3 and, not surprisingly, it looks just like a "clock" !

[2.7] **Birth, Death**, and the bit in-between

For easy reference we reproduce the array from §[2.6], viz.,

A Piece of the infinite Linear Clock-Time

L	T0	T1	T2	T3	T4	T5	T6	T7	T8	T9	......	->
I1	1	1	0	0	0	0	0	0	0	0	......	->
I2	0	1	1	0	0	0	0	0	0	0	......	->
I3	0	0	1	1	0	0	0	0	0	0	......	->
I4	0	0	0	1	1	0	0	0	0	0	......	->
I5	0	0	0	0	1	1	0	0	0	0	......	->
I6	0	0	0	0	0	1	1	0	0	0	......	->
I7	0	0	0	0	0	0	1	1	0	0	......	->
I8	0	0	0	0	0	0	0	1	1	0	......	->
I9	0	0	0	0	0	0	0	0	1	1	......	->

↓ .... ad infinitum .... ->

This array denotes the relation between time-moments and time-intervals.

The interval I_1 is the 1-event represented by the 1-simplex $<T_0\ T_1>$ .. and so on.

The **biological-time** which we experience - as do all living beings - is the experience of a **sequence of events** as we go through the process of **formation, growth,** and **decay**. When we map this experience onto an observed **clock-time** we naturally associate it with **birth, growth, decay** and **death**. The occurrence of **death** is, for the biological entity the **end of time** (as the entity knows it).

So we propose that we should equate **death** with the **end of (clock-) time**, as do the Buddhists ?

The clock-time intervals in between **birth** and **death** is, of course, **life**.

We have traditionally dealt with this **Clock-Time** as **Linear-Time**, and the **Dowker** array for this finite (and linear) subcomplex is the one listed above.

This complex is an example of a continuous **ticking device**, such as the blips put out by the quartz crystal in a computer, or the standard time reference by, say, the **caesium clock** - which is the international standard for the "second" - as already noted. We naturally regard these devices as working away forever - from the past and into the future.

Here we should point out that the devices themselves "know nothing" of these concepts **past** and **future** ; if they were "inumerate humans" their activity might correspond to what is called "jumping on the spot".

This strongly sugests that there is a **personal sense** of **Time** which we experience - what is often called **biological time**, with reference to a **biological clock** - which idea is to provide a pointer for our further discussions.

A Cyclic Piece of the infinite Clock-Time

C	T0	T1	T2	T3	T4	T5	T6	T7	T8	.....	->
I1	1	1	0	0	0	0	0	0	0	.....	->
I2	0	1	1	0	0	0	0	0	0	.....	->
I3	0	0	1	1	0	0	0	0	0	.....	->
I4	0	0	0	1	1	0	0	0	0	.....	->
I5	0	0	0	0	1	1	0	0	0	.....	->
I6	0	0	0	0	0	1	1	0	0	.....	->
I7	0	0	0	0	0	0	1	1	0	.....	->
I8	0	0	0	0	0	0	0	1	1	.....	->
I9	1	0	0	0	0	0	0	0	1	.....	->

We notice that this structure illustrates the concept of a q-loop (v. Appendix-A) since it contains the 1-loop (i.e. a loop which is bounded by 1-simplices/edges) viz.,

$I1 <T0,T1> I2 <T1,T2> I3 <T2,T3> ... ... I8 <T7,T8) I9 <T8, T0> \rightarrow I1$ etc.

whence I9 joins up with I1 again at the clock-moment T0.

[2.8] Komplex-Time (KT) and Clock-Time (CT)

[Note : We use the spelling of complex, viz., Komplex, to remind ourselves that we are dealing with a **simplicial complex** - this avoids confusing the word with the everyday meaning of "complex" - since a simplicial complex is not complicated]

We shall regard this "biological time" as an **example of the time experienced** by any simplicial complex structure via the **events** dynamically unfolding therein, and which generates that intrinsic sense of time.

This built-in intrinsic sense of time we shall call the **Komplex-Time** which is inherent in the structure of those events - which comprise that simplicial complex.

So in Komplex-Time the "possessor" of the complex experiences his/her structure as a sequence of serial or parallel events <u>in that complex</u>, and since these events are graded as p-events the sense of time will also be graded.

If we use the Greek letter τ for Komplex-Time it means that the pattern will be written as
$$\Psi^* \equiv \tau^0 \oplus \tau^1 \oplus \tau^2 \oplus \ldots \oplus \tau^p \oplus \ldots$$
where, for example, τ^p is the pattern of Komplex-Time experienced via the p-Events of the structure.

Life for this being will largely consist in matching (or trying to match) this sequence of p-events with pieces of the standard linear clock-time. When we do this we are, in effect, **blaming the Clock-Time** for our sense of Time "like an everflowing stream ..." as the hymn has it, whereas it is the life-enfolding complex which is to **blame** (if that is not too harsh a word) for what we commonly regard as the **flowing of clock-Time**. And this reversal of roles is neatly expressed in terms of the complex KI(T) and its conjugate KT(I).

Let us take as an example the following **Dowker** complex - one which expresses a structure of Events for some living creature/entity, relating its experience of p-Events E1 ... E7 to its awareness of Clock-Time-Moments T1 ... T8.

K	T1	T2	T3	T4	T5	T6	T7	T8
E1	1	0	1	1	0	0	0	1
E2	1	1	0	0	0	1	1	0
E3	0	0	1	1	1	0	0	1
E4	1	1	1	1	0	0	0	0
E5	0	1	0	1	0	1	0	1
E6	1	0	0	0	1	1	1	0
E7	1	1	1	1	1	0	1	1

This involves the usual approach, discussed in the above, based on the complex KE(T). But what if we adopt the view that **the p-Events determine the sense of Clock-Time** - rather than the other way around ?

Then this amounts to reading the columns (the T's) in terms of the E's. This means that we look at the complex KT(E) as opposed to KE(T). Now the entity which is carrying the p-Events transmits this experience to the awareness of the Clock-Time events.

Assuming this to be the case we are led to the following

Postulate-A
   Komplex-Time is the basic experience of a living existence and this is imposed on the awareness (the progression) of Clock-Time

In the example shown above, and looking at KE(T), we see that (eg) E4 is a σ_3 and is given by E4 = $<T1,T2,T3,T4>$, and since Clock-Time **intervals** only are measured, the 3-Event appears to require the the total time which consists of all the intervals (like T1-T2 etc) to be found in E4. But the total number of edges (pairs of T's, or 1-simplices) in E4 is six - which becomes its Clock-Time experence.

[Note : The number of distinct pairs of n items is the number of ways of choosing pairs out of all n items and without regard to the order of choice, eg. (2,3) is the same as (3,2). This number is $n(n-1)/2$]

Now if we look at KT(E) the p-Event called T1 becomes a σ_4, viz., $<E1,E2,E4,E6,E7>$ - since t T's denote Clock-Time-Moments it means that the entity described by this **Dowker** array identifies T1 by the occurrence of the Events E1, E2, E4, E6, E7. Of course the Clock-Time-Moments need be the naturally assumed one (of a second or less - the ordinary "now") but can be anything like hour, day, week, month, year, century, millennium, or whatever. So experiencing a "moment" like T1 could be like saying "I know it's Thursday because that is when the Events "going shopping", "cleaning the car", "meeting my friend for lunch", and similar Events occur.

[Note : The government only deals in **Clock-Time** and ignores each **personal Komplex-Time**.]

We would, of course, hardly expect Komplex-Time, for a given structure, to coincide with Clock-Time ; it is hardly surprising therefore that **lifetimes** (measured in Clock-Time) differ greatly, since Komplex-Times must dominate the experience. This explains a common experience among human beings, for during our early years - the years of childhood growth and development there are varying numbers of KT-moments included in the outside CT-intervals ; so, looking at the accepted Clock-Time we feel such things as "will next week/year never come ?".

And much later on in the period of adulthood we slowly find the Clock-Time more or less tolerable whilst in old age (when the biological structure is winding down) - the Komplex-Time intervals contain a number of Clock-Time moments - we experience a sense of "how (clock) time seems to fly by".

Any sense of, or desire to, "go back in time" is meaningless when contemplated as a possibility in Clock-Time since Clocks per se have no sense of direction. It must be the stored memory of p-Events in the relevant complex structure which more or less possesses this facility. And when memory is not consciously recalled for this purpose perhaps we get close to the experience via situations known as **deja vu**.

[2.9] The relevance of t-forces.

Referring to Appendix-A where we have introduced the concept of a Π^* and have seen it represented as a graded pattern on a gvien complex viz.,

$$\Pi^* \equiv \Pi^0 \oplus \Pi^1 \oplus \Pi^2 \oplus \ldots \oplus \Pi^n$$

Now if we contemplate **some change** in the complex K - such as the simple case where just one of the vertices is removed - then we will get a new patteern Π^* associated with K. This change of Π^*, say $\delta\Pi^*$, will be experienced throughout K and which will therefore affect any phenomenon which we have associated with the simplices of K, and that means it will affect some or all of the **p-Events** in that K.

We shall therefore call this change, $\Delta\Pi^*$, a set of **t-forces** acting on the **p-events**.

[Note : This has a parallel in **Einsteinian** physics, where for example the gravitational field is due to the warping of the **Euclidean** space-time continuum - provided by the presence of masses : here we regard the Π^* as analogous to the "momentum vector" - changes in which are experienced as "impulse" with its "rate-of-change" being "force".]

The letter "t" in this context not only reminds us of the **topology** of K but also can be used to identify the q-values of the affected simplices/events of the complex K. Thus we can refer to 1-forces (which are the $\Delta\Pi^1$ acting on the 1-simplices/edges) : or, for some particular t-value, to those forces experienced by the t-simplices in K. Social pressures can be seen as t-forces - for example, as a result of the changes in social environment occasioned by moving house, climate change, financial situations, family disturbances, employment ambitions, natural catastrophes, government actions, love affairs etc..

Any such changes will exert t-forces on the p-Events to be found in the relations which surround our daily lives. And since, unlike Physics, we cannot express such forces to some universal basis nevertheless the **ratios** of the various components in $\Delta\Pi^*$ can give us important insights into what is being experienced ; for example where the maximal and minimal forces are to be experienced, but also at what dimensional level (q-value) they are to found - for a high q-level will indicate a more widespread experience of the forces.

[2.10] In total, therefore, we can see our analysis of a complex defined by a **Dowker**-array as consisting in the identification of

      q-simplices & q-components of K
        plus eccentricities, structure vector, stars, loops

      p-events associated with K and

      t-forces generated by changes in patterns on K, or being due
        entirely to changes in the underlying structure (backcloth) of K.

Chapter-3 /PART-4 Various Examples

[3.1] A few of many **Dowker** arrays.

The following arrays are typical of possible **Dowker** arrays defining complexes KY(X) and KX(Y), using the values 0 or 1 - λ_{ij} being the value in the ith row and the jth column.

(a) Y = set of individual People, X = set of named Committees, and
$\lambda_{ij} = 1$ means that person Y_i who sits on committee X_j.

(b) Y = set of kinds of people, X = set of named business enterprises
$\lambda_{ij} = 1$ means that the group Y_i invests cash in business X_j.

(c) Y = set of traffic routes, X = set of types of vehicles, and
$\lambda_{ij} = 1$ means that route Y_i carries vehicles of type X_j.

(d) Y = set of streets in a town, X = retail outlets in the town, and
$\lambda_{ij} = 1$ means that in Y_i we find outlet X_j.

(e) Y = set of educational courses, X = set of named colleges, and
$\lambda_{ij} = 1$ means that course Y_i is offered by college X_j.

(f) Y = set of medical illnesses, X = pathological symptoms, and
$\lambda_{ij} = 1$ means that illness Y_i carries symptom X_j.

(g) Y = set of manufacturers, X = set of industrial techniques, and
$\lambda_i = 1$ means that manufacturer Y_i uses technique X_j.

(h) Y = set of political groups, X = set of policies, and
$\lambda_{ij} = 1$ means that group Y_i adopts policy X_j.

(i) Y = setf of theses, X = setf of intellectual concepts, and
λ_{ij} means that thesis Y_i requires concept X_j.

It appears endless and ubiquitous - indeed insofar as relations are universally occurring we offer **Postulate-B**, viz.,

to every experience of a relation between various sets A and B there is to be found a Dowker complex whose simplices in KA(B) and/or in KB(A) correspond to associated p-Events.

[3.2] Suggestions for patterns of traffic on complexes.

(A) Y = set of commitees, X = set of items on agendas, and
traffic via patterns on KX(Y) = budget allocations.

(B) Y = set of streets in a own, X = setf of retail outlets, and
traffic via patterns on KX(Y) = weekly turnover by the X_j.

(C) Y = set of colleges, X = set of courses offered, and
traffic via patterns on KX(Y) = applications for courses.

(D) Y = setf of illnesses, X = set of available drugs, and
traffic via patterns on KX(Y) = budget spent on X_j as well
as traffic patterns on KY(X) = consulting time needed for Y_i

and so on.

[3.3] Anne Chamberlain : A Study of Behcet's Disease

This study of the medical features of Behcet's disease was undertaken by Anne Chamberlain in 1976 [v. Appendix-E [131]] with a complex covering 227 persons (Yk's) - of whom 32 were patients with the relative symptoms and the rest were relatives and spouses. The number of columns (Xk's) amounted to 71 and represented various significant symptoms of the disease. Of these 32 were probands and the rmaining ones, 227, were relatives.

A coding system of suitable weights was introduced which allowed a number of suitable slicings to be used. This technique made it possible to analyse a number of complexes in order to look for answers to seven questions, viz.,

>Qn 1. Is Behcet's syndrome a disease entity ?
>Qn 2. If so, what features are of the greatest diagnosis importance ?
>Qn 3. Is sacro-iliitis part of the syndrome ?
>Qn 4. Do patients exhibit any links with other sero-negative arthropathies ?
>Qn 5. Do members of the patient's families show features of the disorder ?
>Qn 6. Which features are familial or inherited ?
>Qn 7. Are thes associated with any white blood cell abnormalities ?

Q-analysis gave appropriate answers to these questions.

[3.4] J.H.Johnson : The Q-Analysis of Road Intersections

In his paper [v. Appendix-E [119]] Professor Johnson studied a comprehensive set of road junctions, from the simplest to the complicated. His analysis begins with the structure of a 2-road junction, a narrow junction, a mini-roundabout, and a junction with a central reservation. Each of these is associated with a simplicial complex. He then goes on to define a **flow-pattern** on the complexes, as well as a pattern for **travel times**. These lead to an illustration of t-forces experienced by the system and their transmission throughout the complex - identified by the concept of **q-transmission**. Analyses of two examples of what the author calls **multidimensional spaghetti** junctions emphasise the applicability of the methods. Finally a "real-life" example of a roundabout with six entrances and exits - which had been studied by the Road Research Laboratory (1973) produced a choice of design which happily coincided with that of the research lab, which no doubt did not use Q-Analysis in their deliberations.

[3.5] J.H.Johnson : The Q-Analysis of Road Traffic Systems

This paper [v. Appendix-E [120]] was a follow-up to the previous one and went a great deal further in analysing what might be called a "macro-approach" to any traffic system.

A powerful method for analysing the obvious complexities, in eg a sizeable urban area, is developed by an approach through the concept of **hierarchies**. This was shown to enable (eg traffic flows, journey times, and t-forces, and congestion predictions) q-connectivity ideas to be handled via computers operating on relatively manageable complexes - using hierarchical levels $N, N+1, N+2, N+3 \ldots$. Zoning the urban area via these hierarchical levels provides structures which carry their own appropriate patterns of traffic, and the author demonstrates how these can be mathematically related.

He concludes with "any model (for traffic problems) which ignores the existence of the structural backcloth underlying traffic flows is certain to be inadequate".

[3.6] Beaumont & Gatrell : Examples in Dynamic Geography   (v. Appendix-E [115])

The authors give a clear description of the basic ideas and definitions of Q-Analysis and folow it with a discussion, from the standpoint of a geographer and/or a town planner, of a number of "estates" with associated locations as well as with postulated traffic patterns.

This allows them to illustrate the t-forces which arise from a change of pattern on the backcloth (the **Newtonian** approach) as well as those which arise from the effects of a change in the backcloth itself (the **Einsteinian** approach).

The following examples are given with a detailed discussion and analysis, using the relevant concepts of Q-Analysis.

(a) A typical binary matrix defining a Christallerian central place system based on the marketing principle.

(b) Various analyses, based on consideration of q-connectivities, of Social Area studies.

(c) An in-depth study of some of the work of J.H.Johnson (v. Appendix-E for refs.) chiefly in his contributions to Traffic Flows (literally motor cars etc) in transport systems.

(d) General ideas re Man-Environment relations,   and

(e) Giving insights into what is called Urban Morphology.

[3.7] J.H.Johnson : A Contribution to Artifical Intelligence.   (v. Appendix-E [143])

In this paper Professor Johnson is stressing (without being explicit) the notion, which the present author has long propounded, that **mathematics is a language**. Scholars in Pure Mathematics have been making this language more and more unambiguous over the centuries and the development of what is now called **applied mathematics** is an expression of using that language to discuss the phenomena in the world at large - whether it be in Physics, Chemistry, Economics, Social Studies etc..

In this paper he studies the so-called **complex systems** and convincingly argues the need for using the mathematical notion of **well defined sets of entities** to include those concepts in any serious discussion of the properties of such entities.

This naturally involves the concepts of **hierarchical levels** of observed data and a caveat against the irresponsible use of old-fashioned statistical arguments - where very often the writers appear to be unaware of the significance of **Russell's paradox**.

Chapter-4 /PART-4   Dynamics and Complexes

[4.1] Building a Complex, K

So far we have rather implied that we must start with a given initial complex, eg. by discovering a relevant **Dowker** array, and then proceeding to examine its q-connectivities.

But this cannot always be the case ; "someone behind the scenes" must be creating these structures and maybe challenging us to examine them.

So, building a Complex must be a ubiquitous activity ; how can it be done ? Well, like all building projects, it takes time, and also it almost certainly operates at more than one hierarchical level - we need the "bricks" and then the "mortar" to hold them together and then all the gadgets and things which may be needed to fit it out.

We therefore start with some **time-periods** to accommodate the action. These will usually be pieces of the linear clock-time. Let us denote them by periods P1, P2, P3, ... etc.: there is no need for these periods to be equal.

Since we are building a Complex we must start with at least one **vertex**, say V1, and if we allow for the project to be defined at any hierarchical level (which should be specified) such a vertex at, say, level (N+1) can be a collection of items at level N. For example, if we are building a complex corresponding to, say, a house then at level N the V1 might well be a simple old-fashiioned "brick", whereas at level (N+1) V1 could be about 10,000 such bricks. In any event, having defined the particular hierarchical levels which seem appropriate to the project, we start with, say, two appropriate vertices V1 and V2 - which initially are not connected/related.

Naturally the building of the complex must be controlled by some **agency**, whether **human** or **otherwise**.

This means we start at Stage-1 with these disconnected vertices and proceed to build the **Dowker** arrays in stages as shown in the following self-explanatory form. We demonstrate the process through six stages, as an example. In any particular application not only would the hierarchical structure be specified but also an appreciation of whatever **agent** is involved would be acknowledged.

Stage 1 :

K1	V1	V2	V3	V4	V5	V6	V7	V8
P1	1	0	0	0	0	0	0	0
P2	0	1	0	0	0	0	0	0

Stage 2 :

K2	V1	V2	V3	V4	V5	V6	V7	V8
P1	1	0	0	0	1	0	0	0
P2	0	1	0	0	0	0	0	0
P3	1	0	1	0	1	0	0	0

Stage 3 :

K3	V1	V2	V3	V4	V5	V6	V7	V8
P1	1	0	0	0	0	0	0	0
P2	0	1	0	0	0	0	0	0
P3	1	0	1	0	1	0	0	0
P4	0	1	0	1	0	0	0	0
P5	0	0	0	1	0	1	0	0

Stage 4 :

K4	V1	V2	V3	V4	V5	V6	V7	V8
P1	1	0	0	0	0	0	0	0
P2	0	1	0	0	0	0	0	0
P3	1	0	1	0	1	0	0	0
P4	1	0	0	0	0	0	1	0
P5	0	1	0	1	0	0	0	0
P6	0	0	0	1	0	1	0	1

Stage 5 :

K5	V1	V2	V3	V4	V5	V6	V7	V8
P1	1	0	0	0	0	0	0	0
P2	0	1	0	0	0	0	0	0
P3	1	0	1	0	1	0	0	0
P4	1	0	0	0	0	0	1	0
P5	0	1	0	1	0	0	0	0
P6	0	0	0	1	0	1	0	1
P7	0	1	1	0	0	0	0	0

Stage 6 :

K6	V1	V2	V3	V4	V5	V6	V7	V8
P1	1	0	0	0	0	0	0	0
P2	0	1	0	0	0	0	0	0
P3	1	0	1	0	1	0	0	0
P4	1	0	0	0	0	0	1	0
P5	0	1	0	1	0	0	0	0
P6	0	0	0	1	0	1	0	1
P7	0	1	1	0	0	0	0	0
P8	0	0	1	1	0	0	0	0
P9	0	0	1	0	0	0	1	0

…. and so on …

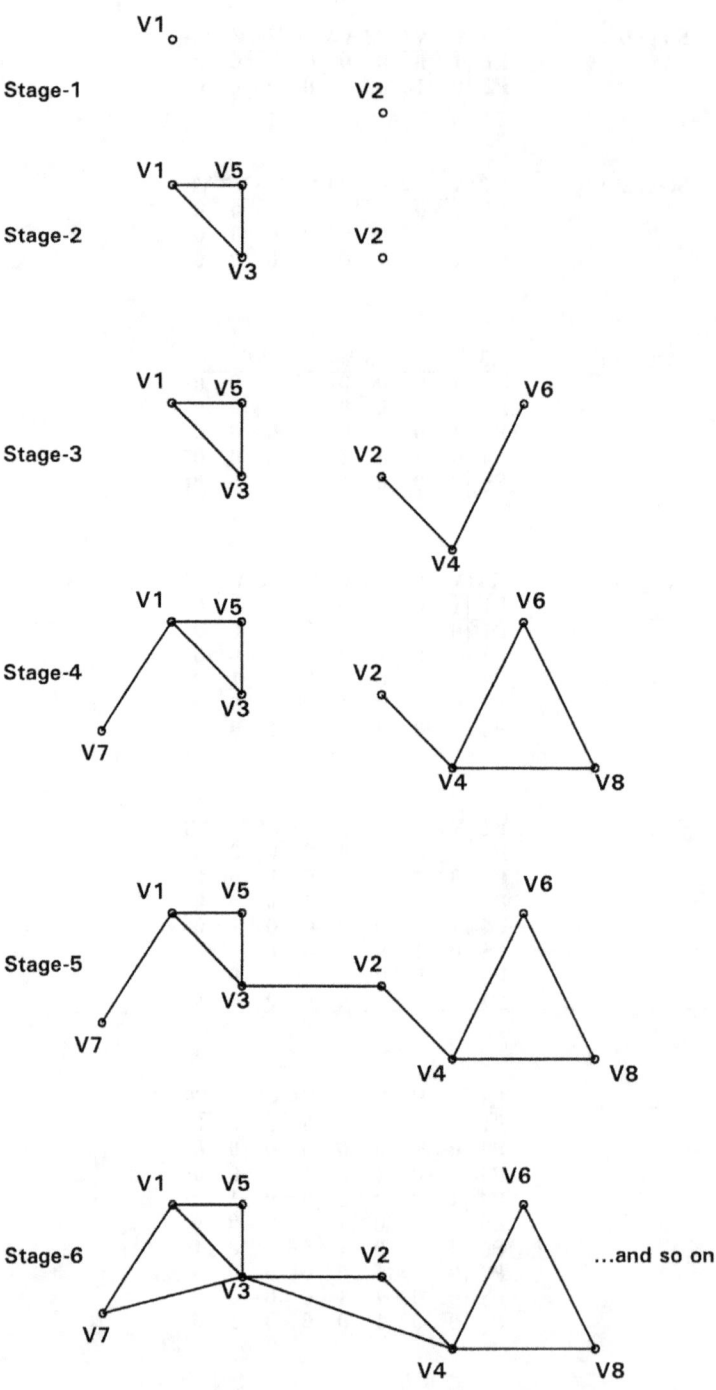

Fig.4 Building a Complex

[4.2] Destroying a Complex - the game of NIM

This game is played with an arbitrary number of piles of matches (say) containing arbitrary numbers of matches in each pile. (v. Appendix-E [142]) We shall illustrate the game with just 5 piles containiing the following numbers of matches, viz.,

```
pile-1   7
pile-2   4
pile-3   8
pile-4   13
pile-5   11
```

There are two players in the game, player-A and player-B, who play alternately ; player-A begins. Each player takes any number (> 0) of matches from any one pile, and the player who takes the <u>last match</u> is the winner.

The analysis depends on expressing these numbers in binary notation.

Thus we have

```
pile-1   07   -->   0111
pile-2   04   -->   0100
pile-3   08   -->   1000
pile-4   13   -->   1101
pile-5   11   -->   1011
```

We now look at these binary digits from the point of view of the <u>columns</u> in which they sit, and we add up the binary bits to give us the following decimal sums, viz. (counting the columns from L to R)

```
col-1 -->  3
col-2 -->  3
col-3 -->  2
col-4 -->  3
```

The fact that <u>at least one column</u> adds up to an **odd** value marks this as an **incorrect** pattern - for the player who is next to play (as we shall see).

If they had <u>all</u> been **even** sums we would call the position a **correct** pattern.

Example-1 :

If there are only two piles with pile-1 = 2 and pile-2 = 2 , then it is a correct position for A (who has left this position for B). Denoting the position by
$$* * \mid * *$$
with binary columns
```
       10
       10     (each column being even)
```
we see that if B takes all of one pile then A takes all of the other ; if B takes just one of one pile then A takes just one of the other pile - and clearly wins.

Example-2 :

When the game ends up with 0 or 1 in every pile, and
it is B's move then it is a correct position, for A, if the
number of 1's is even.

The key theorem in this is :-

If player-B is presented with a correct position by player-A
then any move player-B makes must change the position into
an incorrect position (which he leaves for player-A) and then
player-A can always change it back into a correct position again.

Thus correct play means that player-A must win if he plays
first and the initial position is an incorrect one , which he
then converts into a correct one.
If the initial position is a correct one then player-A must lose
provided player-B plays correctly from then on.

Generally the initial position is an incorrect position and so
A must win if he plays correctly.

Looking at the example shown above we see that it is an incorrect
position, so assuming player-A starts the play he can change it to a
correct position (for himself) when it is player-B's turn to move.

Player-A achieves this by noting whiich columns have an _odd sum_ ; these
are columns 1,2,4 - reading from L to R.

player-A then looks down the piles and finds the _first pile_ whose binary
row has a 1 in column-1. N.B. there will always be one such pile
since column-1 is an odd column.

In this case the appropriate pile turns out to be pile-3, with a
binary number
$$1000 \quad \text{(decimal 8)}$$

Player-A changes this by adjusting the 1st, 2nd and 4th binary digits.
This means that A now changes this row to be

$$0101 \quad \text{(decimal 5)}$$

and he does this by changing

        the 1 in column-1 to 0   (this will make column-1 even)
        the 0 in column-2 to 1   (this will make column-2 even)
        the 0 in column-4 to 1   (this will make column-4 even)

So player-A adjusts pile-3 so as to leave it with just 5 (0101) matches

That is to say he takes 3 (= 8-5) matches from pile-3.

This leaves player-B to face the following correct position, viz.,

        pile-1 07 --> 0111
        pile-2 04 --> 0100
        pile-3 05 --> 0101
        pile-4 13 --> 1101
        pile-5 11 --> 1011

where the column sums are all even, viz.,   2 4 2 4

Whatever player-B now removes (from any one pile) there will be a change in the binary row for that selected pile, and this will alter at least one binary digit in the array of columns - leaving (some) even sums as odd sums. Thus player-B must leave an incorrect position.

[4.3] Examining NIM via Q-Analysis

If we name the piles as
$$P1\ P2\ P3\ P4\ P5$$
and the columns as (bits)
$$B1\ B2\ B3\ B4$$
we can consider the binary array as a typical **Dowker** array and this gives us two complex structures, viz. KP(B) and KB(P). We show the two distinct arrays - Array-1 at the start and Array-2 after Player-A's move.

Array-1	N1	B1	B2	B3	B4	Array-2	N2	B1	B2	B3	B4
	P1	0	1	1	1		P1	0	1	1	1
	P2	0	1	0	0		P2	0	1	0	0
	P3	1	0	0	0		P3	0	1	0	1
	P4	1	1	0	1		P4	1	1	0	1
	P5	1	0	1	1		P5	1	0	1	1

In Array-1, and in KP(B), we are looking at the rows (piles) as simplices and the columns as vertices (bits, from L to R), whilst in KB(P) we are lookong at the columns (bits) as simplices and the rows (piles) as vertices.

In this initial position we have, writing σ_i for the simplices in KB(P),

$$
\begin{array}{ll}
B1 \text{ is a } \sigma_2 & <P3,P4,P5> \\
B2 \text{ is a } \sigma_2 & <P1,P2,P4> \\
B3 \text{ is a } \sigma_1 & <P1,P5> \\
B4 \text{ is a } \sigma_2 & <P1,P4,P5>
\end{array}
$$

In the course of the game these complexes KP(B) and KB(P) are to be demolished.

Now, eg, a σ_2 has 3 vertices - which is why the sum in the corresponding column is an odd number (recall that a σ_q has (q+1) vertices).

So this is an **incorrect position** because at least one of the simplices is of even dimension - and therefore has an odd number of vertices.

Referring to the **Dowkler** Array-1 above, and following the strategy already outlined, to obtain a **correct position** player-A must make sure that <u>all</u> the simplices in KB(P) are of an odd dimension - that is to say one or other of σ_1, σ_3, σ_5 ...

This means that player-A must reduce the initial complex (since he must remove matches) ;

this he does by down-grading some or all of the simplices in the conjugate complex KP(B) - by removing a suitable number of matches from the first pile which is an even dimensional simplex. As we have seen, the first such simplex is a σ_2, viz., P1. The details of this have already been discussed and the new complexes KP(B) and KB(P) arise via Array-2.

So we see that the game involves the destruction of the initial Complex and the winner is the player who gives the final "coup de grace".

[4.4] Ubiquitous nature of **Dowker** complexes.

We have seen in [4.1] above that a complex, under some appropriate agency, can be built into an accompanying complex - over some period of clock-time. In that artificial example the total time taken to get as far as the end of Stage-6 would naturally be the sum of the periods, viz., $P1 + P2 + P3 + P4 + P5 + P6$ and that, indeed, these periods would not neccessarily be equal in the normal clock-time.

This kind of activity, with its accompanying complex structure, can be found in a variety of situations. When the **agency** can be traced to human volition the process will appear in activities like

   (a) manufacturing artifacts - such as machines, computers, tools and implements (various)
   (b) designing and constructing physical buildings, roadways, transport systems, all kinds of structural engineering
   (c) creating a work of art
   (d) pursuing a scientific experiment
   (e) constructing a thesis
   (d) designing a social or political manifesto

and so on.

Such activities will be organised hierarchically from, say,

$$N\text{-level} \to (N+1)\text{-level} \to \ldots (N+k)\text{-level}$$

and at the lowest level the **Dowker** arrays will no doubt be large arrays, whilst the human agent provides the **force** (or **power**) to produce the q-connections needed to build the relevant complexes. But the human agency also includes things like **farming, horticulture, husbandry, fishing** and so involve an agency which is not simply that of human control - either of him/herself or of the living entities engaged in the process.

In particular it is somewhat fascinating to contemplate the reproduction of humanity - and indeed of all mammals.

There must therefore be a kind of **life-agency** which governs the following:

conception -> **embryo** -> **foetus** -> **newborn** -> **growth** -> **achievement** -> **decay** -> **death**

Part of **growth** must include the **mental faculties** as well as **education** of all kinds and of the **bonding** which lies behind **society**.

And the living **flora** and **fauna** complete a universal picture - all described (often at some length) by way of **hierarchically** arranged **Dowker arrays** and **complexes**.

Delving into the micro-organic mechanisms of genetics allows us to go further down into lower hierarchical levels - although with the best will in the world can we hope to find the nub of 266 this living agency? It's rather like the physicists looking (forever) for the ultimate fundamental particle?

Perhaps the extraordinary power of the modern computers has come along just in time for us to be able to visualise their use in unravelling the processes.

Chapter-5 /PART-4   Tactical & Positional Play in Chess

[5.1] Complexes arising in the game of Chess

This ancient game is played on a board of 64 squares which constitute the geometrical space wherein the 32 pieces have the ability to move.

[The initial positions of the pieces are shown in the diagram on the next page - Black in lower case letters and White in upper case letters]

The relationship between a subset of pieces, say P, and any subset of the squares, say S, can obviously be represented by a selection of **Dowker** arrays which are built by placing the 1's where each piece is able to move to each square to defend, capture, occupy, or otherwise aim at. These can be seen as defining simplicial complexes, viz., KP(S) and KS(P).

In the first, KP(S), the rows are named by the selected pieces and the columns are named by the selected squares. In the second, KS(P), the rows are named by the squares and the columns by the pieces. The usual notation is for the squares to be named by the **file & rank** notation ; for example initially the White Queen, Q, is to be found on the square (whose name is) d1 whilst the Black Queen-Bishop, qb, is on square c8 etc. whilst castling is written O-O (kingside) and O-O-O for (queenside).

Now at any stage of the game we can imagine that KP(S) and KS(P) are constructed and then the player whose turn it is has at his disposal two views viz.,

      the view that the pieces have of the squares, using KP(S)   and

      the view that the squares have of the pieces, using KS(P).

In making his/her decision about choosing a move the player must consider the efficacy of placing a piece on a particular square. The purpose of the move can involve several ideas - such as

    (a) defending another piece on a particular square
    (b) capturing an enemy piece on a particular square
    (c) threatening a particular square (or squares) for future occupation
    (d) checking the enemy king
    (e) taking all these into account by looking ahead over a series of moves (by both players)

Where the emphasis is placed, in the choice of a move, on making squares/files/ranks as targets - to create a (long lasting) dominance of significant areas of the chessboard - players are said to be excercising **positional play** ; but when the emphasis is on pieces as targets the play is said to be **tactical**. Naturally master chess-players must have a talent for combining these ideas in a manner which is superior to that of the opponent.

We shall pursue the notion that **tactical play** is best associated with the simplices in KP(S) whilst **positional play** is best associated with the simplices in KS(P).

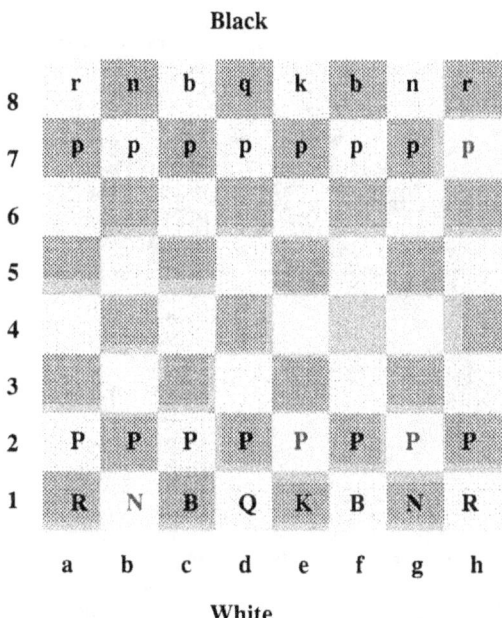

The board is shown here in the intial position - to show the **file & rank** notation for identifying the squares. The White pieces are denoted by the letters P R N B Q K for the pawn, rook, knight, bishop, queen and king ; the Black pieces by the lower case letters p r n b q k.

The relation between the moves, as shown, and the older English notation is as follows.

Older notation	Newer notations
P-K4	P-e4 or e2-e4
Kt-QB3	N-c3 or b1-c3

and so on.

For example, if we use the notation e2-e4 we mean that the piece (whatever it is) on square e2 moves to the square e4 - and if there is a piece on e4 then it is thereby captured.

[5.2] Illustrating **positional play** and **tactical play**

We first show an early example of an all-out attacking (tactical) play in what became known as **The Immortal Game**, viz., **Anderssen v Kieseritsky 1851** which was played in a tournament held at the Great London Exhibition of 1851. We show the final position after **Anderssen's** win, in which he offerred, and **Kieseristky** accepted, 2 pawns, 1 bishop, 2 rooks and a queen having successfully lured the Black pieces away from the Black king's defences.

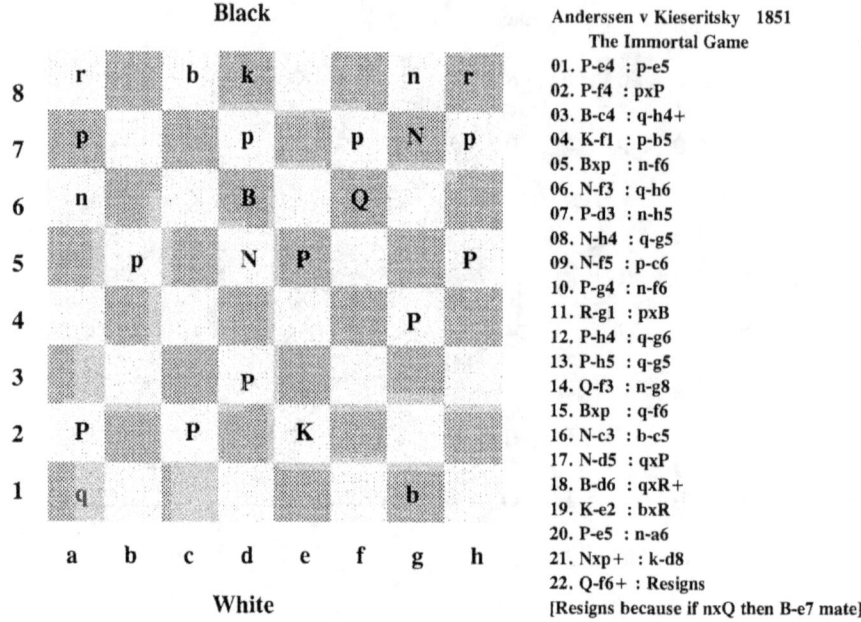

Anderssen v Kieseritsky   1851
The Immortal Game
01. P-e4   : p-e5
02. P-f4   : pxP
03. B-c4   : q-h4+
04. K-f1   : p-b5
05. Bxp   : n-f6
06. N-f3   : q-h6
07. P-d3   : n-h5
08. N-h4   : q-g5
09. N-f5   : p-c6
10. P-g4   : n-f6
11. R-g1   : pxB
12. P-h4   : q-g6
13. P-h5   : q-g5
14. Q-f3   : n-g8
15. Bxp   : q-f6
16. N-c3   : b-c5
17. N-d5   : qxP
18. B-d6   : qxR+
19. K-e2   : bxR
20. P-e5   : n-a6
21. Nxp+   : k-d8
22. Q-f6+ : Resigns
[Resigns because if nxQ then B-e7 mate]

[Position at end of game]

Simplices from KP(S) Complex
QN --> simplex = (b6 c7 e7 f6)
QB --> simplex = (b8 c7 e7 f8)
Q --> simplex = (d8 e7 f7)
KN --> simplex = (e8 e6)
k --> simplex = (c7 e8 e7)
kn --> simplex = (e7 f6 h6)

Simplices from KS(P) Complex
c7 --> simplex = (k qn QB QN)
d8 --> simplex = (Q)
e8 --> simplex = (k KN)
e7 --> simplex = (kn k Q QB QN)

Here we see the Black king's escape squares are c7, e7, e8, but these simplices in KS(P) show that White has pieces attacking them, and this remains the case even after Black's only move of nxQ, for the simplex e7 has the vertex QB with which to deliver checkmate - the Black kn having been moved in the capture - so that it can no longer protect e7.

We note also that, the Black queen, q , and the Black bishop, b , are far away from the scene of the action (having been lured there) and play no part in the defence of the Black king - nor do they offer any immediate threat to the White king, K.

[Note: The data file for the complex in this position is given in Appendix-C §[C.5]]

[5.3] So at any stage of the game we can consider the complexes KP(S) and KS(P), where the set P denotes all or some of the pieces (P's R's B's N's Q K) and the set S denotes all or some of the set of the squares of the board, the selections for P and S being made by the players.

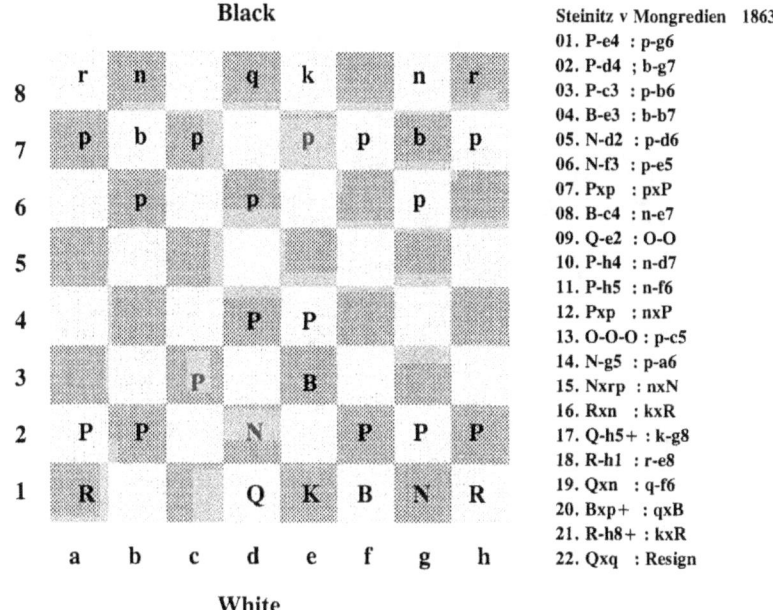

Steinitz v Mongredien 1863
01. P-e4 : p-g6
02. P-d4 : b-g7
03. P-c3 : p-b6
04. B-e3 : b-b7
05. N-d2 : p-d6
06. N-f3 : p-e5
07. Pxp   : pxP
08. B-c4 : n-e7
09. Q-e2 : O-O
10. P-h4 : n-d7
11. P-h5 : n-f6
12. Pxp   : nxP
13. O-O-O : p-c5
14. N-g5 : p-a6
15. Nxrp : nxN
16. Rxn   : kxR
17. Q-h5+ : k-g8
18. R-h1 : r-e8
19. Qxn   : q-f6
20. Bxp+ : qxB
21. R-h8+ : kxR
22. Qxq   : Resign

[Position after 05. N-d2 : p-d6]

Simplices from KP(S) Complex
QN --> simplex = (c3 d2)
QB --> simplex = (c12 d2 d4 f4 g5 h6)
Q --> simplex = (c12 c2 b3 a4 d2 d3 d4 e2 f3 g4 h5)
KB --> simplex = (e2 d3 c4 b5 a6)
qb --> simplex = (a6 c6 d5 e4)
kb --> simplex = (f8 f6 e5 d4 h6)

Simplices from KS(P) Complex
c4 --> simplex = (KB)
c3 --> simplex = (QN)
c2 --> simplex = (Q)
d4 --> simplex = (kb QBP Q QB)
d5 --> simplex = (qb KP)
e4 --> simplex = (qb)
e5 --> simplex = (kb QP)
f3 --> simplex = (Q)
f4 --> simplex = (QB)
f5 --> simplex = (knp KP)

Each move changes both of these complexes - by reduction (when captures occur) and generally by re-arranging them (as manoeuvring takes place) - either effect distorting the geometry of the structures, and thereby may be viewed as the introduction of various t-forces into the game. We summarize these points by claiming that

(i) **tactical play** is based on the complex KP(S), whilst

(ii) **positional play** is based on the complex KS(P).

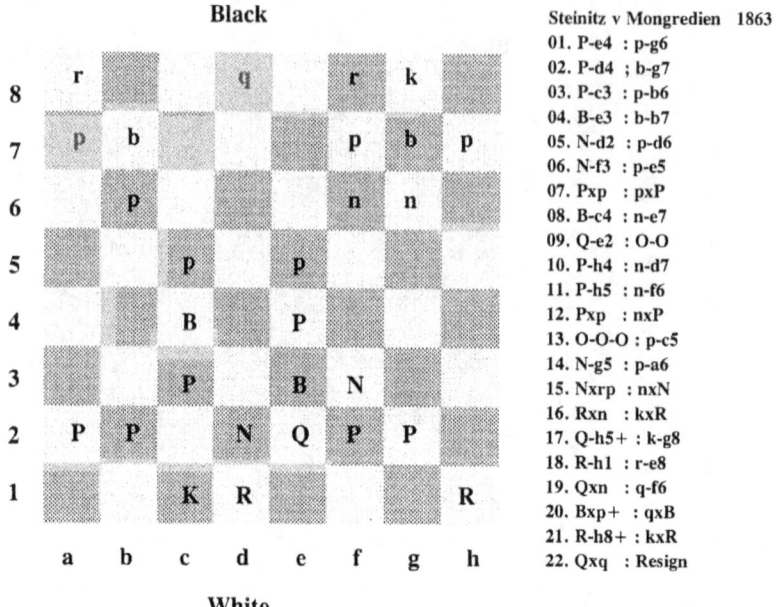

[Position after 13. O-O-O : p-c5]

Simplices from KP(S) Complex

```
QB -->   simplex =   (d4 f2 g5 g6 h6)
QN -->   simplex =   (e4 f3)
Q  -->   simplex =   (c4 f3)
KB -->   simplex =   (d5 f7)
KN -->   simplex =   (e5 g5)
KR -->   simplex =   (h2 h3 h4 h5 h6 h7)
qb -->   simplex =   (d5 e4)
qn -->   simplex =   (h5 h7)
q  -->   simplex =   (f6 g3)
k  -->   simplex =   (h7)
kr -->   simplex =   (f7)
```

Simplices from KS(P) Complex

```
d4 -->   simplex = (kp qbp QBP QB)
d5 -->   simplex = (qb KP KB)
e4 -->   simplex = (qb QN)
e5 -->   simplex = (kn KN)
f5 -->   simplex = (KP)
f7 -->   simplex = (kr KB)
h6 -->   simplex = (kb KR QB)
h7 -->   simplex = (k qn KR)
```

To illustrate these ideas we shall study a few positions from the game :

Wilhelm Steinitz v Mongredien   1863

This means that there is a **Dowker** array which defines the compexes KP(S) and KS(P) and that it changes after every move which is made, either by White or by Black.

The connectivities between the Pieces, as well as that between the Squares can then be explored by the usual Q-Analysis algorithms. Such connectivities arise by way of the simplices which are found in KP(S) or in KS(P).

At four stages of the game we look at these **Dowker** complexes and notice relevant simplices in each of them. These stages are taken at the following points of the game, viz.

Steinitz1 after 05. N-d2 : p-d6 then **Steinitz2** after 13. O-O-O : p-c5
Steinitz3 after 18. R-h1 : r-e8 then **Steinitz4** after 22. Qxq : Resign

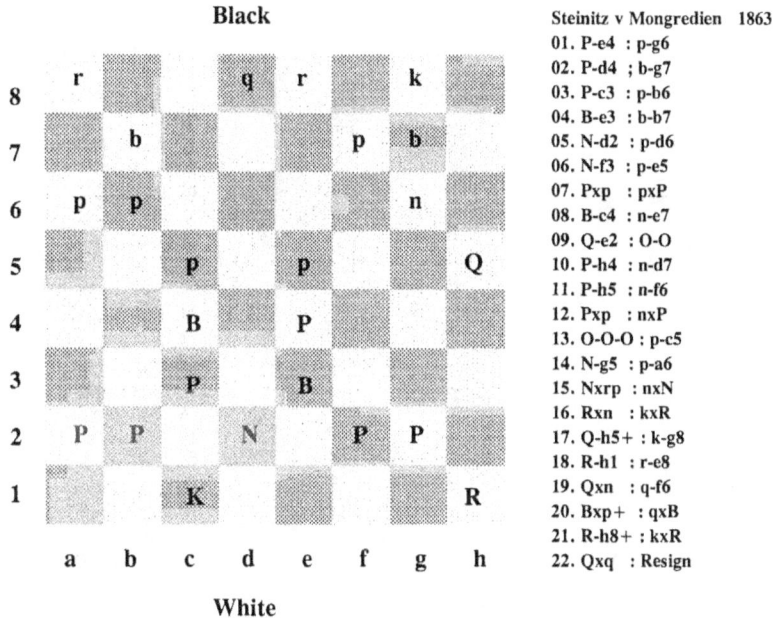

[Position after 18. R-h1 : r-e8]

Steinitz v Mongredien 1863
01. P-e4   : p-g6
02. P-d4   ; b-g7
03. P-c3   : p-b6
04. B-e3   : b-b7
05. N-d2   : p-d6
06. N-f3   : p-e5
07. Pxp   : pxP
08. B-c4   : n-e7
09. Q-e2   : O-O
10. P-h4   : n-d7
11. P-h5   : n-f6
12. Pxp   : nxP
13. O-O-O : p-c5
14. N-g5   : p-a6
15. Nxrp   : nxN
16. Rxn   : kxR
17. Q-h5+ : k-g8
18. R-h1   : r-e8
19. Qxn   : q-f6
20. Bxp+   : qxB
21. R-h8+ : kxR
22. Qxq   : Resign

Simplices from KP(S) Complex
QB --> simplex = (d4 c5 f2)
QN --> simplex = (e4 f3)
Q   --> simplex = (g6 h6 h7 h8)
KB --> simplex = (d5 e6 f6)
QR --> simplex = (g6 h2 h3 h4)
qb --> simplex = (d5 e4)
q   --> simplex = (d2 e7 f6)
k   --> simplex = (f8 h8)
kr --> simplex = (e5 e6 e7)

Simplices from KS(P) Complex
c5 --> simplex = (QB)
d4 --> simplex = (kp qbp QBP QB)
d5 --> simplex = (kb qb KP KB)
e4 --> simplex = (qb QN)
e5 --> simplex = (kr qn)
f4 --> simplex = (kp qn)
f5 --> simplex = (KP)
f7 --> simplex = (kb)
h6 --> simplex = (Q)
h7 --> simplex = (Q)
h8 --> simplex = (k qn Q)

White has set up a defensive/aggressive pawn structure resulting in the positional control of
the centre of the board, viz., squares c4,d4,d5,e4,e5, and supported them by the development of
his QBP, QB, QN. The selected simplices in KS(P) list these and show that Black's response is
limited to the action of his qp and the two bishops. White is now ready for a King-side attack.
Here Black has attacked the centre by advancing his qp but after 07. Pxp : pxP Black's kb
is left blocked by its own kp - where it is doomed to stay for the rest of the game.

266 MATHEMATICAL PHYSICS

The success in holding the centre squares retains White's advantage. Having castled on the queen side he has moved his K to relative safety and deprived Black of any wistful plan to use the open queen's file.

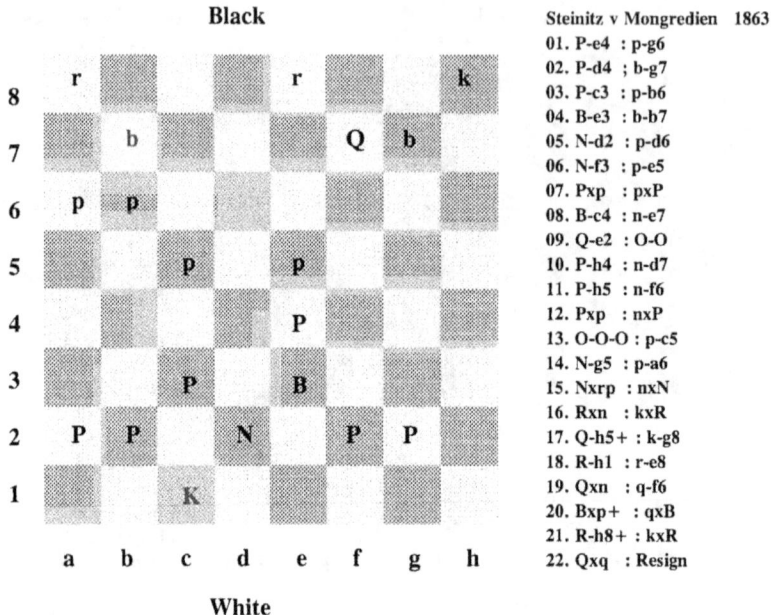

Steinitz v Mongredien 1863
01. P-e4 : p-g6
02. P-d4 ; b-g7
03. P-c3 : p-b6
04. B-e3 : b-b7
05. N-d2 : p-d6
06. N-f3 : p-e5
07. Pxp : pxP
08. B-c4 : n-e7
09. Q-e2 : O-O
10. P-h4 : n-d7
11. P-h5 : n-f6
12. Pxp : nxP
13. O-O-O : p-c5
14. N-g5 : p-a6
15. Nxrp : nxN
16. Rxn : kxR
17. Q-h5+ : k-g8
18. R-h1 : r-e8
19. Qxn : q-f6
20. Bxp+ : qxB
21. R-h8+ : kxR
22. Qxq : Resign

[Position after 22. Qxq : Resign]

Simplices from KP(S) Complex
QB --> simplex = (c5 d4 f3)
QN --> simplex = (c4 e4 f2)
Q --> simplex = (b7 c7 d7 e6 e7 e8 f8 g6 g7 g8 h5)
qb --> simplex = (e4)
k --> simplex = (g8 h7)
kr --> simplex = (e5 e6 e7 f8)

Simplices from KS(P) Complex
d4 --> simplex = (kp QBP QB)
d5 --> simplex = (KP)
e4 --> simplex = (qb QN)
e5 --> simplex = (kr kb)
f5 --> simplex = (KP)
g6 --> simplex = (Q)
g7 --> simplex = (Q)
g8 --> simplex = (k Q)
h5 --> simplex = (Q)

White proceeds to develop his KN to g5 - which announces his intention of tactical play on the king-side.

White has opened the h-file by a sacrifice (15. Nxrp : nxN) whilst Black has no counter from the centre, but hastens to look for reinforcement for the k defences ; seeing danger lurking Black moves his kr hoping that this will give his king a possible escape square.

Notice that White's secured centre squares are still there - although he tries (too late) to prepare a counter on the queen-side (14. ... : p-a6)

The exchanges leading to this position leaves White with the advantage of Queen and two king-side Pawns over Black's two rooks. Black's qb is under attack and the manoeuvrability of the Queen together with the possibilties of a king-side attack via the h-file and the two extra king-side Pawns together with the positional advantage which White still holds in the centre leads to Black's resignation.

Chapter-6 /PART-4 Probabilities and Events

[6.1] A Graded Pattern of Probabilities on Events in a Complex

Suppose we have a **Dowker** complex with an associated set of **p-Events**, that is to say an experience of happenings tied to the various p-simplices found in the complex, then we can identify a set of probabilities attached to these simplices. This means that, since the simplices form a set graded by their dimensions, so the pattern of probabilities associated with the Events will also be **graded** by the same dimensions. [v. Appendix-C §[C.3]]

This pattern will be of the general form

$$\Pi^* \equiv \Pi^0 \oplus \Pi^1 \oplus \Pi^2 \oplus \Pi^3 \oplus \ldots \Pi^n$$

where n is the maximm q-value in the complex.

Π^0 will be the set of probability values assocated with the vertices (0-simplices, σ_1's),

Π^1 will be the set of probability values associated with the edges (1-simplices, σ_1's)

...... until

Π^n will be the probability values associated with the n-simplices (σ_n's)

If we assume an equiprobable property for the Events then, for example, the probability of any p-Event will be calculated as

$$1 \div (\text{number of p-simplices/p-Events in the complex})$$

and these are not necessarily dependent on q-Events, where $q \neq p$; but the sum of all the values on the p-Events must add up to unity - to comply with standard probability theory.

A familiar example of a **non-equiprobable** set of Events is the odds placed, by the bookmaker, on the runners in a horse race ; thus the odds (against a win) of 3-to-1 means that the probability of that particular horse winning is $1 \div (1 + 3)$. It is of course essential for the bookmaker to ensure that the sum of all the probabilities on all the horses in a race adds up to a number less that 1. This caveat also applies to, say, betting on the roulette wheel, where the numbers 1 ... 36 are augmented with the extra 37. Since the betting is only allowed on the 36 numbers the extra number 37 gives the banker a slight but (apparently) adequate advantage.

[6.2] Probabilities in the game of CRAPS

A common example of an equiprobable set of Events is the throw of a single die with its six sides as the Events (N = 6). So, for example, we say that the probability of throwing a 6 is 1/6, etc.. We can obviously regard these Events as forming a particularly simple complex - consisting of six separate vertices (nodes) without any relationship between them ; one throw does not depend on any other throw.

The probability distribution defined on the Events (which we must see as **0-Events**) now becomes a **patttern**, as discussed above, viz.,

$$\Pi^* \equiv \Pi^0$$

all other dimensions being absent.

But the structure of the Event-space becomes a little more interesting when we move to consider the game of throwing two dice, say, die-A and die-B.

Here the events become 1-Events, or 1-simplices σ_1's, in a complex in which the A-scores are always joined with a B-score.

The situation is illustrated by the gambling game known as **Craps** (U.S.A) in which the total sum of a throw is gambled on - under various rules.

If we list the Events as things like $<$A1 B3$>$ as E4 (since the sum would be 1+3) then we can see that the following data file will contain all the results of the throws - as well as many others which in fact cannot occur. The actual possible throws will constitute a **subcomplex** of edges of the complete complex consisting of a single 11-simplex, σ_{11}, as follows.

```
rownames = (A1 A2 A3 A4 A5 A6 B1 B2 B3 B4 B5 B6)
colnames = (E2 E3 E4 E5 E6 E7 E8 E9 E10 E11 E12)
E2  (1 7)
E3  (1 7, 2 8)
E4  (1 9, 2 8, 3 7)
E5  (1 10, 2 9, 3 8, 4 7)
E6  (1 11, 2 10, 3 9, 4 8, 5 7)
E7  (1 12, 2 11, 3 10, 4 9, 5 8, 6 7)
E8  (2 12, 3 11, 4 10, 5 9, 6 8)
E9  (3 6, 4 5, 5 4, 6 3)
E10 (4 12, 5 11, 6 10)
E11 (5 12, 6 11)
E12 (6 12)
```

The σ_{11} has all the vertices A1 A2 ... A6 B1 B2 ... B6 whilst the throws E2 E3 ... E12 form the subcomplex shown above and consists of selected edges (σ_1's) contained therein. For example

(1 7) refers to throw $<$A1 B1$>$ with a score of 2 (E2)
E7 is any one of the pairs (1 12) (2 11) (3 10) ... (6 7) ... and so on.

Since the throws contain a B-score and an A-score there are 6 x 6 = 36 scores with a non-zero probability. So here the prability pattern on the complete σ_{11} has the value of 1/36 on each of the edges listed above. Since there are actually 66 edges in the σ_{10} the remaining 30 edges carry the value of zero. These 30 edges (1-simplices) are those which would occur as pairs of the A's or of the B's - since e.g. (A1 A2) cannot be a throw.

The probability pattern will therefore be

$$\Pi^* \equiv \Pi^1$$

This representation of the throws can clearly be extended to a game in which any number of dice are thrown simultaneously ; for example if 3 dice are thrown the Events would be named

$$E3 \; E4 \; ... \; E18$$

and would consist of a subcomplex of 2-simplices, σ^2, triangles in a single simplex σ_{17} - the vertices being A1 A2 ... A6 B1 B2 ... B6 C1 C2 ... C6, and the most likely score to be thrown will be 10 - with a total of 27 ways of scoring that number. Since the total number of possible scores is 216 ($= 6^3$) this means that we allot E10 the probability of $27 \div 216$ ($= 1/8$).

The other possibilities are easily calculated, although it is a bit tedious ; the system is the basis of the **game of over ten**.

[6.3] Loaded Dice in CRAPS

If, say, die A is "loaded" - so as always to turn up a 3, then the **Dowker** complex, CRAPS, is drastically altered, with σ_{10} changing to σ_6 with the rows being

A3 B1 B2 B3 B4 B5 B6   and the Events/Throw-sums being

E31 E32 E33 E34 E35 E36   [1]

Now the 1-Events are identified by the scores

4 5 6 7 8 9

and the probability pattern on CRAPS, say Σ^1, is simply

$\Sigma^1 = 1/6$ on each of the six Events in σ_6 and 0 on the remaining edges.

There has therefore been introduced, into the game itself, a **1-force** experienced in the supporting simplex σ_{10} - because the 1-Events E31 ... E36 have an altered value whilst the other Ers are deleted.

The magnitude of this 1-force could well be measured by

$$\Delta\Pi = \Sigma^1 - \Pi^1 \quad \text{(taken over the six new Events only)}$$

with values $(1/6 - 1/36)$ on each of the existing Events in [1] and $-1/36$ on the others. This latter value is an impossible value for any probability, and it indicates that such Events <u>do not occur</u> in the game.

So the 1-force values experienced in the game are $+5/36$ on each Event listed in [1].

[Note: In the non-loaded case there are no t-forces in the game]

A similar argument applies to any case such as "loading A so that it scores a 2 or a 4" (but no others).

In this case we would have the scores as

$$3\ 4\ 5\ 6\ 7\ 8 \quad \text{for E21 E22 ... E26} \quad \text{and}$$

$$5\ 6\ 7\ 8\ 9\ 10 \quad \text{for E41 E42 ... E46}$$

The probabilities for the scores will now be

1/12	for the score 3
1/12	for the score 4
2/12	for the score 5
2/12	for the score 6
2/12	for the score 7
2/12	for the score 8
1/12	for the score 9
1/12	for the score 10

Denoting this probability pattern by Ξ^1 the 1-force experienced by the Events

has the values

$$\Delta\Pi = \Xi^1 - \Pi^1 \quad \text{on the 12 possible Events}$$

That is to say

$$1/12 - 1/36, \text{ or } 2/36 \text{ on E21 E22 E45 E46} \quad \text{and}$$
$$2/12 - 1/36, \text{ or } 6/36 \text{ on E23 E24 E25 E26 E41 E42 E43 E44}$$

relative to the unloaded CRAPS.

[6.4] Some Basic Probabilities using a **Dowker Complex**

Suppose we have a **Dowker Complex** with its associated set of p-Events - for example :

K	X1	X2	X3	X4	X5	X6	X7	X8
Y1	1	1	0	0	0	1	0	1
Y2	0	0	1	1	0	1	1	1
Y3	1	0	1	1	1	0	1	0
Y4	1	0	0	1	0	1	0	1
Y5	0	0	0	1	1	1	1	0
Y6	0	1	1	0	0	1	0	1
Y7	1	0	0	0	1	1	1	0

then we can associate the 0-Events of KY(X) with the **column names** X1, X2, ..., X8

since these are to be the vertices of KY(X) and **are also** the Events via which all

the other p-Events occur. Now let us suppose that we know the probabilities attributed

to these 0-Events - so that

$$\text{prob}(X1) = p_1 \ \text{prob}(X) = p_2 \ .... \ \text{prob}(X8) = p_8$$

together with the proviso that $\text{prob}(\text{any } X) > 0$ and

$$p_1 + p_2 + p_3 + ... + p_8 = 1$$

Of course when we have an equiprobable system of Events each p_r, in this case would

be 1/8.

This knowledge, together with the **Dowker Complex**, allows us to find the probabilities of all the other Events (the Y's), as follows.

For example, take the case of the 2-Event represented by Y2, viz.,

$$Y2 = <X3, X4, X6, X7, X8>$$

This Event occurs when the combination $X3 \cap X4 \cap X6 \cap X7 \cap X8$ occurs, and so the

$$\text{prob}(Y2) = \text{product } p_3 \cdot p_4 \cdot p_6 \cdot p_7 \cdot p_8 \qquad (A)$$

and similarly for the others.

In standard probability theory it is found useful to have the notion of **conditional** probability. This is written, for example, in the form $\text{prob}(A|B)$ which stands for the probability of Event-A given the Event-B. In the above example we might consider $\text{prob}(Y2|Y5)$ - that is to say, assuming Y5 occurs what is the probability of Y2 occurring, or what is the probability of Y2 **relative** to the Event Y5, and this means that we look for the **intersection** of the rows Y2 and Y5 (given by $Y2 \cap Y5$). So we are really looking at the simplex which is a **face** shared by Y5 and Y2.

This is $\quad Y2 \cap Y5 = <X4, X6> \quad$ and that has a probability $\quad p_4 \cdot p_6$ when it stands alone. But to get the value of $\text{prob}(Y2|Y5)$ we must ensure that the Event Y5 has actually occurrred ; the probability of this occurring is simply $\text{prob}(Y5)$ and so we must write $\qquad \text{prob}(Y2 \cap Y5) = \text{prob}(Y5)\text{prob}(Y2|Y5)$

the rule being usually written in the form

$$\text{prob}(A|B) = \text{prob}(A \cap B) \div \text{prob}(B) \qquad (B)$$

We can proceed to the other basic theorem in the theory, viz., **Bayes' Theorem**.

This applies in situations where there is a **partition** of the 0-Events by way of a selection of the p-Events . That means that we have a set of the Y's which are pairwise disjoint but whose union encompasses the whole set of X's. We illustrate this by the following simple case.

K	X1	X2	X3	X4	X5	X6	X7	X8	X9	X10
Y1	1	0	0	0	0	0	0	0	0	1
Y2	0	1	1	0	0	0	0	0	0	0
Y3	0	0	0	1	0	1	1	1	0	0
Y4	0	0	0	0	1	0	0	0	1	0
Z5	1	0	0	1	1	0	1	0	1	1

Here we see that such a partition is provided by the Events Y1, Y2, Y3, Y4 - and any other Event is typified by Z5, and it follows that such an Event (as Z5) can be written as

$$Z5 = (Z5 \cap Y1) \cup (Z5 \cap Y2) \cup (Z5 \cap Y3) \cup (Z5 \cap Y4) \qquad \text{which is also}$$

$$Z5 = (Y1 \cap Z5) \cup (Y2 \cap Z5) \cup (Y3 \cap Z5) \cup (Y4 \cap Z5)$$

since eg $Z5 \cap Y1 \equiv (Y1 \cap Z5)$ - and so

$$\text{prob}(Z5) = \text{prob}(Y1 \cap Z5) + \text{prob}(Y2 \cap Z5) + \text{prob}(Y3 \cap Z5) + \text{prob}(Y4 \cap Z5)$$

and this gives

$$\text{prob}(Z5) = \text{prob}(Y1)\text{prob}(Z5 \mid Y1) + \text{similar} \ldots + \text{prob}(Y4)\text{prob}(Z5 \mid Y4) \quad (D)$$

Given the Event-Z5 **Bayes' Theorem** provides us with the conditional probability of Event-Y_i relative to this Event-Z5 - that is to say $\text{prob}(Y_i \mid Z5)$.

We can therefore write **Bayes'** result, viz.

$$\text{prob}(Y_i \mid Z5) = \text{prob}(Y_i \cap Z5) \div \text{prob}(Z5) \text{ which finally becomes}$$

$$\text{prob}(Y_i \mid Z5) = \text{prob}(Y_i)\text{prob}(Z5 \cap Y_i) \div \text{prob}(Z5) \quad \textbf{Bayes' Theorem}$$

where $\text{prob}(Z5)$ is given by (D).

[6.5] The Time Significance of the Grading

It is inevitable that we should take into account the Clock-Time we experience in waiting for an Event to occur. This is because we are accustomed to regarding the probabilty of an Event as in some way giving us an idea of "when it is likely to occur" ; we constantly **ask the question** "what are the chances of it (some p-Event) occurring in the next hour/day/week ...".

A low value attributed to the p-Event tells us it is relatively unlikely ; a high(er) value says it is relatively likely.

In this sense the probability we attach to a p-Event is the **answer to such questions**.

The answer does not have the precision found in, say, physics - except when the probability is zero when the answer equals "it will not occur", or when the value is 1 when the answer is "it will certainly occur".

If we refer the edges (the σ_1's) in the complex to linear Clock-Time (the vertices being the Clock-Moments) then we must associate a p-Event (a σ_p) with the time it takes for the simplex to be manifest.

Writing n for $(p+1)$, the number of Time-Moments involved, then the time needed to set up the σ_p will be that involved in identifying the sequence of pairs of 0-Events to be found in the vertices of a given p-Event. We can read off the σ_p as an Event in the **Dowker** array - for example in (see our earlier Chapter-2, §[2.8])

K	T1	T2	T3	T4	T5	T6	T7	T8
E1	1	0	1	1	0	0	0	1
E2	1	1	0	0	0	1	1	0
E3	0	0	1	1	1	0	0	1
E4	1	1	1	1	0	0	0	0
E5	0	1	0	1	0	1	0	1
E6	1	0	0	0	1	1	1	0
E7	1	1	1	1	1	0	1	1

So the Clock-Time needed to identify any particular σ_p will be the number of edges/intervals found in the p-Event. Since any of the p-Events is to be found in this **Dowker** array we can see not only the value of p but also which Clock-Time-Moments are involved. Then the answer will be $n(n-1)/2$ intervals ($n = p+1$).

Of course the Time-Interval which is being used for the calculation depends upon the circumstances and the choice of the Clock-Time-Moments(eg. microseconds, hours, days, weeks ... etc.).

[6.5] An **Unexpected Event** versus A **Surprise Event**

If we have a set of Events with known probabilities, then if there exists an Event with zero probability it is natural to suppose that it will be an **Unexpected Event** ; if there is at least one probability which is non-zero, but very small, it is natural to regard it as an **Unlikely Event**, for example the horse running at 100-1 ?

But what if an Event occurs which is not in the list of what we thought of as possible Events ? We shall regard this as a **Surprise Event**.

Since the probability of an Event constitutes an answer to a question, such as "will it occur", we can only suppose that

a SURPRISE is the answer to a QUESTION which has not been asked.

For example, suppose Player-A and Player-B are playing CRAPS, when B (by sleight of hand) produces a throw of 14 - by substituting an 8-sided die into the game - surely it would come as a Surprise to Player-A ?

Perhaps we can see this situation arising where the **Dowker complex** which contains the defined Events has more than one component at the (q=0)-level. For while the questions are being asked with regard to one 0-component, the Surprise Event is part of the other distinct 0-component.

APPENDIX-A   SOME ALGEBRAS

[Note : The reader will appreciate that the following notes are meant only to be a sketchy introduction to the subject - on which there is a growing literature. The reward to be found in a serious course of study will be an appreciation for a rich panorama of disciplined abstract intellectual concepts.
Applying them to study Physics is yet another reward.
If needed the reader may care to consult a few of the references.
Appendix-E [21], [35], [39], [40], [41], [42], [67], [95]]

[A.0]   An important underlying concept in modern algebra is that of **FIELD**.

A **Field**, **F**, is a set of symbols **(a b c ...)** together with two modes of combination, viz., + and x (preferably $*$) (or $\oplus$ and $\otimes$, when we wish to stress the general nature of the operations) together with the following Rules.

<u>Closure</u>

C1   **a** + **b** is a member of **F** whenever **a** and **b** are members of **F**.

C2   **a**x**b** (or simply **ab**) $\in$ **F**. { $\in$ denotes "is a member of"}

<u>Additive Properties</u>

A1   **a** + **b** = **b** + **a**   (Commutative law of addition)

A2   **a** + (**b** + **c**) = (**a** +**b**) + **c** = **a** + **b** +**c**   (Associative law)

A3   There exists an <u>additive identity</u> viz. **0** such that
a + **0** = **a**   for all **a** in **F**.

A4   For any **a** in **F** there exists an <u>additive inverse</u> viz. **-a** such that **a** + (**-a**) = **0**.

<u>Multiplicative Properties</u>

M1   **ab** = **ba**   (Commutative law of multiplication)

M2   **a**(**bc**) = (**ab**)**c** = **abc**   (Associative law)

M3   There exists a <u>multiplicative identity</u> viz. **1** (sometimes **e**) such that $a * 1 = 1 * a$

M4   For any **a** $\neq$ **0** there exists a <u>multiplicative inverse</u> viz. a^{-1} such that $a * a^{-1} = a^{-1} * a = 1$.

When **C1** and **C2** apply we say that **F** is <u>closed</u> under + and $*$.

When conditions **A2,A3,A4** (or equally **M2,M3,M4**) apply the symbols are said to define a **GROUP**, **G**. If, in addition, **A1** (or **M1**) applies then the **Group** is described as **Abelian** (after the mathematician **Abel**).

In addition to the above laws **C1** ... **M4** a field **F** possesses the

<u>Distributive Property</u>, viz.

D1 $a(b+c) = ab + ac$

[A.1] These Rules are not haphazard or capricious, but represent the structural properties of the more familiar "elementary algebra" - which itself is a reflection of common arithmetic.

Thus the **rational numbers** (of the form **p/q** where **p,q** are integers $q \neq 0$) denoted by $\Re$, form a field - as is easily checked - as do also the real numbers, denoted by $\Re^{\#}$; these latter consisting of all the rationals together with the so-called **irrationals**.

But the **integers** (denoted by $\mathbf{J} = \{0, \pm 1, \pm 2, \pm 3, ...\}$) do not form a **Field**, there being no multiplicative inverse.

But if **x, y** are in $\Re$ (or in $\Re^{\#}$) then all numbers of the form $x+iy$, where $i^2 = -1$ form a **Field** - these being the Complex numbers, denoted by $\mathbb{C}$. In this $\mathbb{C}$ we must note that **equality** is to be interpreted via

$x_1 + iy_1 = x_2 + iy_2$ if and only if

$x_1 = x_2$ and $y_1 = y_2$

and + and * obey the well known equations

$(x_1 + iy_1) + (x_2 + iy_2) = (x_1 + x_2) + i(y_1 + y_2)$

$(x_1 + iy_1) * (x_2 + iy_2) = (x_1 x_2 - y_1 y_2) +$
$\qquad i(x_1 y_2 + x_2 y_1)$

If all **C1** ... **M4** apply except **M1** the structure is called a **skewField** - shortened to **sField**.

An example of a such an sField is the set M_n of all non-singular $n * n$ matrices with the usual definitions of + and * , viz.

If the elements of M_n are c_{ij} (members of $\Re$, $\Re^{\#}$ or $\mathbb{C}$) then the zero matrix is such that $c_{ij} = 0$, for all **i,j**, and the unit matrix (the multiplicative identity) is $\mathbf{I} = (c_{ij})$ with $c_{ij} = 1$ when $i = j$ and $c_{ij} = 0$ whenever $i /= j$.

Also $H(h_{ij}) + K(k_{ij}) = M(c_{ij})$

if $c_{ij} = h_{ij} + k_{ij}$ for all **i,j** and

$\qquad H(h_{ij}) * K(k_{ij}) = M(c_{ij})$

if $c_{ij} = \sigma_r (h_{ir} * k_{rj})$ summing for $r = 1 ... n$.

It is easily seen that, in general, $H * K \neq K * H$.

But if the apparent **Field** fails to exhibit a multiplicative inverse property then the structure is called a **RING, R**.

To be precise:

A set $\mathbf{R} = \{a \; ; \; \oplus \; ; \; \otimes\}$ is an **algebraic ring** if it is

(i) a **group** under $\oplus$

(ii) is closed and associative under $\otimes$

(iii) and if $\otimes$ is distributive over $\oplus$.

We notice that in a **Ring** it is not necessary to have the rules **A1,M1,M3,M4**.

Then, for example, the integers, **J** form a ring (not a **Field** because **M4** fails).

Expressed in words, we can say that a **Ring** becomes a **Field** when it contains

(i) a multiplicative identity (a unit)

(ii) multiplicative inverses for every non-zero element, and

(iii) is commutative over $\otimes$.

[A.2] Vector Spaces and Modules

A set of quantities $V\{x,y,...\}$ is said to form a **Vector Space over a Field F**, whenever binary operations $\oplus$ and $\odot$ are defined by

V1 $x, y \in V$ implies $x \oplus y \in V$

V2 $x \in V$ and $\alpha \in F$ implies $\alpha \odot x \in V$.

[Note: it is common to replace $a \odot b$ by the simple ab]

When in addition **A1**, **A2** apply, and when the following distributive rules occur:

D2 $\alpha(x \oplus y) = \alpha x \oplus \alpha y \quad \alpha \in F$

D3 $(\alpha \oplus \beta)x = x(\alpha \oplus \beta) = \alpha x \oplus \beta x \quad \alpha, \beta \in F$.

the members of **F** are commonly referred to as **scalars** whilst the members of **V** are simply called **vectors**.

If, instead, the **scalars** are taken from a **Ring**, then the structure is called a **module** over the **Ring**.

Furthermore, if each member of **V** can be uniquely expressed as a **linear combination** of independent vectors $\quad u_1, u_2, u_3 \ldots u_n$

(the coefficients being in **F**) then the Vector Space is said to be of **dimension n** and the set $\{u_r\}$ is said to form a **Basis** for **V**.

[Note: **independent** means that none of the u_r can be expressed in terms of any of the others]

Example 1

The set of all polynomials of degree 5 (eg) in an undefined **x**, over the field $\Re^{\#}$, is a vector space of dimension 6 - and the vectors

$$x^0, x^1, x^2, x^3, x^4, X^5 \quad \text{form a Basis.}$$

Example 2

All solutions of a given **n**th order differential equation, over the field $\mathbb{C}$, form a vector space. The **Basis** is any set of independent solutions, say,

$$u_1(x), u_2(x), \ldots u_n(x)$$

the dimension being **n**.

Example 3

All $n * n$ matrices, with elements in $\mathbb{C}$, over the field $\mathbb{C}$, form a Vector Space. A Basis is formed by the n^2 matrices M_{rs} where $r,s = 1 \ldots n$ and where (eg) $M_{rs} = (m_{rs})$ and $m_{rs} = 1$ whilst $m_{ij} = 0$ whenever $i,j \neq r,s$.

[A.3] When the elements **x, y, ...** of a Vector Space can be combined by a second binary operation, say, $\otimes$ in such a way that

the **associative law** $x(y \otimes z) = (x \otimes y) \otimes z = x \otimes y \otimes z$ holds

then the resulting structure $\{V ; \oplus ; \otimes ; F)$ is called an **ALGEBRA**.

An **Algebra, A,** is **commutative** if $x \otimes y = y \otimes x$ for all **x,y** in **V**.

An Algebra, **A**, is a **Division Algebra** if each **x** ($\neq$ **0**) possesses an **inverse**.

[Note: It is only in a Division Algebra that we can solve for **x** an equation like

$\alpha \otimes x = 1$ - since we need the inverse a^{-1} to obtain $x = a^{-1} \otimes 1$.]

Generally it follows that an **Algebra, A** consists of $\{M ; R\}$; that is to say, is a **Module, M**, over a **Ring, R**, - when **M** is itself a **Ring**.

In particular it is clear that **M** and **R** may both be **Fields** - even the same **Field**.

An important property of, for example $\Re$, $\Re^{\#}$, **J**, is that of being **ordered** - that is to say we can speak of the elements being 1st, 2nd, 3rd .. etc.. In mathematics this is usually denoted by the signs " > " (for after) and " < " (for before), and this relation has the properties that :-

For all members **x,y** ... in an **Ordered Field, F** we can write

(i) **x > 0** or **x < 0** or **0 > x**   where **0** is the additive zero in **F**

(ii) if **x > 0** and **y > 0** then **x+y > 0** and **xy > 0**

(iii) **x > y** if and only if **x-y > 0**.

This property of being ordered cannot be found in all Fields ; in particular it can be shown that, for example, the Complex Numbers cannot be ordered, but we shall see that this concept of **order** can play an essential part in many aspects of Physics [v.Appendix-B].

Now the property of being a **division algebra** includes the property of there being no right- or left- divisors of zero. This means that the equation $x \otimes y = 0$ has no solutions other than $x = 0$ or $y = 0$.

In the well-known **vector algebra** (which is <u>not</u> a **division algebra**) [and which was emphasised by the mathematician <u>Heaviside</u>] $a * b = 0$ need not imply that either $a = 0$ or $b = 0$ (they might be parallel vectors).

Example-4

Elementary Algebra is formed by the **Field** of reals, $\Re^\#$ (which behaves like a **vector space** over the same **Field** $\Re^\#$) and it also forms a **Field** under its own operators $\oplus = +$ and $\otimes = *$. This algebra is **commutative, associative** and is a **division algebra**.

Example 5

Complex Algebra, $\mathbb{C}$, is a commutative associative division algebra and forms a **vector space** (with Basis (**1 i**)) over the reals $\Re^\#$: its dimension = 2 and every element can be written in the form $x + iy$ where $i^2 = -1$.

[A.4] Example 6

Quaternion Algebra, $\mathbb{Q}$, is a **non-commutative associative division algebra** over the reals $\Re^\#$ (or equally over the rationals $\Re$). The Basis of its **vector spce** is usually denoted by (**1, i, j, k**) and is of dimension = 4. Each element can be written uniquely in the form

$$q = q_0 + q_1 i + q_2 j + q_3 k \quad \text{(each } q_r \in \Re^\#)$$

in which we have the rules (**Hamilton's** notation) :

$1^2 = 1$, $i^2 = -1$, $j^2 = -1$, $k^2 = -1$

and $ij = k$, $jk = i$, $ki = j$

whilst $ji = -k$, $kj = -i$, $ik = -j$.

What is now known as **vector algebra** consists of the properties of the **vector** part of the **quaternions**.

The product of two quaternions is given by

$$(\alpha + ai + bj + ck) * (\beta + Ai + Bj + Ck)$$
$$= (\alpha\beta - aA - bB - cC) +$$
$$i(bC - cB) + j(cA - aC) + k(aB - Ba)$$

which reminds us of the **vector algebra** notation (where $\wedge$ is the "product").

If we write a quaternion as (eg) $\alpha + \underline{A}$ where $\underline{A}$ is the "vector" part of the quaternion, then we get (what might be more familiar to the reader)

$$(\alpha + \underline{A}) * (\beta + \underline{B}) = (\alpha\beta - \underline{A}.\underline{B}) + \underline{A} \wedge \underline{B}.$$

Each quaternion $q = \alpha + ai + bj + ck$ possesses a **conjugate quaternion**, say q^* where $q^* = \alpha - ai - bj - ck$ or $\alpha - \underline{A}$

and the product of these defines the Norm of q, written $\|q\|$, via

$$\|q\| = |q|^2 = \alpha^2 + a^2 + b^2 + c^2$$

Then the multiplicative inverse of q is defined as

$$q^{-1} = q^* \div \|q\|$$

Since, in $\Re^\#$, $\|q\|$ will be positive this shows that if we take the **quaternions** as an algebra over $\mathbb{C}$ (instead of $\Re^\#$) - where the norm can vanish - then it will cease to be a **division algebra** - although <u>some elements</u> can still have inverses.

But it can be shown that :

There are only 3 associative division algebras over the reals $\Re^\#$, viz.

Elementary Algebra $\Re^\#$, **Complex Algebra** $\mathbb{C}$, and **Quaternion Algebra** $\mathbb{Q}$.

The first two of these are sub-algebras of $\mathbb{Q}$.

[This helps to explain why these Algebras have played such a big part in Physics, where the need to unravel a $\Gamma(1\text{-}1)$ depends on the ability to <u>divide</u>].

It is also the case that an associative algebra with a finite basis can be represented by an isomorphic image in an algebra of matrices. [v. Appendix-B [39].

An illustrative example can be provided by considering the base units

$\{1, i, j, k\}$ of $\mathbb{Q}$ - by taking their respective corresponding (2 x 2) matrices as

$$\begin{bmatrix} 1 & 0 \\ 0 & 1 \end{bmatrix} \quad \begin{bmatrix} i & 0 \\ 0 & -i \end{bmatrix} \quad \begin{bmatrix} 0 & 1 \\ -1 & 0 \end{bmatrix} \quad \begin{bmatrix} 0 & i \\ i & 0 \end{bmatrix}$$

The product combinations of these are easily verified as copying those of pairs of $\{1, i, j, k\}$.

[A.5] Example-7

Exterior Algebra, Λ^n, is a Vector Space **L** (of dimension **n**) over the reals $\Re^\#$ and possesses a "product" - denoted by $\wedge$ (read "wedge") - which is distributive over $+$, associative and anti-commutative ; that is to say

if α and $\beta \in \Lambda$ then $\alpha \wedge \beta = -\beta \wedge \alpha$, so that it follows that

$\alpha \wedge \alpha = 0$ for all $\alpha \in \Lambda$.

The dimension of the vector space can be any $n \in J^+$ (where J^+ denotes all the non-negative integers) but in this thesis we shall only need those cases in which **n** has the values 1, 2, 3, 4 . More generally, if we write a Basis for **L** as $\{e_1, e_2, e_3 \ldots e_n\}$ then the algebra can be seen as a **graded algebra** in that we can write

$\Lambda L = \Lambda^0 \cup \Lambda^1 \cup \Lambda^2 \cup \Lambda^3 \ldots \cup \Lambda^n$ where, eg,

All sequences like $(e_1, e_2, e_3, \ldots e_p\}$ will form a possible Basis for

Λ^p , $p <= n$ and so it follows that the dimension of such a Λ^p will equal the number of ways of selecting a **p**-subset out of **n** items - which is the usual ${}^nC_p$.

[A.6] Clifford Algebras

Take a Vector Space, V, spanned by the unit vectors $\{e_r, r = 1 \ldots n\}$ so that every vector in V is a linear combination of these - eg

$\underline{x} = x_1 e_1 + x_2 e_2 \ldots x_n e_n$

then impose a **quadratic groundform** on V by requiring the conditions

$e_r.e_s = -e_s.e_r$ whenever $r \neq s$

and $e_r.e_r = k$ a constant in $\Re^\#$

For example, when $n = 3$ we have

$x^2 = (x_1 e_1 + x_2 e_2 + x_3 e_3)^2$

will reduce to $(x_1^2 + x_2^2 + x_3^2)k$

We then have three important examples, viz.,

(i) When $k = -1$ we get **Quaternion algebra**

(ii) When $k = 0$ we get **Exterior algebra**

(iii) When $k = 1$ we get "ordinary" **Clifford algebra**, ℭ

$$\mathfrak{G} \equiv \mathfrak{G}^0 \oplus \mathfrak{G}^1 \oplus \mathfrak{G}^2 \oplus \ldots \oplus \mathfrak{G}^n$$

$\mathfrak{G}^0$ being simply $\mathfrak{R}^{\#}$, and $\mathfrak{G}^r$ being spanned by all r-products of the form

$$e_1 e_2 e_3 \ldots e_r$$

Thus the dimension of $\mathfrak{G}^r$ is ${}^n C_r$ and since ${}^n C_r = {}^n C_{n-r}$

there is a natural correspondence between vectors in $\mathfrak{G}^r$ and $\mathfrak{G}^{n-r}$

The one is usually referred to as the "pseudo"-of the other ; thus the single element in $\mathfrak{G}^n$ is a pseudo-scalar (corresponding to the scalar in $\mathfrak{G}^0$).

Various other properties of $\mathfrak{G}(V)$ follow [v. Ref 67].

[A.7] Calculus and Field Theories

The invention of the **differential and integral calculus**, by both **Newton** and **Leibnitz**, (although that of **Leibnitz** has given us the modern notation) has enabled the modern development of **Field Theories** - by allowing Physicists to deal with measures **at a point** in a **continuum** (the mathematician's field $\mathfrak{R}^{\#}$). It means that when we are faced with a "space" containing a function (field) defined at every point of it we can express changes in it in a **vanishing neighbourhood**, and from this we can deduce properties of the measures over a finite (and macroscopically observed) region of the space.

This process is expressed by the **derivative**, df, and the **integral**, $\int$df, of the calculus. It must be assumed, of course, that the functions dealt with are suitably continuous and differentiable.

So in dealing with, eg **force fields, velocity fields, potential fields**, etc., the correspondences which are relevant **at a specified point** in the continuum will be specified by things like dx, dt, df ..., that is to say by **differentials** and the **laws** in the theoretical Physics will be expressed via subsequent **differential equations**. To obtain macroscopic "laws" from these will involve specific solutions of these as well as the **inverse process** of **integration**.

We shall see that these concepts are discussed in eg. Appendix-B as well as in most of the chapters in this thesis.

[A.8]
The **exterior derivative** of a p-form, ϖ, is defined so as to take it into a (p+1)-form.

If ϖ_1 is a p-form and ϖ_2 is a q-form, we define the operator Δ by

(1) $\Delta(\varpi_1 \wedge \varpi_2) = (\Delta\varpi_1) \wedge \varpi_2 + (-1)^p \varpi_1 \wedge \Delta\varpi_2$

together with the requirements

(2) $\quad \Delta^2 = \Delta\Delta = $ Z(ero) operator $= 0_{op} = 0$, reducing the operand to zero,

whilst for a differentiable function $f(\underline{x})$, a 0-form, we require

(3) $\quad \Delta f = \Sigma_i\, (\partial f/\partial x_i)\, dx_i \qquad i = 1 \ldots n.$

We notice that this case gives **grad**f, in **Euclidean** 3-space.

Example-8 We also notice that when $f \equiv x$ then $Dx = dx$ (the usual differential) and also that since $\Lambda^0 \wedge \Lambda^p$ is contained in Λ^p we can treat $f(\underline{x})\, dx$ as $f(x) \wedge dx$ and applying (1) gives, eg.

$$\Delta(f \wedge dx) = -(\partial f/\partial x)dx \wedge dy + (\partial f/\partial z)dz \wedge dx$$

In the case of Λ^1, where $\varpi = udx + vdy + wdz$ (which often occurs with a Work Function in a vector field) we shall get

$$\Delta\varpi = (\partial v/\partial z - \partial w/\partial y)\, dz \wedge dy + \text{similar}$$

exhibiting the **curl** of the vector with components **u,v,w**.

Condition (3) then expresses the vector relation, viz., **curl grad** $f = 0$

Example-9 Taking the 2-form $\varpi = \mathbf{P}dy \wedge dz + \mathbf{Q}dz \wedge dx + \mathbf{R}dx \wedge dy$

gives us $\quad \Delta\varpi = (\partial P/\partial x + \partial Q/\partial y + \partial R/\partial z)\, dx \wedge dy \wedge dz$

- the **divergence** of the vector $(\mathbf{P,Q,R})$.

In this case condition (3) states the vector identity $\quad$ **div curl** $\underline{\mathbf{F}} = 0$.

APPENDIX-B Some GEOMETRIES

[B.0] As a part of the **language** of mathematics, geometry is an **abstract exercise** which exists independently of any physical representation of the concepts.
For this reason, as mathematicians commonly require, drawing pictures to represent these concepts is merely a tool (albeit often a very useful one) to help the imagination in its perambulations.

Consequently we shall see how the introduction of **algebraic geometry** is an essential step in the development of the subject, and in this connection we need to say something about the introduction of what are called **co-ordinates** ; and this needs some analysis of what are called the **real numbers**, $\Re^{\#}$, and their relationship to the **rational numbers**, $\Re$.

Numbers like $\sqrt{2}$ are meant to be solutions of algebraic equations - in this case a solution of the equation $x^2 = 2$. Although this cannot be a rational number (that is, a number of the form p/q where p and q are integers but with q $/=$ 0) it can nevertheless be approximated to any desired degree of accuracy by rational numbers - usually expressed as decimals with a finite number of digits after the decimal point.
This situation was made precise by the mathematician **Dedekind** who defined, (eg) $\sqrt{2}$, as the "Dedekind section" - being the pair of **sets of rationals**, viz., all those rationals **k**, where $k^2 < 2$ together with all those rationals **k** where $k^2 > 2$.
But this means that $\sqrt{2}$ is to be found in the **power set**, $P(\Re)$; that is to say the **set of all subsets** of $\Re$. It cannot therefore be of the same nature as one of the rationals. Things of the form $a + b\sqrt{2}$ is clearly a member of a **vector spce** with Basis $\{1 \sqrt{2}\}$. And this is why if we contemplate an equality like
$a + b\sqrt{2} = c + d\sqrt{2}$ we needs must conclude that $a = c$ and $b = d$.
The same things follow when we consider the so-called **complex numbers**, $\mathbb{C}$, being members of a vector space with Basis $\{1\ i\}$, where $i^2 = -1$.
Confusing "numbers" like $\sqrt{2}$ with "rationals" breaks the syntactical logic - which was elegantly exhibited by **Bertrand Russell** in his famous **Barber Paradox**. We cannot ask questions of $\sqrt{2}$ by regarding them as members of the same set as the set of rationals. All this does not preclude the mathematician's use of irrationals (c.f. his use of complex numbers) but it certainly places a big question mark beside the Physicist's use of them as the result of his "observations".

It is therefore legitimate to distinguish

>   (i) the mathematician's continuum, say $\Re^{\#}$, which consists of the rationals together with all the irrationals, and

>   (ii) the physicist's continuum, say $\Re$, which excludes the irrationals.

Other irrational numbers, such π or **e**, are also known as **transcendental numbers** being derived from functional (as opposed to algebraic) equations. No experimentalist in a laboratory is going to **measure** (eg) π - because he/she can only observe rationals. So the **circle cannot be squared**! - "getting close" is not the same logical thing. Nor can he measure $\sqrt{2}$ by the hypoteneuse of a rt-angled triangle with sides 1 and 1 - he can only "get close" by numbers in $\Re$.

[B.1] After the great achievements by the Ancient Greeks (typified by **Pythagoras, Euclid** and **Appollonius** etc) there was another startling advance by the Europeans in the 18th and 19th centuries. Pionering work by (eg) **Poncelet, Von Staudt, Cauchy, Laguerre, Gauss** and others too numerous to mention (v. Appendix-E, refs [2], [3], [4], [19]) has shown that the traditional dependence and assumption of **Euclidean Geometry** is more an accident of history than an absolute necessity for our understanding of the natural world.

Poncelet, in particular, laid the foundations of **Projective Geometry** whilst Von Staudt showed that (i) **Projective Geometry** is more fundamental than **Euclidean Geometry**, and (ii) the attribution of numbers as co-ordinates in a geometry does not necessarily require the concept of **distance** - as is always assumed in **Euclidean Geometry** (which uses the **Pythagorean** metric).

Assigning co-ordinates to points and lines only needs a mapping, say,

$\gamma : \Re \times \Re \to \mathbf{P}^2$ or $\gamma : \Re \times \Re \times \Re \to \mathbf{P}^3$

and this is dealt with in the modern **algebraic geometry**.

Projective Geometry, $\mathbf{P}^2$, (in 2 dimensions) consists of entities called **points** and **lines** which satisfy the **propositions of incidence**, viz.,

>   (i) any two distinct lines define a unique point

>   (ii) any two distinct points define a unique line.

Projective Geometry, $\mathbf{P}^3$ (in 3-dimensions) consists of entities called **points, lines** and **planes** which satisfy the extended **propositions of incidence**, viz.,

>   (i) any two distinct lines define a unique point

(ii) any two distinct points define a unique line

(iii) a line and a plane define a unique point

(iv) a line and a point define a unique plane

(v) any two distinct planes define a unique line.

It follows, for example, that any three non-collinear points define a plane.

Projectiv Geometry, P^1 consists of entities called **points** and are defined by a single co-ordinate, x, (or the pair (x,z) in gomogeneous co-ordinates).

It is worth noticing that, since $kP \equiv P$ for $k \in \Re^{\#}$, the "points at infinity" viz., $+\infty$ and $-\infty$ coincide [v. §[B.3]].

We shall find most results, but not all, which are relevant to this study can be expressed in P^2.

A noticeable absence in a projective geometry is the concept of **parallelism** - so <u>any</u> two distinct lines meet in a point.

Notice, too, that if we complain about this - because of what we think we observe - then we have really drifted into the realm of **Physics** (where <u>observation</u> is supposed to be experienced). In spite of this, and because the business of **observing** has yet to be analysed, we shall see that the properties of P^n are highly relevant to a **Physics**.

An important feature of a projective geometry is the <u>absence of any idea of distance</u>. This concept is properly relegated to structures such as **Euclidean Geometry** or its extension into **Riemannian Geometry** - which we shall discuss later.

So **Projective Geometry** is said to be basically **non-metrical**.

If in fact we allow the idea of parallelism to be built into our P^n then we call the resultant structure an **Affine Geometry** - but we shall not need this in our present thesis.

It also follows from the propositions of incidence that there is a natural **duality** to be found among theorems. For example, in P^2, a theorem about lines and points gives rise to a **dual theorem** about points and lines : that is to say, we need only replace "point" with "line" - and "line" with "point".

In a similar way, in P^3, we need only replace "point" with "plane" and "plane" with "point" to obtain a **dual result** : in this case "line" is **self-dual** (v. figs B.1, B.2).

[B2] As its name implies **Projective Geometry** is centred on the idea of **projection**, and this is a generalisation of the idea of **perspectivity** - one figure being in **perspective** with respect to another if the joins of corresponding points are concurrent at, say **V**, the **centre of perspective**. The dual of this being when the meets of corresponding lines are collinear on, say the line v; this being the **axis of perspective**.

A series of perspectives (via distinct centres V_r) is called a **projectivity** (v. fig B.3).

We are particularly interested in any **invariants** in P^2 which are preserved under this process of projection, and the most important of these is that of **cross-ratio** of any 4 points (or lines) related by a projectivity.

For example (v. fig B.3) the 4 points A_1, B_1, C_1, D_1 on the line p_1 have an associated cross-ratio written as $(A_1 \; B_1 \; C_1 \; D_1)$ and, however we characterise it algebraically, this is to be equal to the cross-ratio of (eg) $(A_2 \; B_2 \; C_2 \; D_2)$ on the line p_2, via a perspectivity from V_1, and thence equal to the cross-ratio of the range $(A_3 \; B_3 \; C_3 \; D_3)$, via the perspectivity from V_2.

When we define parameters with which to co-ordinatise the **rays** we also equate their cross-ratio with the above - so the duality holds.

A significant property of **projectively related** pencils of lines, that is to say those which have equal cross-ratio values, is the following.

Given two vertices S_1 and S_2 with associated pencils of lines, (a_1, b_1, c_1, d_1) and (a_2, b_2, c_2, d_2), which are projectively related, then the intersection points of corresponding lines (rays) viz., $P_a \; (= (a_1 a_2))$, P_b, P_c, and P_d all lie on a "point-conic". [ref Appendix-E [2] [3] [7] [43] [44]].

The dual of this defines an envelope curve which is a "line-conic".

[B3] Moving into **algebraic geometry** we introduce "co-ordinates" as follows. [v. Fig B.4]

Let **X,Y,Z** constitute a **triangle of reference** and let co-ordinates be such that the point **X** has co-ordinates $(1,0,0)$, the point **Y** has co-ordinates $(0,1,0)$ and the point **Z** has co-ordinates $(0,0,1)$. It is also postulated that, for any non-zero $k \in \Re^{\#}$, the symbolic points **kP** and **P** are the same point - that is to say $(kx, ky, kz) \equiv (x, y, z)$.

This means that it is the **ratios** of the co-ordinates which are significant. This also means we are free specify one point in the plane as the so-called **unit point**, $U(1,1,1)$.

There is a clear relation with the traditional system of Cartesian co-ordinates (x,y) by supposing that $z = 1$. So comparing these systems $P(x,y)$ corresponds to $(x,y,1)$ in $\mathbf{P}^2$. In metrical geometry these co-ordinates have traditionally been **areal co-ordinates** or **trilinear co-ordinates** (v. refs Appendix-E : [43], [44]).

Any **line**, λ say, in $\mathbf{P}^2$ is defined by the **homogeneous equation** $ax + by + cz = 0$ and is said to have **line co-ordinates** **(a, b, c)** - and contains the **point (x, y, z)**.

Lines which, in traditional language, are called **parallel** will have eqations like $ax + by + cz = 0$ and $ax + by + c'z = 0$. Consequently they will meet where these equations are simultaneously true - which is where $(c - c')z = 0$. If the lines are distinct it follows that $z = 0$. Since in Cartesian terms this means that they meet at $(x \div 0, y \div 0)$ we call this a **point at infinity** and the line $z = 0$ as the **line at infinity** (denoting it by Λ_∞). [These concepts are required by the propositions of incidence, since any two lines must meet in a point].

We also see that by projection any point on a line can be projected onto Λ_∞ - [v. Fig B.5 following] - so such a point is accessible in a commonsense way, since if we can project a finite point to infinity then we can reverse the procedure.

[These matters are fully discussed in any of the following references in Appendix-E : [2], [3], [4], [43], [44]]

Following the work of **von Staudt**, in particular, we can demonstrate how the introduction of algebraic "co-ordinates" in the projective plane $\mathbf{P}^2$ can give rise, by simple constructions which do not involve any concept of distance, representations of all the usual algebraic operations. A full discussion of this point will be found in **H.F.Baker's** excellent work [v. Appendix-C Ref. [2]] - and we demonstrate this in Figs. B.01 & B.02.

However we choose a mapping γ (v. supra) to describe co-ordinates we define the **cross-ratio (A B C D)**, of points **A,B,C,D** on a line λ, as the value of **(a b c d)** - where **a** etc. are the co-ordinates attributed to the point **A** etc. - and this is to have the value $(a-b)(c-d) \div (a-d)(c-b)$

Clearly we must ensure that the co-ordinate values are taken from a **Field** - so using $\mathfrak{R}^\#$ becomes almost (but not quite) essential, $\mathfrak{R}$ would serve equally well.

Although it is not possible to deduce Pappus' theorem from the propositions of incidence alone, yet it can be proved in algebraic geometry - although in that case it requires that the co-ordinate symbols be **multiplicatively commutative**. [v. Refs Appendix-E [2],[43]]

Another point of interest lies in the fact that Desargue's theorem (v. Figs B.6 and B.7) can only be proved in $\mathbf{P}^3$ (from the propositions of incidence) - and in algebraic geometry the proof requires that the symbols be **multiplicatively associative**.

[B4] Homographies, $\Gamma(1\text{-}1)$, Harmonic ranges, and Involutions

If one range of points on a line λ_1 are projectively related to another range of points on a line λ_2 then it is clear that the separate points are in a one-to-one correspondence. We denote such a correspondence by the symbol $\Gamma(1\text{-}1)$ and in algebraic geometry this becomes an algebraic expression which is linear in each variable. Thus if **x** denotes the co-ordinate of any point on λ_1 whilst **y** denotes the same thing for the **corresponding point**, on λ_2, in the projectivity, the expression (called a **homograpy**) is

$$axy + bx + cy + d = 0 \qquad \text{with } a,b,c,d,0 \in \Re^{\#}$$

where $ac\text{-}bd \neq 0$ (otherwise the expression factorises and is degenerate).

The points **A,C** are called **mates** in the homography - as are the points **B,D**.

It is not difficult to prove that, under such an expression, as a $\Gamma(1\text{-}1)$, the cross-ratios

$(x_1 x_2 x_3 x_4)$ and $(y_1 y_2 y_3 y_4)$ are equal in $\Re^{\#}$.

So, in **algebraic geometry** we have an **invariant** (viz., cross-ratio) under such a $\Gamma(1\text{-}1)$.

In a given $\Gamma(1\text{-}1)$ although an **x** might correspond to a particular **y** it will not necessarily follow that the particular **y** corresponds to the original **x**. In other words it is not always **symmetric** homography. This will occur whenever $b \neq c$.

In the event that (i) the ranges lie on the same line λ
(or, dually, rays lie on the same point **V**)

and (ii) the homography is symmetrical, i.e. $b = c$,

the homography is called an **involution** (on the range of points, or on the pencil of rays).

When the cross-ratio $(abcd) = -1$ we call the range (or pencil) **harmonic** so that, under a projectivity a harmonic range (or pencil of rays) maps into another harmonic range.

[The name "harmonic" derives from the following fact :
Taking **A** as an origin and mapping points **B,C,D** into co-ordinates b,c,d we can show that the harmonic condition $(\mathbf{ABCD}) = -1$ gives the relation $2/c = 1/b + 1/d$ - showing that b,c,d are members of a harmonic series].

In an **involution** we require that the **interchange of mates A,C** (or that of **B,D**) leaves the cross-ratio be unaltered - because of the symmetry. That is to say that, in an **involution**

$$(\mathbf{ABCD}) = (\mathbf{CBAD}) = (\mathbf{ADCB}).$$

and the homography then looks like

$$axy + b(x + y) + c = 0 \qquad \text{the symbols being in (say) } \Re^{\#}.$$

Special cases arise by taking various values for the coefficients **a,b,c** (v. earlier chapters).

In an **involution** an important role is played by the existence of its **double points**, that is to say those points for which $x = y$. These will be given by the equation

$$at^2 + 2btx + c = 0$$

and if we are given its roots, as α and β, then the equations

$$a\alpha^2 + 2b\alpha + c = 0 \quad \text{and} \quad a\beta^2 + 2b\beta + c = 0$$

enable us to solve for the ratios b/a and c/a - thus defining the involution.

So an involution is determined by its two double points.

Also, if we take an origin O such that points **A,C** have co-ordinates -**a**, +**a**

whilst **B,D** have co-ordinates **b,d**, it follows from (-a b a d) = -1

that $bd = a^2$ or in terms of metrical geometry $OB.OD = OC^2$.

This situation is illustrated in the case of **inverse points** with respect to a circle.

Figures B.8 and B.9 illustrate some of these ideas.

B and **D** are defined as inverse points with respect to the circle σ if

$OB.OD = OC^2 = $ (radius)^2, and consequently **B,D** are mates in the involution defined

by the double points **C** and **C"**. It is also a theorem of Euclidean Geometry that

any circle through the points **B** and **D** cut the circle σ orthogonally.

[Note: All these concepts and definitions (except the ideas of parallelism and the line
at infinity, Λ_∞) can still apply in a metrical geometry such as **Euclidean geometry**.
But in that case the factors in the cross-ratio expression, viz., $(x_1 - x_2)$ etc.
will be interpreted as **distances** or **angles**. In many textbooks this is how they are introduced.
It is only in ,eg, $\mathbf{P}^2$ or $\mathbf{P}^3$, that "distance" is not needed.]

[B5] Introducing a metric into $\mathbf{P}^2$

We can show that $\mathbf{P}^2$ (and $\mathbf{P}^3$) contains the well-known **Euclidean Geometry** by showing

how a **metric** (concept of **distance** or **interval**) can be introduced.

First we need to list the properties of such a thing. It is in fact a mapping of $\mathbf{P} \times \mathbf{P}$ into

the reals $\mathfrak{R}^\#$, viz., $d : \mathbf{P} \times \mathbf{P} \to \mathfrak{R}^\#$ satisfying the following conditions.

(i) $d(\mathbf{P},\mathbf{P}) = 0$

(ii) $d(\mathbf{P},\mathbf{Q}) > 0$ and $d(\mathbf{P},\mathbf{Q}) = -d(\mathbf{Q},\mathbf{P})$ whenever $\mathbf{P} \neq \mathbf{Q}$

(iii) $d(\mathbf{P},\mathbf{R}) \leq d(\mathbf{P},\mathbf{Q}) + d(\mathbf{Q},\mathbf{R})$ for all triads **P,Q,R**.
equality occurring when the points are collinear.

Such a set of properties of the mapping "d" is called a **metric** on the geometry.

If **O,U** are points on a line λ and **P,Q** are any two other points on λ **Laguerre** defined

the distance $d(\mathbf{P},\mathbf{Q})$ as $(1/2i)\ln(\mathbf{POQU})$. Since it can easily be shown that, generally,

$$(\mathbf{POQU}) * (\mathbf{QORU}) = (\mathbf{PORU})$$

we deduce that this gives $d(\mathbf{P},\mathbf{Q}) + d(\mathbf{Q},\mathbf{R}) = d(\mathbf{P},\mathbf{R}) + n\pi$, for any $n \in \mathbf{J}^+$ and we must appreciate that "distance" between points **P,Q** in the plane is no longer an absolute property of the two points, as in Euclidean Geometry, but depends on the choice of what are usually called **absolute points** (following **Cayley**) .

These points of reference are found by asserting a **basic, or absolute, conic** (often a circle) whose intercepts on a line joining **P,Q**, say **I** and **J**, and which allow us to specify the cross-ratio needed in the above definition of distance. [It is sometimes convenient to take the points **I,J** as the two points intercepted (by every circle) on the line at infinity, Λ_∞.]

In a similar way, if **V** is a vertex, **o,u** two arbitrarily chosen rays through **V** and **p,q** two other rays through **V**, the **angle** between **p** and **q** can be defined by the expression $\angle$(**p,q**) = (1/2i) ln (**poqu**). The lines **o,u** are identified as the two tangents from **V** to the chosen **absolute conic**.

[Further discussion of this topic can be found in Appendix-E [2],[43],[44],[45]]

But suffice it to say that, following the work of Laguerre and Cayley, but we can show that metric notions can be introduced into P^2 (and P^3) which result in a geometry mirroring the well-known **Euclidean Geometry**.

[A fuller discussion of the this topic can be found in Appendix-E [2], [4], [43] [44]]

[B6] <u>Euclidean Geometry</u> possesses a **metric** which is usually called the **Pythagorean Metric** - since it imposes the famous **Pythagoras' Theorem** on the space as a "natural" measure of distance in the plane. In this space the incremental distance between the point (**x,y**) and (**x+dx,y+dy**) is written as

$$ds^2 = dx^2 + dy^2$$

where eg dx^2 means $(dx)^2$.

This naturally assumes that the concept of a "right-angled Δ" is admitted and that the real numbers $\Re^\#$ are to be used in the geometry.

This metric, together with the notion of **parallelism**, constitutes the essence of **Euclidean Geometry**. Other concepts and properties of P^2 (or P^3) are accepted which do not involve any of these special additives.

In this geometry **distance, angle**, and **congruence** are **invariants** under the general group of **linear transformations**.

[Note : We assume that the reader is knowledgeable about the common theorems of this Euclidean Geometry].

Following the notation and argument found in Ref. Appendix-D [2] we can show that this

Euclidean metric is compatible with the structure of projective geometry.

[B7] Defining **distance/interval** in the projective space P^2 is achieved by taking the **absolute conic** S_0 in any one of the following forms :

(a) $x^2 + y^2 + z^2 = 0$ giving elliptical metrical geometry

(b) $-x^2 - y^2 + z^2 = 0$ giving hyperbolic metrical geometry

(c) $z^2 = 0$ giving parabolic metrical geometry

[The case (c) is the form of a degenerate parabola]

It is the case (c) which contains the familiar **Euclidean metrical geometry** with the **Pythagorean metric**.

Points on S_0 in elliptical geometry cannot have co-ordinates in $\Re^\#$, since (0,0,0) is excluded.

Points on S_0 in hyperbolic geometry can have co-ordinates in $\Re^\#$.

Points on S_0 in parabolic geometry have co-ordinates $(x,y,0)$ where **x** and **y** are in $\mathbb{C}$.

Interval in Elliptical geometry is obtained by writing $S \equiv x^2 + y^2 + z^2$ and by denoting $x_p x_q + y_p y_q + z_p z_q$ by S_{pq}.

If now $P_1(x_1,y_1,z_1)$ and $P_2(x_2,y_2,z_2)$ are two points with real homogeneous co-ordinates in the plane, and using the algebraic inequality

$$S_{11}S_{22} - (S_{12})^2 \geq 0$$

then it follows that $0 < (S_{12})^2/S_{11}S_{22} \leq 1$

Hence there is a real number θ such that

$$\cos \theta = S_{12}/\sqrt{(S_{11}S_{22})} \quad \text{and} \quad 0 \leq \theta < \pi.$$

If **I,J** are on P_1P_2 (and also on $S = 0$) and if $I = \lambda_1 P_1 + \lambda_2 P_2$ and $J = \mu_1 P_1 + \mu_2 P_2$ then (by **Joachimstahl's** equation) λ_1/λ_2 and μ_1/μ_2 are the roots of

$$\lambda_1^2 S_{11} + 2\lambda_1 \lambda_2 S_{12} + \lambda_2^2 S_{22} = 0$$

This gives the values

$$\lambda_1/\lambda_2 , \mu_1/\mu_2 = [-\cos \theta \pm i \sin \theta]\sqrt{(S_{22}/S_{11})}$$

whence dist $(P_1 P_2) = k \log (\infty \; \lambda_1/\lambda_2 \; 0 \; \mu_1/\mu_2) = k \log e^{2i\theta}$

and this gives dist $(P_1 P_2) = ik (2\theta + 2n\pi)$

If we want the distance beteen two real points to be a real number then we take $k = 1/2i$.

so that $\quad\text{dist}(P_1P_2) = \theta + 2n\pi \quad (\mu)$

where $\quad \theta$ is given by $\quad \cos\theta = S_{12}/\sqrt{(S_{11}S_{22})}$

Then to obtain the metric in parabolic geometry we use a limiting process on the conic
$S \equiv R^2(x^2 + y^2) + z^2$ as $R \to 0$.

Writing $k = 1/2iR$ and $d = \text{dist}(P_1P_2)$ in the elliptic case we get

$$-\sin^2(Rd) = [(S_{12})^2 - S_{11}S_{22}]/S_{11}S_{22}$$

and this is

$$-\sin^2(Rd) = [-R^2D + R^4A]/[(z_1z_2)^2 + R^2B + R^4C]$$

where A,B,C are polynomials in the co-ordinates of P_1 and P_2 and where

$$D = (z_1x_2 - z_2x_1)^2 + (z_1y_2 - z_2y_1)^2$$

Since, as $R \to 0$, $\lim [\sin^2(Rd)]/R^2d^2 = 1$ we have

$\text{dist}^2(P_1P_2) = d^2 = d^2 \lim [\sin^2(Rd)]/R^2d^2 = \lim [\sin^2(Rd)]/R^2$

giving $\quad \text{dist}^2(P_1P_2) = D/(z_1z_2)^2$

Now (x_r,z_r) and (y_r,z_r), with $r = 1,2$ are homogeneous co-ordinates in P^2,

so putting $z_1 = z_2 = 1$ we get the **Euclidean** case wherein

$$\text{dist}^2(P_1P_2) = (x_2 - x_1)^2 + (y_2 - y_1)^2$$

the usual **Pythagorean** metric.

[B8] Riemmannian Geometry, owing its inspiration to the work of **Gauss**, is an

n-dimensional structure with a generalised (Euclidean) metric of the form

$$ds^2 = \Sigma_r \Sigma_s g_{rs} d x_r x_s$$

where the summations are over the values $r,s = 1,\ldots n$.

It is easy to see that this form of a metric follows from the idea of changing from a set of perpendicular axes into a set of **oblique axes**. For [v. Fig B.10] that there will always be a **linear transformation** from the one set to the other. So, for example, in a simple 2-space, to change from co-ordinates (u,v) to (x,y) via, say

$u = ax + by \quad$ and $\quad v = hx + ky$

means that the **Euclidean metric**

$$ds^2 = du^2 + dv^2 \to (adx + bdy)^2 + (hdx + kdy)^2$$

which gives a metric like $\quad g_{rs}dx^rdx^s \quad$ (i.e. **Riemannian**)

Its importance lies in its possible application in a space of **n**-dimensions.

Its interest for Physicists (following the work of **Einstein**) has naturally been associated with the case of **n** = 4. In this context we can notice the metrics which were proposed by **Lorentz-Einstein** and by **Minkowski**.

The former being $\quad ds^2 = dx_1^2 + dx_2^2 + dx_3^2 - (cdt)^2$

and the latter being $\quad ds^2 = (cdt)^2 - dx_1^2 - dx_2^2 - dx_3^2$.

The importance of the **Minkowski** metric being in that it ensures ds^2 is always ≥ 0 this being compatible with the experimental measure that no velocity can be greater than c.

[v. Refs Apendix-E [6] [22]].

[B9] Hilbert Space is an example of a space spanned by a vector basis of **infinite dimensions**. The basis consists of a **countable** set of unit vectors, say $\{\hat{e}_r\}$, so that any vector in the geometry is of the form

$$\mathbf{X} \equiv x_1\hat{e}_1 + x_2\hat{e}_2 \ldots + x_n\hat{e}_n + \ldots \infty$$

where the components x_i are complex numbers, in $\mathbb{C}$.

We must also insist on **norm (X)** being **finite**. This means that **Hilbert Space** consists of formal sums, such as **X**, being subject to the convergence of $\Sigma_n |x_n|^2$.

Linear operators in such a space will therefore be repersented by **infinite matrices**, and these matrices will differ in many important respects from that of **finite matrices**. For example, finding the **eigenvalues** of such an operator will involve the solution, if there is one, of an equation of **infinite degree**.

Similarly the product of two matrices, **AB**, will only exist if the infinite series

$$c_{\alpha\beta} = \Sigma_g a_{\alpha\gamma} b_{\beta\gamma}$$

is absolutely convergent. Even then it is not necessarily true that **BA** exists.

[B10] We shall introduce **Dirac Space** in chapter-16 of Part-1.

Algebraic Sum $a + b$ in P^2 : without using a Metric

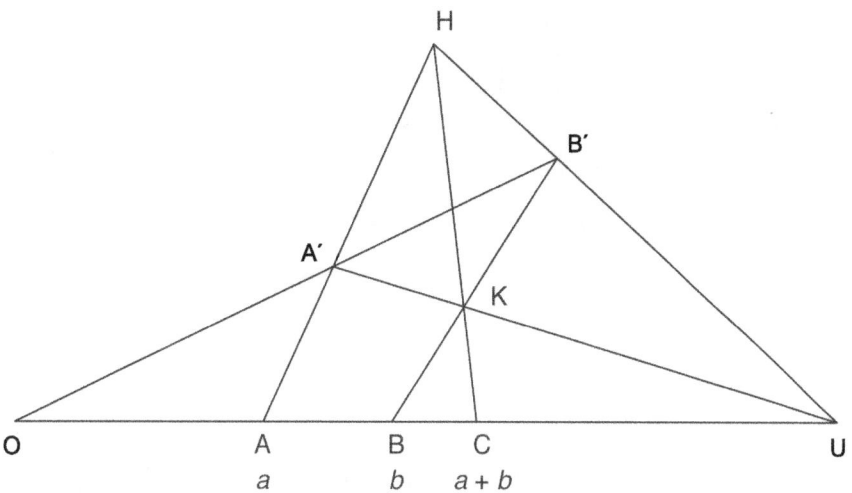

Given: Line OABU, point H, line OA'B' to meet HA, HU.
Join A'U, BB' to meet in K. Let C= HK∩OU.
To Show That, with A = O + aU & B = O + bU then C = O + (a + b)U.
Take H = A + hA' = O + aU + hA' ⇒ H - aU = O + hA'
Since H - aU is a point on HU and O + hA' is a point on OA'
so B' = OA'∩HU is O + hA'
But then B' - = O + hA' - (O + bU) = hA' - bU which = K
Then H - K = A + bU which defines point C
so C = O + aU + bU = O + (a + b)U

Fig B.01

Algebraic PRODUCT ab in P^2 : without using a METRIC

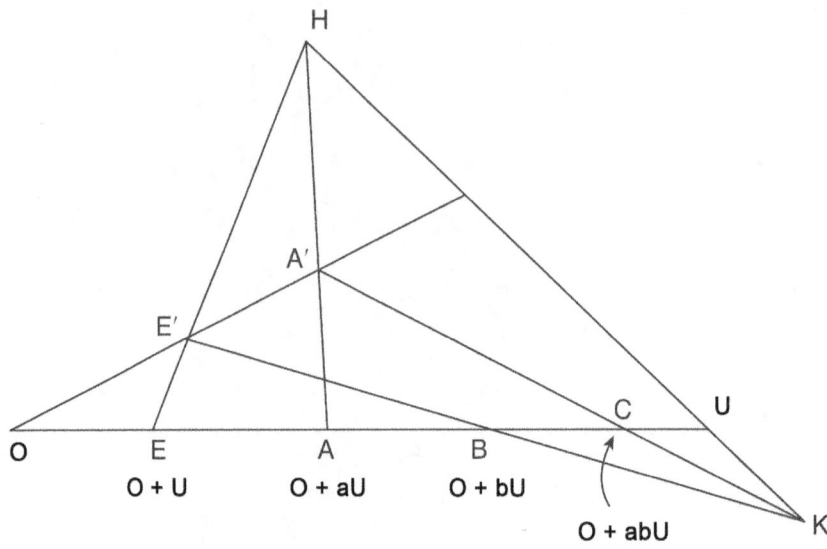

Given: Line OE ABH; E = unit point = $O + U$ + aU, B = $O + bU$
H is any external point; Join HE, HA, HU; Any line OE'A' as shown.
OHU as Δ of ref., Take A' = $O + aU + hH$ & E'=OA∩HE is a linear combination of O and A'; hence we can write E' = $aO + aH + hH$
Then K = HU∩E'B can be K = $aO + aU + bH - a(O + bU)$
So K = $(a - ab)U + hH$ and since C = A'K∩OU it can be written as C = $O + aU + hH - \{(a - ab)U + hH\}$ = $O + abU$
Hence we can identify C as a point with "value" of ab

Fig B.02

APPENDICES **297**

Pappus Theorem

points A_1, B_1, C_1 on line p_1
A_2, B_2, C_2 on line p_2

$\Rightarrow$ meets of lines A_1B_2, A_2B_1 (= X)
A_1C_2, A_2C_1 (= Y)
B_1C_2, B_2C_1 (= Z)
are collinear (on line p_3)

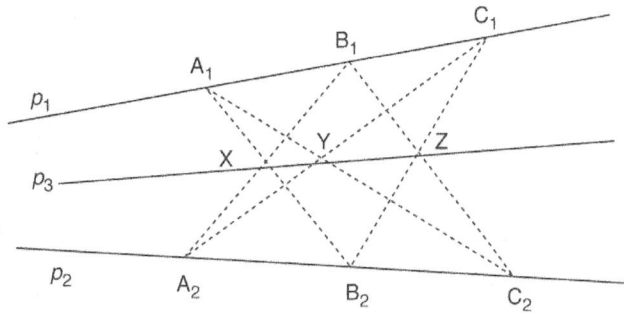

Fig B.1

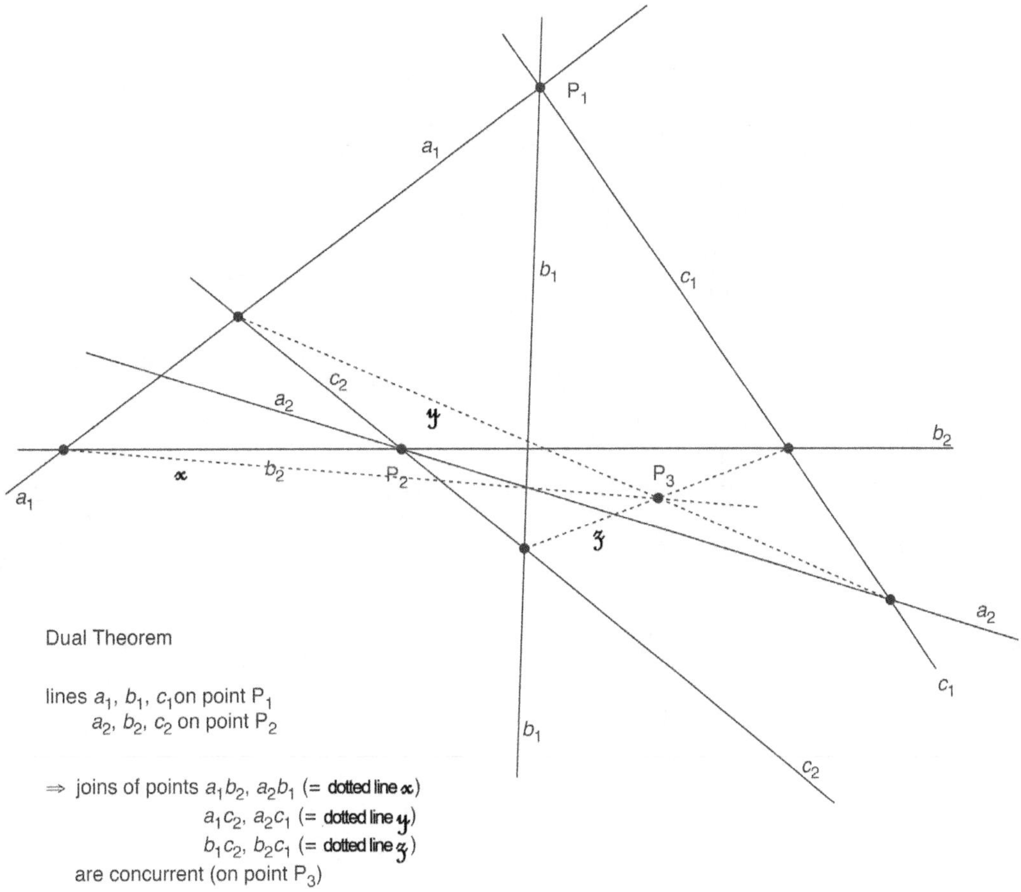

Dual Theorem

lines a_1, b_1, c_1 on point P_1
a_2, b_2, c_2 on point P_2

$\Rightarrow$ joins of points a_1b_2, a_2b_1 (= dotted line x)
a_1c_2, a_2c_1 (= dotted line y)
b_1c_2, b_2c_1 (= dotted line z)
are concurrent (on point P_3)

Fig B.2

298 MATHEMATICAL PHYSICS

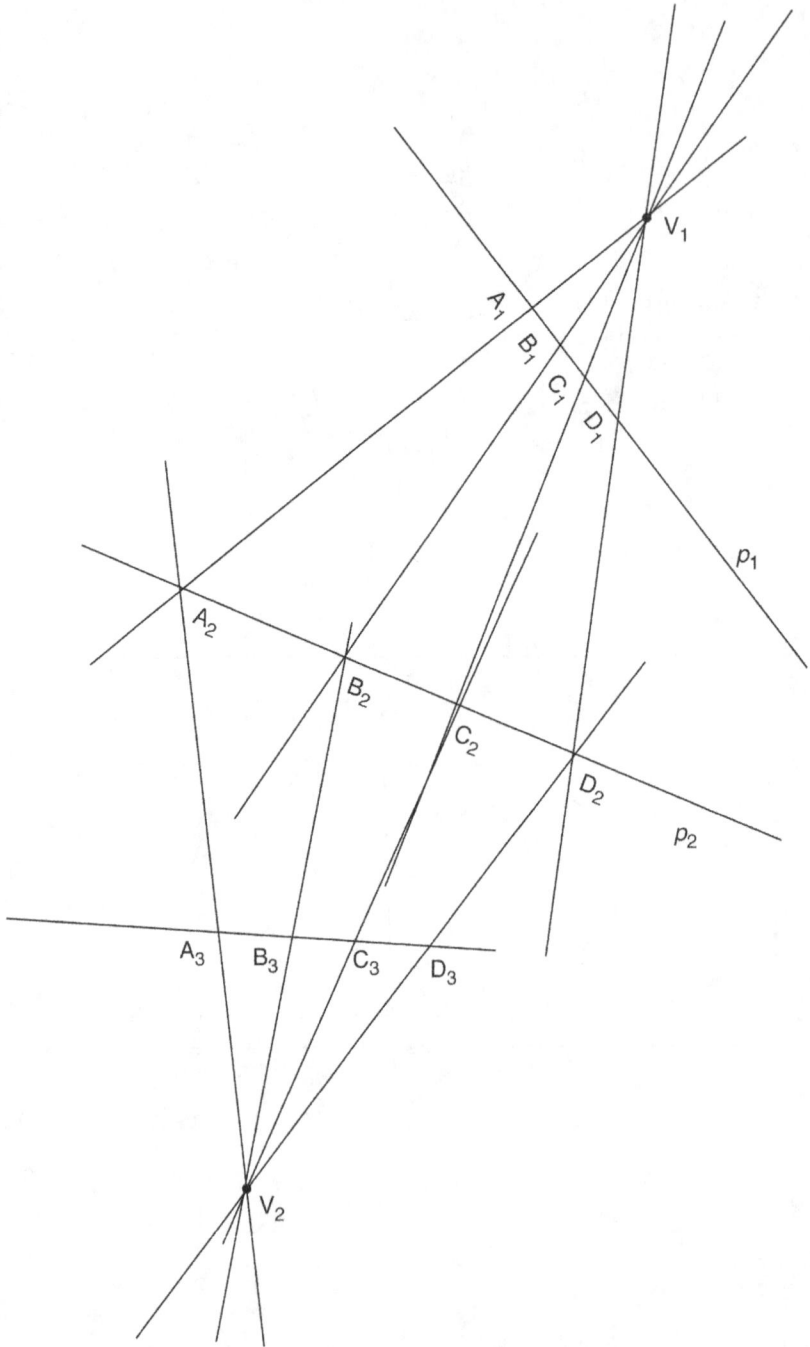

Fig B.3

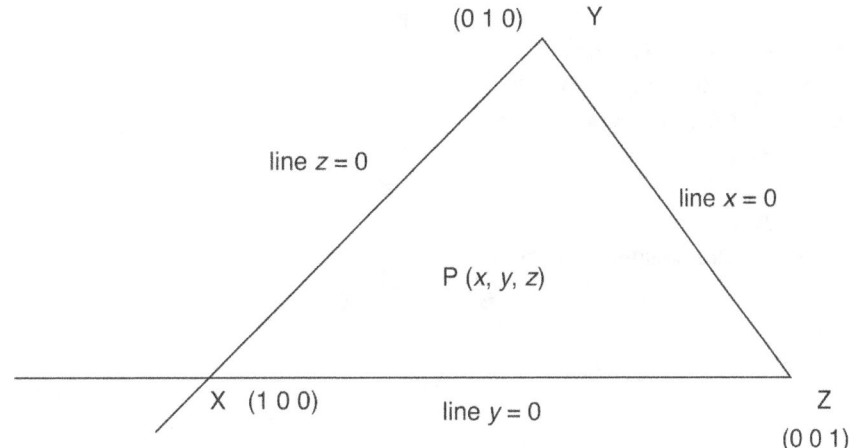

Fig B.4

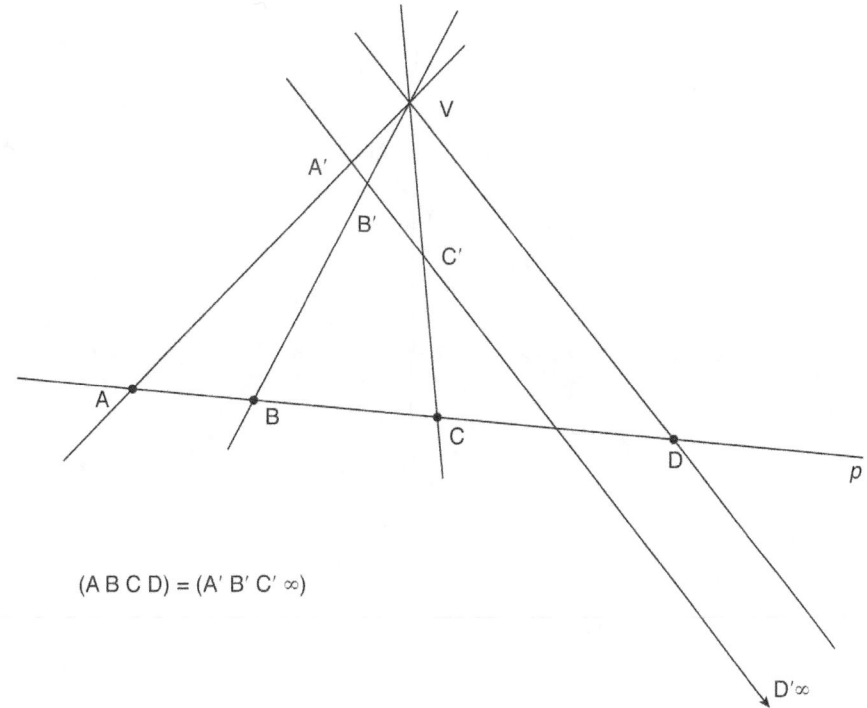

(A B C D) = (A' B' C' ∞)

Fig B.5

DESARGUE'S THEOREM

If the triangles $A_1B_1C_1$ and $A_2B_2C_2$ are in perspective from V then the Meets of corresponding sides, viz
A_1B_1, A_2B_2 in X
B_1C_1, B_2C_2 in Y
C_1A_1, C_2A_2 in Z
then X, Y, Z are collinear on line p

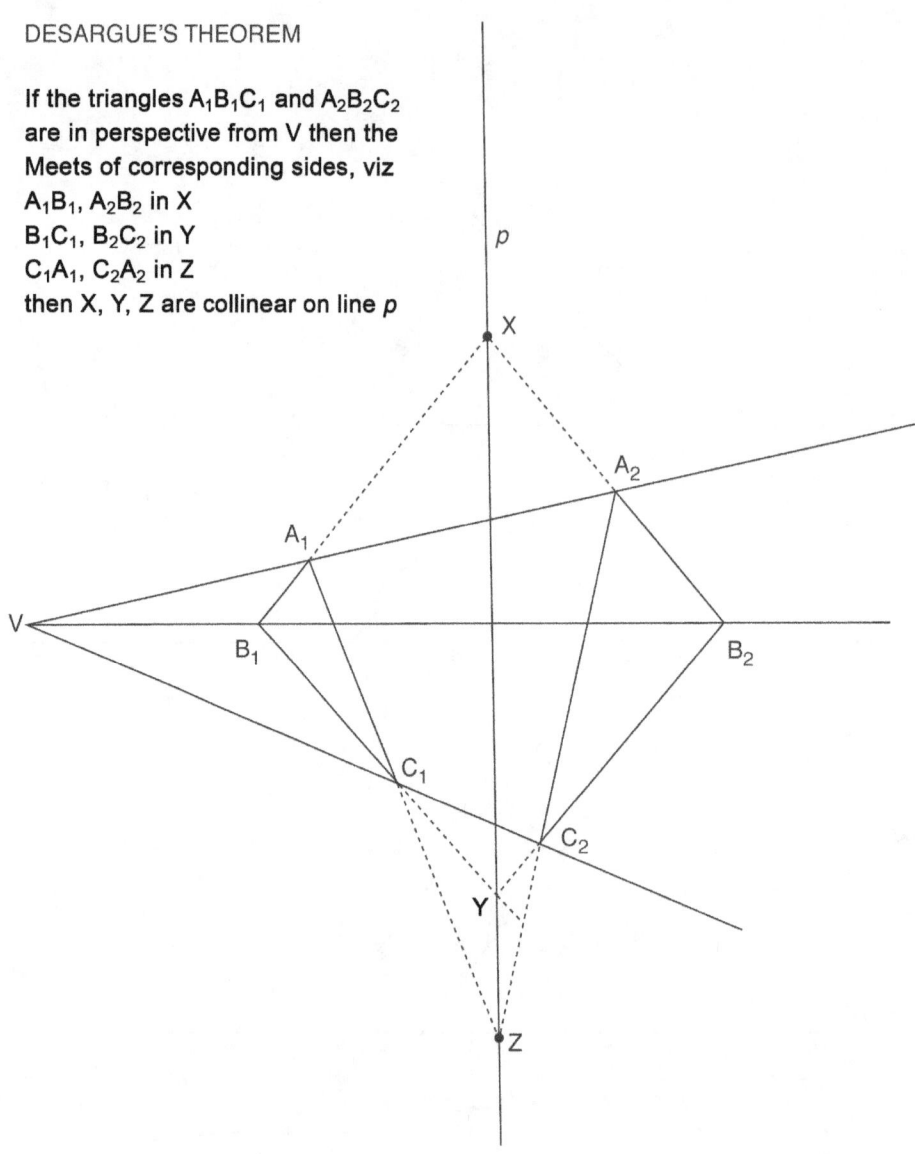

Fig B.6

DUAL of DESARGUE'S THEOREM (is actually the converse)

If Trilaterals $a_1b_1c_1$, $a_2b_2c_2$ are such that the meets of corresponding sides are collinear, on p, then the joins of corresponding vertices are concurrent: i.e. x, y, z meet at V.

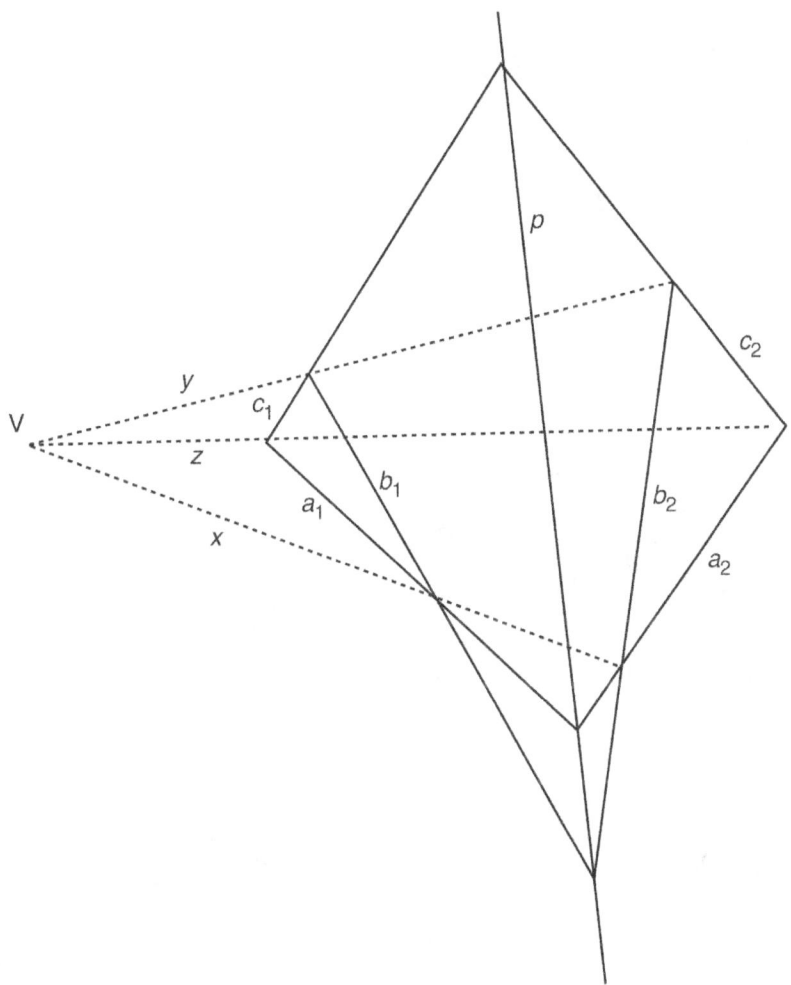

Fig B.7

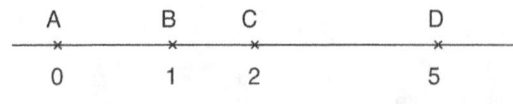

$$(ABCD) = (0\ 1\ 2\ 5) = \frac{(-1)(-3)}{(-5)(1)} = -\frac{3}{5}$$

"Harmonic Range"

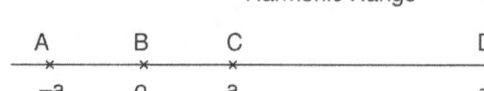

$$(ABCD) = (-a\ o\ a\ \infty) = \frac{(-a)(a-\infty)}{(-a-\infty)(a)} = -1$$

Fig B.8

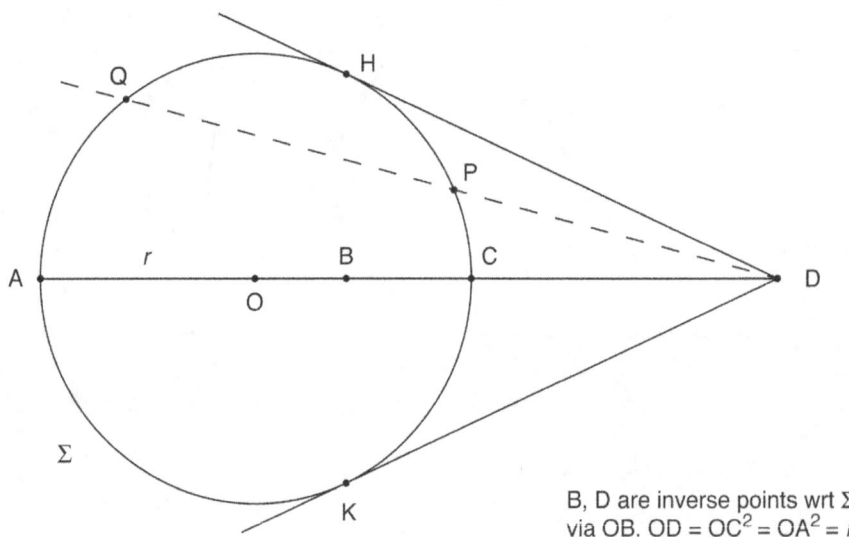

B, D are inverse points wrt Σ via OB. $OD = OC^2 = OA^2 = r^2$

$\Rightarrow$ B, D are mates in involution on defined double points H, K. P, Q is another pair of mates.

Fig B.9

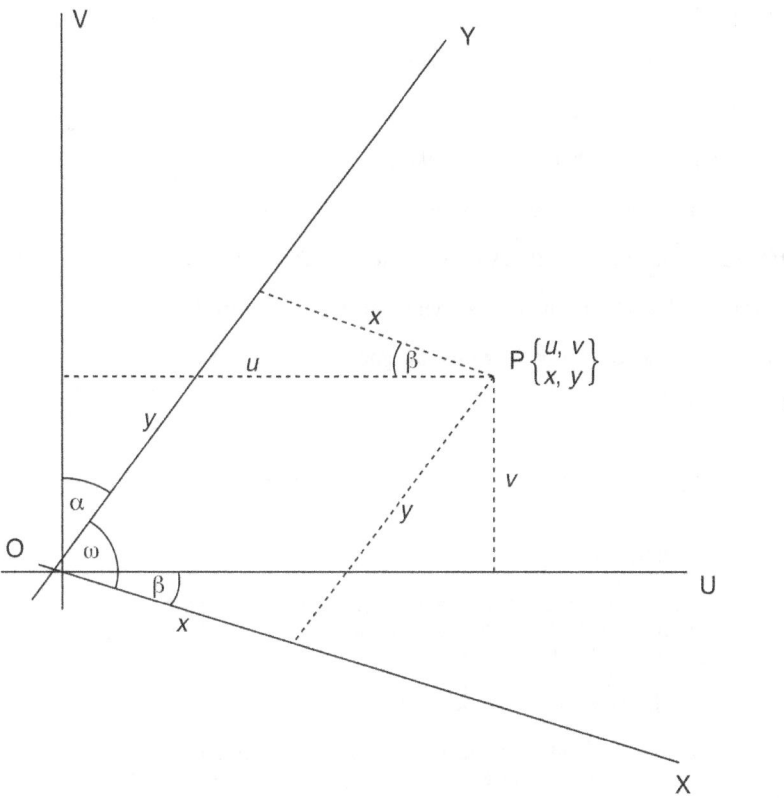

$$u = x\cos\beta + y\sin\alpha \qquad v = -x\sin\beta + y\cos\alpha$$

$$u^2 + v^2 = x^2 + 2xy\sin(\alpha - \beta) + y^2 = x^2 + 2xy\cos\omega + y^2$$

Fig B.10

Appendix-C Outlines for Q-Software

[C.0] The author has written software for q-analysis in Common Lisp

- before the modern approach with object-oriented languages -

but other writers have used different computer languages.

Because of this we give only a bare outline of the flow of any

language - only for those who would like a few simple suggestions

for calculating the q-components, etc, associated with any **Dowker** array.

Implied references to "functions" in the following will normally correspond to

"methods" in a more modern approach, whilst the complexes themselves will be

"instances" of objects.

[C.1] The <u>file containing the data</u> for the complex will normally consist of :-

```
*title
*rows     (= number of rows)
*cols     (= number of columns)
*type     (= 1 means the rows-data numbers are ordinals - fixing the
              columns where the 1's are to be found)
*rownames (names of the rows, eg Y1 Y2 ...)
*colnames (names of the columns, eg X1 X2 ...)
*rows-data in each row as a list/vector (eg 1 5 9 10 ...)
```

The format of the *rows-data need only contain the column-numbers where the 1's are to be found : the example means that the 1's in this Dowker-array are found in columns 1 5 9 etc.

This is because the q-values of the shared faces are simply numbers like :-

(length of the intersection of row1 and row2) minus 1

(Note : Examples of this format can be found in the following §[C5])

The simplest way ahead is probably to form the <u>shared face matrix</u> or *sfmat.

This is a <u>symmetric square matrix</u> with both rows and columns named (for KY(X))

as, say, Y1 Y2 ...etc.

For the analysis of KX(Y) these would be X1 X2 ... .

This matrix contains all the information needed to find, say,

    topq values for each row
    bottomq values for each row
    q-values of any shared faces

The topq value for each Yk is simply its diagonal element in *sfmat.

The bottomq value for any particular Yk is simply the value of the

maximum of values found in the kth row of *sfmat - but ignoring the diagonal values.

The q-components range over q-values from maxq to 0, and for any q-value *sfmat will provide the Yk's which have topq-values equal to this q. Connection with other Yk's sharing such a q-face can then be tracked through this *sfmat. A recursive function is helpful in collecting up the separate simplices which form a q-component.

[C.2] <u>Other properties</u> have been found to be of interest in some projects, viz.,

(a) Eccentricity - which can be defined as [topq - bottomq] ÷ [bottomq + 1]

(b) Structure Vector - which collects up the number of different pieces or separate components which occur at any q-value, giving an overall macroview of the complex.

(c) Identifying the possible loops (<u>q-loops</u>) or <u>holes</u> in the structure.
a 1-hole might be formed by the sides of a hollow triangle whilst
a 2-hole might be formed by the four 2-simplices around a hollow tetrahedron
and so on. Such a <u>hole</u> will form an obstacle to an entity which is
moving through the structure. In that sense it can be thought of as
an <u>object</u> in the world of the complex. These "holes" bear some relationship
to the generators of the Homology Groups studied by **Dowker** in his paper,
(v. Appendix-B [5]).

(d) Listing the **q-stars** for any given simplex. These are all the immediate neighbours of the given simplex - i.e. those which share a q-face with it.

[C.3] The Concept of a Pattern on a complex

If we start with a complex, KY(X), we can consider any set of values associated with the vertices $\{X_i\}$ as a **pattern** (strictly speaking a **0-pattern**) on the complex.

So if we are considering a complex KY(X) the 0-pattern will

be a set of values on the column names : in the case of a complex KX(Y)

a 0-pattern will be a set of values on the row names.

Such a pattern, in the simplest case, will be denoted by the symbol

Π^0 - which will be values on the 0-simplices (vertices)

In a similar way we can consider a pattern, say, Π^k - which will be

a set of values on the k-simplices.

The most general occurrence of a pattern on a complex will therefore be a

collection of sets like Π^k, and we shall refer to this as a **graded pattern**

it being graded by the q-values of K. We shall write such a collection as a Π^*

where $\Pi^* = \Pi^0 \oplus \Pi^1 \oplus \Pi^2 \oplus ... \oplus \Pi^k \oplus ... \oplus \Pi^n$ [C]

it being understood that the underlying complex K is of <u>finite dimension</u>, n.

[C.4] The concept of an **anti-complex** $\overline{K}$

Given a **Dowker** complex K, with its rows a mixture of 0's and 1's its anti-complex $\overline{K}$ is the complex defined by a new array formed from the given one by replacing all the 0's by 1's and all the 1's by 0's.

It follows that if we think of the union of K and $\overline{K}$, viz. $K \cup \overline{K}$, as the superimposition of K and $\overline{K}$ we will have a simplex full of 1's, that is to say a defining simplex, σ_N, where N is the number of columns in the original K.

This σ_N will be repeated M-times, where M is the number of rows in either K or $\overline{K}$.

As a simple example - writing H for $\overline{K}$:

A Complex-K

K	X1	X2	X3	X4	X5	X6	X7	X8	X9	X10	X11
Y1	1	1	1	1	0	0	0	0	0	0	0
Y2	1	1	0	0	1	0	0	0	0	0	0
Y3	0	0	0	0	1	1	0	0	0	0	0
Y4	0	0	0	0	0	0	1	1	1	0	0
Y5	0	0	0	0	0	0	1	0	1	0	0
Y6	0	0	0	0	0	1	0	0	1	0	
Y7	0	0	0	0	0	0	1	0	0	1	

It's Anti-Complex-H

H	X1	X2	X3	X4	X5	X6	X7	X8	X9	X10	X11
Y1	0	0	0	0	1	1	1	1	1	1	1
Y2	0	0	1	1	0	1	1	1	1	1	1
Y3	1	1	1	1	0	0	1	1	1	1	1
Y4	1	1	1	1	1	1	0	0	0	1	1
Y5	1	1	1	1	1	1	0	1	0	1	1
Y6	1	1	1	1	1	1	0	1	1	0	1
Y7	1	1	1	1	1	1	1	0	1	1	0

[C.5] Data Files used for Q-Analysis of Chessboard positions in Chapter-5.

These refer to **The Immortal Game**, also to stages 1,2,3,4 in **Steinitz** v **Mongredien**.

(1) The data file for chessboard complex after 22. Q-f6+ : Resign

```
"The Immortal Game  1851"
(11 25 1)
("QN" "QB" "Q" "KN" "KP" "qn" "qb" "q" "k" "kb" "kn")
 "b1" "b2" "b6" "b7" "b8" "c1" "c3" "c5" "c7"
 "d1" "d4" "d6" "d8" "e3" "e5" "e6" "e7" "e8"
 "f1" "f2" "f6" "f7" "f8" "g1" "h6")
(3 9 17 21)
(5 9 17 23)
```

(13 17 22)
(18 16)
(12 21)
(5 8 9)
(4)
(1 6 10 19 24 2 7 11 15)
(9 18 17)
(20 14 11 8 3)
(17 21 25)

(2) The data file for chessboard complex after 05. N-d2 : p-d6

"Steinitz1.lsp"
(16 29)
("QN" "QB" "Q" "K" "KB" "QBP" "QP" "KP"
 "qn" "qb" "q" "k" "kb" "kn" "qnp" "knp")
("a4" "a5" "a6"
 "b3" "b4" "b5"
 "c12" "c2" "c3" "c4" "c5" "c6" "c8"
 "d2" "d3" "d4" "d5"
 "e2" "e4" "e5"
 "f3" "f4" "f5" "f6" "f8"
 "g4" "g5"
 "h5" "h6")
(9 14)
(7 14 16 22 27 29)
(7 8 4 1 14 15 16 18 21 26 28)
(14 18)
(18 15 10 6 3)
(5 16)
(11 20)
(17 23)
(3 12)
(3 12 17 19)
(13)
(25)
(25 24 20 16 29)
(24 29)
(2 11)
(23 28)

(3) The data file for chessboard complex after 13. O-O-O : p-c5

"Steinitz2.lsp"
(21 29 1)
("QR" "QN" "QB" "Q" "K" "KB" "KN" "KR"
 "QBP" "KP" "KBP"
 "qb" "qn" "q" "k" "kb" "kn" "kr"
 "qbp" "kp" "krp")
("b4" "c4" "d1" "d2" "d4" "d5"
 "e1" "e3" "e4" "e5" "f1" "f2"
 "f3" "f5" "f6" "f7" "f8" "g1"
 "g3" "g5" "g6" "h1" "h2" "h3"
 "h4" "h5" "h6" "h7" "h8")
(7 11 18 22)
(9 13)
(5 12 20 21 27)
(2 13)
(3 4)
(6 16)
(10 20)
(23 24 25 26 27 28)
(1 5)

(6 14)
(8 19)
(6 9)
(26 28)
(15 19)
(28)
(15 27)
(10 29)
(16)
(5)
(5)
(21)

(4) The data file for chessboard complex after 18. R-h1 : r-e8

"Steinitz3.lsp"
(17 25 1)
("QR" "QN" "QB" "Q" "K" "KB" "QBP" "KP" "KBP"
 "qn" "qb" "q" "k" "kb" "kr" "qbp" "kp")
("b4" "c5" "d2" "d4" "d5" "e3" "e4" "e5" "e6" "e7"
 "f2" "f3" "f4" "f5" "f6" "f7" "f8" "g6"
 "h2" "h3" "h4" "h5" "h6" "h7" "h8")
(18 19 20 21)
(7 12)
(4 2 11)
(18 23 24 25)
(3)
(5 9 15)
(1 4)
(5 14)
(6)
(8 13 25)
(5 7)
(3 10 15)
(17 25)
(5 9 16)
(8 9 10)
(4)
(4 13)

(5) The data file for chessboard complex after 22. Qxq : Resign

"Steinitz4.lsp"
(12 27 1)
("QN" "QB" "Q" "K" "QBP" "KP" "KBP"
 "qb" "k" "kb" "kr" "kp")
("b7" "c4" "c5" "c7" "d2" "d4" "d5" "d7"
 "e3" "e4" "e5" "e6" "e7" "e8"
 "f2" "f3" "f4" "f5" "f6" "f8" "g6" "g7"
 "g8" "h5" "h6" "h7" "h8")
(2 10 15)
(3 6 16)
(1 4 8 12 13 14 20 21 22 23 24)
(5)
(6)
(7 18)
(9)
(10)
(23 26)
(11 19 20 25 27)
(11 12 13 20)
(6 17)

Appendix-D Representing a Complex in Exterior Algebra

[D.1] The basis of vectors

To remind ourselves of the basics of Exterior Algebra, which has been introduced as an example in Appendix-A, we begin with a finite dimensional vector space V which has a basis of independent vectors (serving as "axes" in the Space), viz.,

$$\{u_1\ u_2\ u_3\ u_4\ ....\ u_n\}$$

By imposing a product on the u_i we can transform V into an algebra, which possesses two binary operations "+" and "×".

This product is to be the **exterior product** (often called the **wedge product**) and its chief characteristic is its **antisymmetry**. Thus the wedge product of two elements u_1 and u_2, written as $u_1 \wedge u_2$, behaves as

$$u_1 \wedge u_2 = - u_2 \wedge u_1$$ which gives rise to the property of **nilpotency**, viz., $\quad u_i \wedge u_i = 0 \quad$ for all such u_i in V.

We also require that $\wedge$ shall be distributive over + so that, for example,

$$(a_1 u_1 + a_2 u_2) \wedge (b_1 u_1 + b_2 u_2) = (a_1 b_2 - a_2 b_1)\ u_1 \wedge u_2$$

This wedge product gives us a vector space with all the products $u_i \wedge u_j$ ($i<j$) as a basis.

We denote this modified vector space by $\Lambda^2 V$, the superscript 2 denoting the pairs of its basis vectors.

In a similar way we can form the wedge spaces

$$\Lambda^3 V,\ \Lambda^4 V,\ \Lambda^5 V\ ....\ \Lambda^n V$$

taking three's, four's etc of the original base vectors, the last member having a single base element $\quad u_1 \wedge u_2 \wedge u_3 .... \wedge u_n$

[D.2] The whole series of such wedge spaces is written as ΛV and amounts to a **graded** setf of spaces. IF F is the field of numbers from we take any of the coefficients a, b, c, etc then we can conveniently let this be introduced as $\Lambda^0 V$ and by using V itself as an obvious $\Lambda^1 V$ we get the graded algebra

$$\Lambda V = \Lambda^0 V \oplus \Lambda^1 V \oplus \Lambda^2 V \oplus\ .....\ \oplus \Lambda^n V$$

Since the wedge product is associative in the sense that

$$(u_i \wedge u_j) \wedge u_k = u_i \wedge (u_j \wedge u_k)$$

the algebra is associative.

Example

Take a 3-dimensional vector space V, over the reals $\Re^{\#}$, with basis $\{u_1\ u_2\ u_3\}$

Then the exterior alebra $(\Lambda V, +, \wedge)$ is graded via

$$\Lambda V = \Lambda^0 V \oplus \Lambda^1 V \oplus \Lambda^2 V \oplus \Lambda^3 V$$

with $\Lambda^0 V \equiv \Re^{\#}$.

The product of two elements of $\Lambda^1 V$ lies in $\Lambda^2 V$, for example writing

$$A = (a_1 u_1 + a_2 u_2 + a_3 u_3)$$
$$B = (b_1 u_1 + b_2 u_2 + a_3 u_3)$$
$$C = (c_1 u_1 + c_2 u_2 + c_3 u_3)$$

we can check the product $A \wedge B \wedge C$ is $\delta(a,b,c)(u_1 u_2 u_3)$

where δ is the determinant with

row-1 = $a_1\ a_2\ a_3$
row-2 = $b_1\ b_2\ b_3$
row-3 = $c_1\ c_2\ c_3$

This determinant is an example a **pseudoscalar**, it being the one (and only) component of a vector in the 1-dimensional space $\Lambda^3 V$ - just as $\Lambda^0 V$ (which equals $\Re^{\#}$) is a 1-dimensional space (of scalars) - the only difference being that the sign of the pseudoscalar depends on the order of the base $u_1 \wedge u_2 \wedge u_3$ of $\Lambda^3 V$.

[D.3] The general case

In the general case, when V is an n-dimensional vector space, we notice that the dimension of the space $\Lambda^p V$ is ^nC_p - the number of p-combinations of n unlike things. This means that $\dim \Lambda^p V = \dim \Lambda^{n-p} V$ since $^nC_p = {}^nC_{n-p}$.
In this case $\Lambda^n V$ is the space of pseudoscalars.

The general element **x** in ΛV will be the sum of elements out of the separate vector spaces $\Lambda^p V$, so we can write

$$\mathbf{x} = \mathbf{x}_0 + \mathbf{x}_1 + \mathbf{x}_2 \ldots + \mathbf{x}_p + \ldots + \mathbf{x}_n$$

where $\mathbf{x}_p \in \Lambda^p V$. The product rule then ensures that

$$\mathbf{x}_p \wedge \mathbf{x}_q \in \Lambda^{p+q} V \quad \text{provided } p+q < n.$$

Since we know that for every $\mathbf{x} \in \Lambda V$ $\mathbf{x} \wedge \mathbf{x} = 0$ (nilpotency) then if we consider a polynomial in $\Lambda^1 V$ it can only be of the form

$$a_0 + a_1 \mathbf{x} \qquad a_0, a_1 \text{ being } \Re^{\#}$$

and when $\mathbf{x} \neq \mathbf{y}$ a polynomial in $\Lambda^2 V$ will look like

$$a_0 + a_1 u_1 + a_{ij} u_i u_j + \ldots + a_{1\ldots n} u_1 u_2 u_3 \ldots u_n \qquad (D)$$

where we have written, eg, $u_i u_j$ for $u_i \wedge u_j$, the $\wedge$ being understood.

[D.4] Representing a **Dowker** complex

A **Dowker** complex of dimension q (= n-1) has n verices and (q+1)-subsets of them can be associated with those of a q-simplex, σ_q, - no vertex being repeated in its description. This corresponds exactly to a base vector in $\Lambda^q V$, as described above ; it also has the added bonus of allowing us to introduce an **orientation** into the various simplices.

In fact the exterior algebra contains a **Dowker** complex (of finite dimension) by way of a polynomial such as (D) above - by taking each coefficient as unity, 1.

Furthermore (D) will represnt a **graded pattern** on the underlying complex - by specifying the values of the coefficients $a_{ijk...}$ therein.

Further discussion of these matters will be found in Apendix-E [118].

Appendix-E REFERENCES & BIBLIOGRAPHY

[1] Atkin, R.H. *Abstract Physics, Nuovo Cimento Serie X vol-38* 1965

[2] Baker, H.F. *Principles of Geometry Vols 1,2,3* CUP 1929

[3] Semple & Kneebone *Algebraic Projective Geometry* OUP 1952

[4] Maxwell, E.A. *General Homogeneous Co-ordinates* CUP 1948

[5] Eddington, A. *Fundamental Theory* CUP 1953

[6] Eddington, A. *The Mathematical Theory of Relativity* CUP 1952

[7] Salmon, G. *A Treatise on the Conic Sections* LONGMANS 1900

[8] Fox, C. *The Calculus of Variations* OUP 1950

[9] Whittaker, E.T. *Analytical Dynamics* DOVER 1936

[10] Jenkins & White *Fundamentals of Optics* McGRAW-HILL 1951

[11] Stewart, C.A. *Advanced Calculus* METHUEN 1940

[12] Atkin, R.H. *Classical Dynamics* HEINEMANN 1959

[13] Atkin, R.H. *Theoretical Electromagnetism* HEINEMANN 1962

[14] Trigg, G.L. *Quantum Mechanics* VAN NOSTRAND 1964

[15] Smythe, W.R. *Static & Dynamical Electricity* McGRAW-HILL 1950

[16] Milne, E.A. *Vectorial Mechanics* METHHUEN 1948

[17] Turnbull, H.W. *Determinants, Matrices & Invariants* BLACKIE 1929

[18] Whittaker, E.A. *History of Theories of Aether & Electricity* NELSON 1951

[19] Kline, M. *Mathematical Thought from Ancient to Modern Times* OUP 1972

[20] Pedersen & Pihl *Early Physics & Astronomy* MACDONALD & JANES 1974

[21] Dickson, L.E. *Algebras and their Arithmetics* DOVER 1923

[22] Joos, G. *Theoretical Physics* BLACKIE & SON 1944

[23] Roberts, J.K. *Heat and Thermodynamics* BLACKIE & SON 1943

[24] Born, M. *Atomic Physics* BLACKIE & SON 1945

[25] Ramsey, A.S. *A Treatise on Hydrodynamics* BELL 1949

[26] Jacobson, N. *Lectures on Abstract Algebra, Vols 1,2,3* VAN NOSTRAND 1951

[27] Blumenthal, L.M. *A Modern View of Geometry* FREEMAN 1961

[28] Adler, Bazin & Schiffer *Introduction to General Relativity* McGRAW-HILL 1965

[29] Nelson & Parker *Advanced Level Physics* HEINEMANN 1983

[30] Steffens, H.J. *Joule and the Concept of Energy* DAWSON 1979

[31] Sternberg & Smith *The Theory of Potential* UNIVERSITY OF TORONTO 1964

[32] Mott & Sneddon *Wave Mechanics & its Applications* OUP 1948

[33] Massey, W.S. *Algebraic Topology, and Introduction* Harcourt, Brace & World 1967

[34] Atkin, R.H. *Mathematics and Wave Mechanics* HEINEMANN 1956

[35] Tait, P.G. *Elementary Theory of Quaternions* CUP 1890

[36] Phillips, E.G. *Functions of a Complex Variable* OLIVER & BOYD 1949

[37] Whittaker & Watson *A Course of Modern Analysis* CUP 1952

[38] Baer, R. *Linear Algebra and Projective Geometry* AP 1952

[39] Littlewood, D.E. *A University Algebra* HEINEMANN 1950

[40] Flanders, H. *Differential Forms* AP 1963

[41] Helgason, S. *Differential Geometry and Symmetric Spaces* AP 1962

[42] Chevalley, C. *Fundamental Concepts of Algebra* AP 1956

[43] Faulkner, T.E. *Projective Geometry* OLIVER & BOYD 1949

[44] O'Hara & Ward *An Introduction to Projective Geometry* OUP 1937

[45] Robinson, G. de B. *Foundations of Geometry* University of Toronto Press 1950

[46] Ramsey, A.S. *Statics* CUP 1934

[47] Weyl, H. *Space-Time-Matter* DOVER 1950

[48] Fox, C. *Calculus of Variations* OUP 1950

[49] Coulson, C.A. *Waves* OLIVER and BOYD 1949

[50] MacRobert, T.M. *Spherical Harmonics* METHUEN 1947

[51] Ramsay, A.S. *Hydrostatics* CUP 1947

[52] Noakes, G.R. *Textbook of Light* MacMILLAN 1960

[53] Widder, D.V. *The Laplace Transform* Princeton University Press 1946

[54] Baker, B.B. & Copson, E.T. *Huyghens' Principle* OUP 1949

[55] Levi-Civita, T. *The Absolute Differential Calculus* BLACKIE & Son 1950

[56] Weatherburn, C.E. *Riemannian Geometry & the Tensor Calculus* CUP 1950

[57] Dryer, J.L.E. *Tycho Brahe* DOVER 1963

[58] McConnell, A.J. *Applications of the Absolute Differential Calculus* BLACKIE & Son 1951

[59] Todd, J.A. *Projective and Analytical Geometry* PITMAN & Sons 1947

[60] Carslaw, H.S. & Jaeger, J.C. *Operational Methods in Applied Mathematics* OUP 1949

[61] Noakes, G.R. *A Textbook of Heat* MacMILLAN 1960

[62] Atkin, R.H. *Cohomology in Physics* Int. J. Man-Machine Studies 1972

[63] Schouten, J.A. *Tensor Analysis for Physicists* OUP 1951

[64] Dennery & Krzywicki   *Mathematics for Physicists*   HARPER & ROW   1967

[65] Jeans, J.H.   *Electricity and Magmetism*   CUP   1951

[66] Dirac, P.A.M.   *The Principles of Quantum Mechanics*   OUP   1947

[67] Chevally, C.   *The Construction & Study of Certain Important Algebras*
         Maths Society of Japan   1955

[68] Lawden, D.F.   *An Introduction to Tensor Calculus and Relativity*   METHUEN   1967

[69] Dowker, C.H.   *Homology Groups of Relations*   Ann. Math. [56, 84]   1952

[70] McClachlan, N.W.   *Bessel Functions for Engineers*   O.U.P   1934

[71] Bohr, N   *Phil. Mag., vol-26, 1, 476, 857*   1913

[72] Heisenberg, G.   *Zeitz. f. Phys., vol-33, 879*   1925

[73] Dirac, P.   *Proc. Roy. Soc. A, vol-109, 642*   1925

[74] De Broglie, L.   *Phil. Mag. vol-47, 446*   1926

[75] Schrodinger, E.   *Collected Papers on Wave Mechanics*   BLACKIE & Son   1928

[76] Hamburger, H.L. & Grinshaw, M.E.   *Linear Transformations*   CUP   1951

[77] March, A.   *Quantum Mechanics of Particles and Wave Fields*   WILEY   1951

[78] Crowther, J.A.   *Ions, Electrons, and Ionizing Radiations*   ARNOLD   1945

[79] Lagrange, J.L.   *Mecanique Analytique*   1788

[80] Hammilton, W.R.   *Phil. Trans. 1834 ; ibid 1835*

[81] Hamilton, W.R.   *Brit. Ass. Report 1834 ; Phil. Trans. 1835*

[82] Hamilton, W.R.   *Trans. R. Irish Acad. vol XV 1828, XVI 1830, XVII 1837*

[83] Forsyth, A.R.   *Concomitants of Differential Forms in Four Variables*
         Proc. Roy. Soc. Ed.   1921-2, p.147 et seq

[84] Meschkowski, H.   *Noneuclidean Geometry*   AP   1964

[85] de Rham, G.   *Varietes Differentiables*   HERMAN Paris   1960

[86] Sze-Tsen Hu   *Homology Theory*   HOLDEN-DAY   1970

[87] Hocking, J.G., Young, G.S.   *Topology*   ADDISON-WESLEY   1961

[88] Eilenberg, S & Steenrod, N.   *Foundations of Algebraic Topology*   Princeton U.P.   1952

[89] Whitehead, J.H.C.   *Combinatorial Homotopy*   Bull. Am. Maths. Soc
         Vol 55 (1949) pp 213-245

[90] Mansfield, M.   *Introduction to Topology*   Van NOSTRAND   1963

[91] Kowalsky, H.J.   *Topological Spaces*   AP   1965

[92] Atkin, R.H.   *Mathematical Structure in Human Affairs*   Heinemann   1974

[93] Čech, E.   *Theorie generale de l'homologie dans une espace quelconque*   Fund. Math. 1932

[94] Hilton, P.J. & Wylie, S.   *Homology Theory*   CUP   1962

[95] Baumslag, B. & Chandler, B.   *Group Theory*   McGRAW-HILL   1968

[96] Goldberg, S.I.   *Curvature and Homology*   AP   1962

[97] Hodge, W.V.D.   *The Theory & Applications of Harmonic Integrals*   CUP   1952

[98] Nishijima, K.   *Fundamental Particles*   BENJAMIN   1964

[99] Roman, P.   *Theory of Elementary Particles*   NORTH HOLLAND   1964

[100] Swartz, C.E.   *The Fundamental Particles*   ADDISON-WESLEY   1965

[101] National Research Council (US)   *Elementary Particle Physics*
      NATIONAL ACADEMY PRESS (US)   1986

[102] Weinberg, S.   *The Discovery of Subatomic Particles*   CUP   2003

[103] Veltmen, M.   *Facts & Mysteries in Elementary Particle Physics*
      VE-WORLD SCIENTIFIC   2003

[104] Wittgenstein, L.   *Tractatus Logico Philosophicus*   RUTLEDGE & KEGAN PAUL   1961

....

REFERENCES & BIBLIOGRAPHY for *PART4*

[105] Stoll, Robert R.
      Sets, Logic and Axiomatic Theories
      W.H.Freeman & Co.   1961

[106] Halmos, Paul, R.   *Naive Set Theory*
      Van Nostrand Company   1960

[107] Maher, L.P.
      Finite Sets : Theory, Counting & Applications
      Merrill, Charles, E.   1968

[108] Lipschutz, S.   *Set Theory and related topics*
      Schaum Publishing Co.   1964

[109] Dowker, C.H.   *Homology Groups of Relations*   Ann. Math. [56, 84]   1952

[110] Atkin, R.H. Johnson J.H. Mancini V.
      An Analysis of Urban Structure using concepts of algebraic topology
      Environment and Planning B   1971

[111] Atkin R.H. Wittten I.H.
      Multidimensional structure in the game of chess
      Int. J. Man-Machine Studies   1972

[112] Beaumont J.R Gatrell A.C.
      An Introduction to Q-Analysis
      Concepts & Techniques in Modern Geography   No. 34   1982

[113] Atkin R.H. Hartston W.R. Witten I.H.
      Fred Champ, positional-chess analyst
      Int. J. Man-Machine Studies   1976

[114] Atkin R.H.

An Approach to Structure in Architectural and Urban Design I,II,III
Environment and Planning B   1974, 1974, 1975")

[115] Atkin R.H.
An Algebra for Patterns on a Complex I,II
Int. J. Man-Machine Studies   1974, 1976

[116] Johnson J.H.
Q-Analysis of Road Intersections
Int. J. Man-Machine Studies 1976

[117] Johnson J.H.
Q-Analysis of Road Traffic Systems
Environment & Planning B   1980

[118] Johnson J.H.
A Study of Road Transport
Research Report XI, Maths Dept. Essex University 1977

[119] Johnson J.H. & Wanmali S.
Q-Analysis of Period Market Systems
Geographical Analysis   1981

[120] Beaumont J.R. & Beaumont C.D.
Comparatife Study of multivariate structure of towns
Management Centre, Aston University   1981

[121] Chapman G.P.
Q-Analysis
Quantitative Geography : a British View
Eds   Wrigly N. & Bennett R.J.   1981

[122] Gatrell A.C & Gould P.
A Structural analysis of the Game :
Liverpool F.C v. Manchester United, Cup Final 1977
Social Networks 1977

[123] Atkin R.H.
Time as a Pattern on a Multidimensional Structure
Journal of Social and Biological Sciences   1978

[124] Atkin R.H.
Q-Analysis : a hard language for the soft sciences
Futures   1978

[125] Atkin R.H.
A Theory of Surprises
Special Issue on Q-Analysis - Environment and Planning B   1981

[126] Atkin R.H.
Developing a Hard Science for the Study of Human Society
Environment and Population   ed John B. Calhoun
Praeger 1983

[127] Melville B.
Notes on the Civil Applications of Mathematics
Int. J. of Man-Machine Studies (1976) vol 8

[128] Chamberlain M. Anne
Study of Behcet's Disease by Q-Analysis
Int. J. of Man-Machine Studies 1976

[129] *Special Issue on Q-Analysis*
Environment & Planning B   1981

[130] Nijenhuis & Wilf
Combinatorial Algorithms
Academic Press 1975

[131] Strehler B.L.
Time, Cells, and Aging
Academic Press 1963

[132] Koestler A.
The Roots of Coincidence
Pan Books Ltd. 1974

[133] Dunne J.W.
An Experiment with Time
Hampton Roads Publishing Co.

[134] Atkin R.H.
Multidimensional Man
Penguine Books 1981

[135] Hodge S.E., Seed M.L.
Statistics and Probability
Blackie & Son Ltd 1972

[136] Lindley D.V.
Introduction to Probability & Statistics Pts 1,2
C.U.P. 1969

[137] Halmos Paul R.
Boolean Algebras
Van Nostrand Inc. 1963

[138] Clarke M.R.B
Avances in Computer Chess
Edinburgh University Press 1977

[139] Hardy G.H & Wright E.M.
The Theory of Numbers
O.U.P 1971

[140] Johnson J.H.
Representation, Knowlege Elicitation and Mathematical Science
A Contribution to Artificial Intelligence in Mathematics, published by Institute of Mathematics and its Applications 1993

[141] Laguerre
Nouv. Annal. xii, 1853 and *Oeuvres 1905*

Index

A
Algebras 275
  Elementary 3, 6
  Clifford 281
  Complex 4, 6
  Exterior 281
  Matrix 278
  Quaternion, È33 4, 6
  Sedenion 115
absolute conic 291
  points 12
adiabatic 66
affine space 97
d'Alembert 54, 56, 92
Ampere 81
anti-particle 218, 219
Appollonius 6, 285
Archimedes 38

B
Baker, H.F. 9
barber paradox 284
base-element 14
Bayes' theorem 272, 273
Bernoulli 90, 149
Betti number 237
Bohr 124, 134
boson 222, 223

C
Cantor 230
Cauchy 69
Cayley 12, 13
centre of gravity 00
chess 260 et seq, 306, 307, 308
clock-time 228, 240
cochain 181

co-homology 156
  ring of 182
chain group 168, 178
  graded 239
complex 163, 252
  Cech 167
  CW 166
  Dowker 164
  simplicial 163
  in $L(V) 165
conic 12
co-ordinates 6
CROP 151, 152, 153
cross-ratio 11, 148, 287, 288

D
Dedekind 289
Descartes 6
De Broglie 142
De Rham 165, 195
Desargue 25
dipole 77
Dirac space 141 et seq
  bra, ket 210
division algebra 280
double points 289
Dowker 233, 236, 271
duality 22, 39, 51, 179

E
Eddington 115 et seq
eigenvalues 61, 62
Einstein 46, 50, 105, 107, 209
  v. relativity 00
energy
  potential 41
  kinetic 48

envelope 27
Euclid, geometry 6, 9, 13, 292
Euler 26, 88
E-M field 113, 114
exterior derivative 282

F
factor group 168
Faraday 81
field 275, 282
Forsyth 108, 109

G
generating elements 15
gravitation 49, 50, 105, 107
graded
    complex 239
    pattern 305

H
hadron 222
Hamilton 54, 57, 60, 125, 190, 191
Hamiltonian operator 211
harmonic
    range 10
    pencil 10
    oscillator 136, 146
Hausdorf space 186
Heaviside 46
Heisenberg 185
Hermite 134 et seq
Hodge operator 199, 200
holes in space 157, 158, 159 169
homeomorphism 160
homography 10, 16, 69, 114, 194
homology 156, 177, 237
Huygens 20, 21, 54, 61, 94

I
infinity 9, 288
invariants 17, 110
involution 11, 289

K
Kelvin 67
Kepler 50
Komplex-time 243, 245
Kunneth formula 207

L
Lagrange equations 55, 56, 57
Laguerre metric 13, 120, 145
Laurent series 186, 187
Laplace 69, 71, 91, 187
Legendre 145
Leibnitz 27
leptons 222, 224
line at infinity 9
linear operator 141, 142
linear-time 240
Lorentz operator $N 49
Lorentz-Einstein equations 46

M
mappings 15
Maxwell equations 83, 95, 97, 104, 198
meash 177
mesons 224
metacentre 88
metric 7, 12, 23
    Euclidean/Pythagorean 292
    Lorentz-Einstein 48
    Riemannian 192
    projective 28, 290
module 277

N
nerve of a covering 167
Newton 23, 27, 46, 48, 50, 51, 105, 107, 153
nim (game) 255
nilpotency 168, 180
norm 280

O

optics, geometrical	17, 19
Oersted	81

P

Pappus' theorem	25
partition, of a set	234, 239
patterns, on a complex	305
perpective	287
p-events	240, 244
physics, definition	23
Pfaff	58, 103
Planck	124, 143, 150
Poincare	46
poin	147
Poisson	52, 78, 134 et seq, 188, 189
Poncelet	6, 285
positron	218
potential	
function	41
barrier	68
Poynting vector	86, 112, 113
projective	
geometry P~2~	6
metric	147, 292
proposition of incidence	6, 285
Pythagoras	7

Q

q-analysis	239
q-chain	239
q-component	247
q-connection	239
q-loop	305
q-simplex	236
quark	224

R

reflexive measures	18, 21
relation	233, 234
reflexive	234
symetric	234
transitive	234
equivalence	234
relativity	
special	46
general	107
Riemann	50, 69
Riemann-Christoffel	70
Riemann metric	105, 192
ring, in an algebra	277
Ritz	122, 125
Rutherford	124

S

scales	14, 15
Schwarzchild	107, 109
Schwarz-Christoffel	70
Schrodinger	143, 144, 146
shared-face matrix	304
singularity	157
spin	213, 214
stationary states	144
surprise event	274

T

tensors	100
t-forces	236, 247
times (various)	243, 244, 246
traffic on a complex	248
triangle of reference	8

U

unitary	
matrix	139
operator	139

V

Venn diagrams	231, 232
Von Staudt	6, 9

W

work function, W	243
Weyl	109

X
X-rays as base-element 157

Y
Young's slits 94, 119

www.ingramcontent.com/pod-product-compliance
Lightning Source LLC
Chambersburg PA
CBHW080423230426
43662CB00015B/2191